APPLICATIONS

APPLICATIONS

Editor

Ennio Fini

scitus
academics

Applications

Edited by **Ennio Fini**

This Book is Printed on Acid-Free Paper. ♾

ISBN: 9781681179575

Library of Congress Control Number: 2017949102

SCITUS Academics LLC,

616, Corporate Way, Suite 2, 4766,

Valley Cottage, NY 10989

www.scitusacademics.com

Printed in United States of America.

Preface

Mathematics is a methodical application of matter. Mathematicians seek out patterns and use them to formulate new assumptions. Mathematicians determine the truth or falsity of assumptions by mathematical proof. When mathematical structures are good models of real phenomena, then mathematical reasoning can provide insight or forecast about nature. Through the use of abstraction and logic, mathematics developed from counting, calculation, measurement, and the systematic study of the shapes and motions of physical objects. Practical mathematics has been a human activity from as far back as written records exists. Mathematics makes our life orderly and prevents chaos. Certain qualities that are nurtured by mathematics are power of reasoning, creativity, abstract or spatial thinking, critical thinking, problem-solving ability and even effective communication skills. Mathematics is the frame of all creations, be it a cook or a farmer, a carpenter or a mechanic, a shopkeeper or a doctor, an engineer or a scientist, a musician or a magician, everyone needs mathematics in their day-to-day life.

This book discusses various new issues and developments in different areas of mathematics and its applications. It presents topics that treat mathematical analysis and its numerous applications. The book is devoted to the mathematical treatment of questions arising in physics, chemistry, biology, and engineering, particularly those that stress analytical aspects and novel problems and their solutions. This book will of interest to mathematicians, scientists and engineers providing original research studies in all aspects of pure mathematics, applied mathematics as well as applications of mathematics in chemistry, biology, physics, engineering, geophysics, computing and finance.

Table of Contents

CHAPTER 1

Numerical Simulation of Flow around Diamond-Shaped Obstacles at Low to Moderate Reynolds Numbers

Seyed Reza Djeddi[1], Ali Masoudi[1], Parviz Ghadimi[1]

[1]*Department of Marine Technology, Amirkabir University of Technology, Tehran, Iran*

ABSTRACT

In this paper, viscous fluid flow over an unconventional diamond-shaped obstacle in a confined channel is simulated in low to moderate Reynolds numbers. The diamond-shaped obstacle is altered geometrically in order to represent different blockage coefficients based on the channel height and different aspect ratios based on the length to height ratios of the obstacle. An in-house finite difference Navier-Stokes solver using staggered grid arrangement and Chorins projection method is developed for the simulation of the laminar viscous flow. The numerical solver is validated against numerical results that are presented in the literature for the flow over rectangular cylinders and good agreement is observed. Grid resolution has been studied within a mesh convergence test and as a result, suitable grid dimension is achieved. A series of simulations have been carried out for each set of geometry and configuration in order to find the critical Reynolds number for each case in which the vortex shedding will occur. Therefore, simulations are divided into two groups of steady and unsteady flows. In the case of unsteady flow, non-dimensional Strouhal Number (St) is investigated and results prove the dependency of St on the blockage coefficient and aspect ratio. It is shown that the Strouhal number will increase with the rise of blockage ratio and the local maximum of *St* will occur at lower *Re* for geometries with lower aspect ratios (bluff bodies) than geometries with higher aspect ratios, i.e. with more streamlined bodies.

Keywords: flow around obstacle, diamond-shaped, finite difference, Chorin's projection method, Strouhal number

1. INTRODUCTION

Flow around bluff and streamlined bodies has attracted many scientists and researchers for long period of time. The importance of this problem can be sought in its vast applications in engineering problems. Separation and vortex shedding are the common phenomena which happen at relatively low Reynolds numbers as a result of high positive pressure gradients. Vortex shedding causes fluctuation of pressure distribution on the body surface. Due to the vortex-induced forces, the body oscillates with a definite frequency. This may cause damages to structures such as suspended bridges and offshore oil platforms. Despite of many studies which have been done on flow around circular cylinders, rectangular shaped obstacles have been studied less experimentally and numerically, however, these shapes have many applications in analysis of aerodynamics of structures and fluid-structure interaction (FSI) problems. In the problem of flow around a confined cylinder in addition to the common parameters such as Reynolds number, another parameter called blockage coefficient (which is defined as the ratio of square side length to the channel height) should be considered [1].

Based on Reynolds number, different regimes of flow can be observed for the problem of flow around confined rectangular cylinder [2]. At very low Reynolds numbers ($Re \ll 1$), viscous forces dominate the flow. In this creeping flow, no separation occurs. With the increase of Reynolds number, flow separates at trailing edge and forms a recirculation region which consists of two symmetric vortices. Size of recirculation zone increases with an increase of Reynolds number and by reaching the critical Reynolds number, Von Kármán vortex street with periodic vortex shedding happens. With further increase of Reynolds number beyond this critical value, separation will occur in the leading edge, but the range of *Re* for this phenomenon has not been studied clearly and only the range of 100-150 is reported [2, 3].

Okajima [3] has investigated the Strouhal number for aspect ratios (ratio of cylinder's length to height) of 1 to 4 and a range of Reynolds numbers and stated the commence of periodic vortex motion at Reynolds number of 70 leading to an upper limit of critical Reynolds number at $Re \approx 70$. A smaller value of critical Reynolds number $Re \approx 54$ was determined by Klekar and Patankar [4] based on a stability analysis of the flow. Davis et al. [5] interpreted flow around a rectangular cylinder for a wide range of Reynolds numbers and aspect ratios of 1/4 and 1/6. Two-dimensional study of this problem for Reynolds numbers of 90 to 1200 and aspect ratios of 1/8 and 1/4 has been done by Mukhopadhyay et al. [6]. Also, Suzuki et al. [7] has investigated this problem for a range of Reynolds numbers from 56.3 to 225 and aspect ratios of 1/20 to 1/5. Sohankar et al. [8] have studied the effects of blockage coefficient on an unsteady 2-D laminar flow at different angles of incident for Reynolds numbers of 100 and 200. Moreover, experimental and numerical analyses have been carried out by Nakamura et al. [9] for low Reynolds numbers. Experiments have been carried out at a low speed wind tunnel and the numerical analysis was based on the finite difference method of 2-D Navier-Stokes equations. The aspect ratio of the tested rectangular cylinders ranged from 3 to 16 in the experiments and 3 to 10 in the numerical computations.

Contrary to the flow around circular cylinder in which Reynolds number of 180 is introduced as the starting point for the formation of 3D structures in the wake, there has not been an accurate and comprehensive study on this phenomenon for flow around rectangular cylinders. As a hint, Reynolds number of 300 has been introduced by Franke [2] as starting point of this phenomenon. Breuer et al. [1] stated that this criterion is slightly beyond the limit of 2D simulation and the disturbances which occur in the periodic nature of 2D vortex shedding, has proved this theory. They studied flow around rectangular cylinder for the blockage coefficient of 1/8 and by two different methods of finite volume method (FVM) and lattice-boltzmann method (LBM). The experimental studies of Dutta et al [10] has verified the results of Breuer et al [1] and showed that beyond the Reynolds number of 200 aspect ratio of cylinder has to be taken into account due to the formation of 3D structures. One of the most recent studies in this field has been carried out by Berrone et al. [11] by means of adaptive finite element method (AFEM) and FVM for aspect ratios of 1 to 5 and Reynolds numbers of 100, 400 and 1000.

For higher Reynolds numbers, numerical modelings have been done by different methods like large eddy simulation (LES) and high expensive computational method of direct numerical simulation (DNS). As an example, Bruno et al. [12] studied flow around a cylinder at Re = 40,000 by means of a finite volume discretization method. John and Rang [13] applied finite element method to the large eddy simulations of the flow past a square cylinder at Re = 22,000. Other studies have been done by Xin liang et al. [14] and Kogaki et al. [15].

Dutta et al. [10] have experimentally studied flow past a square cylinder placed at an angle of incidence using particle image velocimetry, hot wire anemometry, and flow visualization. Data for four cylinder orientations θ=0, 22.5, 30, and 45° (diamond-shaped) and two aspect ratios of 16 and 28 have been reported. Reynolds numbers ranging 200 to 410 have been chosen. Results are presented in terms of drag coefficient, Strouhal number, time averaged velocity, stream traces, turbulence intensity, powerspectra, and vorticity field. A minimum of drag coefficient and maximum of Strouhal number is reported at a cylinder orientation of θ=22.5°. The effect of the aspect ratio has been studied and the results proved the formation of 3D structures and entrance to turbulent zone for $Re > 200$.

For the current problem of flow around diamond-shaped obstacle another investigation has been done by W. Regulski and Szumbarski [16]. They applied Lattice-Boltzmann approach and studied flow around circular and diamond-shaped obstacle in the blockage coefficient of 0.4 with the use of a Spectral Element Method (SEM) in the regime of low and moderate Reynolds numbers ranging from 10 to 200. The velocity profiles for the channel are reported and Strouhal number of the Von Kármán vortex street behind the cylinder and the pressure drop is calculated for different Reynolds numbers.

In the present study, 2D laminar flow around a confined diamond-shaped obstacle has been investigated using a finite difference Navier-Stokes solver. Five different geometries have been studied in a range of different Reynolds numbers starting from fully viscous or creeping flow at Re = 1, subcritical Reynolds numbers, critical Reynolds number Re_{cr} and super critical Reynolds numbers up to the limit of 2D simulations, i.e. $Re \leq 200$.

Mesh dependency study has been carried out for a set of four different grid resolutions which includes minimum cell sizes of Δh = 0.05, 0.01, 0.005, 0.001 *mm*.

Validation of results is carried out using the present data for the problem of flow around rectangular cylinder in a range of subcritical Re and blockage coefficient of 1/8 [1].

Governing equations and the numerical procedure are discussed in section 2. In section 3, the computational domain, geometry configurations, boundary conditions along with mesh dependency and validation tests are presented. Section 4 includes the numerical results from flow simulation in both steady and unsteady regimes. Investigations on the critical Reynolds number and the St in the regime of unsteady flow are presented in sections 4.2 and 4.2.1, respectively. A discussion and conclusion on the obtained results are offered in section 5.

2. DESCRIPTION OF NUMERICAL METHOD

Numerical representation of fluid flow dynamic is sought through a set of governing formula known as Navier-Stokes equations. The laminar viscous flow in 2-dimensional framework is written in form of two momentum equations and a continuity equation which satisfies the conservation of mass in the computational domain and within the fluid. The non-dimensional governing equations are the momentum equations as in

$$\frac{\delta u_i}{\delta t} + \frac{\delta\left(u_i \cdot u_j\right)}{\delta x_j} = -\frac{\delta p}{\delta x_i} + \frac{1}{\mathrm{Re}}\left(\frac{\delta^2 u_i}{\delta x_i \cdot \delta x_j}\right) + g_i \qquad i = 1, 2 \tag{1}$$

and the continuity equation for the conservation of mass as in

$$\frac{\delta u_i}{\delta x_j} = 0 \tag{2}$$

in which u_i are the velocities, p is the pressure and g_i is the external forces acting on the fluid. In the non-dimensional form of equations as stated above, *Re* is the Reynolds number that can be calculated using the fluid properties, velocity and a characteristic length as follows:

$$\mathrm{Re} = \frac{D^* U}{\nu} \tag{3}$$

Here, ν is the dynamic viscosity of the fluid. In order to overcome the problem of oscillating pressure field, a staggered grid arrangement is used in which the pressure data is located at the cell centers and the u_1 and u_2 velocities are placed on the vertical and horizontal cell faces, respectively. This type of variable positioning has an infamous alternative called co-located grids [18] for which u_1, u_2 and p are all evaluated at cell centers and a finite volume-based discretization employs special interpolation schemes to determine the flux across the cell edges. It is well-known that a naïve co-located grid arrangement leads to a decoupled (checkerboard) pressure field and oscillations in solutions. Therefore, in the present study, the staggered grid arrangement is chosen for the positioning of the primitive variables. An equidistant orthogonal grid with the help of a flagging technique is used in order to distinguish obstacle cells from fluid cells in which the governing equations should be solved.

Using a finite difference discretization scheme, the momentum equation can be discretized in both space and time. An Euler explicit (forward differencing) temporal discretization along with a chorin's projection method, originally introduced by Chorin [19] is used to solve the time-dependant (transient) incompressible fluid-flow. Using this formulation, intermediate velocities u_i^* are calculated explicitly using the momentum equation while ignoring the pressure gradient in the right-hand-side of the equation.

$$\frac{u_i^* - u_i^n}{\Delta t} = -\frac{\delta(u_i^n \cdot u_j^n)}{\delta x} + \frac{1}{\text{Re}}\left(\frac{\delta^2 u_i^n}{\delta x_i \cdot \delta x_j}\right) + g_i^n \tag{4}$$

in which u^n is the velocity at the nth time level and the intermediate velocities, u^*, do not satisfy the continuity equation. By taking the divergence-free (continuity) condition, a Poisson's equation for p^{n+1} is derived as in

$$\frac{\delta^2 p^{n+1}}{\delta x_i^2} = \frac{1}{\Delta t} \times \frac{\delta}{\delta x_i} \cdot u_i^* \tag{5}$$

This equation is solved using the following boundary condition for the pressure at the $(n+1)^{th}$ time level.

$$\frac{p^{n+1}}{\delta n} = 0 \quad on\ \delta\Omega \tag{6}$$

in which Ω is the domain boundary. After solving the Poisson's equation, the projection step is performed in which the velocities at the next time level are calculated using the following equation.

$$u_i^{n+1} = u_i^* - \Delta t \frac{\delta p^{n+1}}{\delta x_i} \tag{7}$$

One of the main advantages of the Chorin's projection method is that the calculated velocity field is forced to satisfy the continuity equation in each time step.

In the present study, an in-house computer code is developed using the above mentioned methodology and algorithm. A schematic of the generated staggered grid and the location of the primitive variables are depicted in Figure 1.

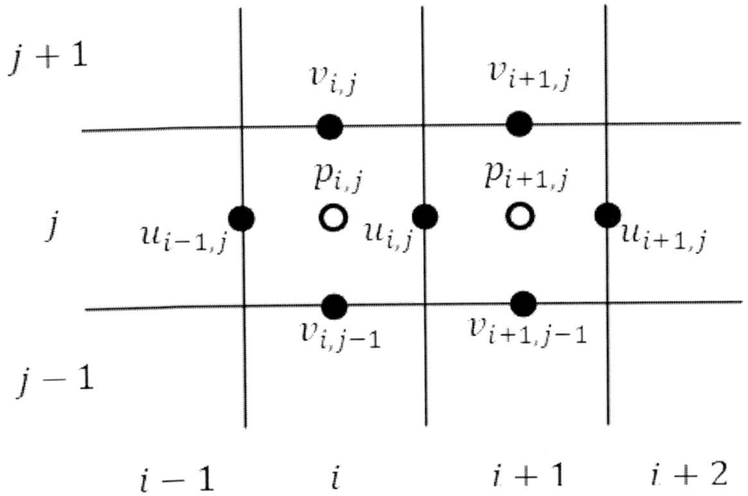

Figure 1. Staggered grid and the location of primitive variables inside the computational domain

Fluid and Obstacle cells are defined in the initialization step and the obstacle cells that serve as boundaries are flagged based on their location with respect to the adjacent fluid cells.

In the present study, study of the flow around an obstacle in a confined channel, is aimed. Therefore, an inlet/outlet configuration has to be implemented on the upstream and downstream boundaries, respectively. The inflow boundary condition at the upstream uses an explicitly given inlet velocity while the outflow boundary condition at the other end of the channel is set in a way that the normal derivatives of both velocity components will vanish. The latter boundary condition means that the total velocity will not change in the direction normal to the outflow boundary and therefore can act as a non-reflective exit boundary. On the other hand, the no-slip wall boundary condition will force the continuous velocities to vanish in order to satisfy the no-slip condition in a way that all the tangential velocities for the locations lying directly on the boundary of a non-moving wall will have zero value.

The computer code uses a successive over-relaxation (SOR) iterative method to solve the Poisson's equation for the next time level pressure values based on the intermediate velocities. The iterative loop will help the conservation of mass to be satisfied.

Although, the solution algorithm is easily implemented and the orthogonal equidistant grid leads to many advantages in storage and simplicity of computer code, the explicit time integration that was discussed above will impose some restrictions on time step size that has to be used for the numerical solution of the equations. This restriction is due to the stability reasons and based on the Courant-Friedrichs-Levi (CFL) criterion, in order to preserve a maximum courant number (C_N) of 1.0, the following restriction of

$$\delta t < \frac{\delta x_i}{|u_{i,\max}|} \tag{8}$$

has to be fulfilled which is based on the courant number definition as follows:

$$C_N = \max\left(\frac{u_{i,\max}\,\delta t}{\delta x_i}\right) \qquad i = 1, 2 \tag{9}$$

in which, $u_{i,\max}$ denotes the maximum velocity in x_i direction.

In the solution algorithm, first the initial and boundary conditions are set. Then, the intermediate velocities will be calculated using the time-explicit finite difference discretized momentum equations in the absence of pressure gradients. The intermediate velocities will then be used to solve the Poisson pressure equation by SOR method. Afterwards, the pressure gradients at the new time level will be used to calculate velocities at the n^{n+1} time level through a projection step. The algorithm will continue until a desired final time is reached.

3. DETAILS OF THE TEST CASE

3.1. Geometry of the Computational Domain

In this section, the computational domain and different geometries and configurations of the obstacle are presented. The diamond-shaped obstacle used for these studies has 5 different shapes based on the aspect ratio and the respective blockage ratio. In the standard form, hereafter referred to as ST, the obstacle has the shape of a rhombus or an equilateral quadrilateral with equal four sides. It can also be seen as a shifted rectangular cylinder which has been rotated 45 degrees. The diameter in ST shape is set in a way that satisfies the blockage ratio of 0.2. The obstacle is then stretched in both vertical and horizontal directions by multiplying each of the two diagonals by 1.5 and 2.0 in a way that 4 different geometries with different aspect ratios can be generated. Spanwise and streamwise stretched geometries are called H1, H2 and L1, L2, respectively. Different geometries and the computational domain are depicted in Figure 2.

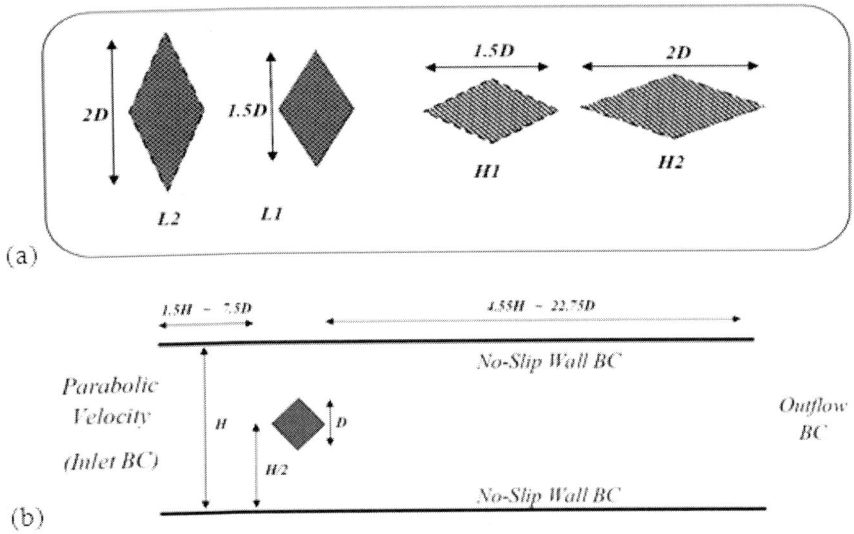

Figure 2. Geometries of different obstacles (a), and the computational domain (b)

It can be seen in Figure 2 that the diamond-shaped obstacle is placed in the middle of a confined channel with equal distances from both side-walls. An inlet boundary is placed at a distant 1.5H, upstream of the obstacle with H being the height of the channel. With the inradius (radius of the incircle) of the diamond obstacle being 0.2H, it can be inferred that the inlet boundary is far enough from the obstacle based on the respective hydraulic diameter (D) of the obstacle, with the distance being close to 8D. The outflow or exit boundary is also placed far away from the obstacle with the distance of 4.55H or almost 23D. Geometrical details of the 5 different obstacles are presented in Table 1. The aspect ratio is defined as the ratio of streamwise length to the spanwise length.

Table 1. Geometrical details of five different obstacles

Obstacle Type	Spanwise Diagonal	Streamwise Diagonal	Aspect Ratio (AR)	Blockage Ratio (BR)
ST	0.2H	0.2H	1.0	0.2
H1	0.3H	0.2H	0.66	0.3
H2	0.4H	0.2H	0.5	0.4
L1	0.2H	0.3H	1.5	0.2
L2	0.2H	0.4H	2.0	0.2

3.2. Boundary Conditions

General boundary conditions for the solution of the Navier-Stokes equations have been presented in section 2. Stokes's no-slip boundary condition is applied to the side-walls of the channel in which the tangential velocity on the nodes lying directly on the wall is set to zero or the relevant wall velocity (for the case of moving wall). Here, the fixed no-slip wall condition requires $u_{\tan g} = 0$ on the wall boundary nodes.

For the inlet boundary, a parabolic velocity profile is prescribed at the entrance of the channel in order to simulate a fully developed laminar flow inside the confined channel and upstream of the diamond cylinder. For the sake of simplicity in the calculation of the flow's Reynolds number, the maximum velocity of the parabolic profile is set to be $u_{\max} = 1.5\frac{m}{s}$ in order to achieve the similar uniform velocity of $u_{uniform} = 1.0\frac{m}{s}$.

The outlet boundary condition for the flow inside a channel with inflow/outflow boundaries at each end, needs careful attention in terms of numerical representation. However, due to the large computational domain that is set downstream of the diamond cylinder, the solution of the flow in vicinity of the obstacle will not be affected. Nevertheless, the zero gradient of the normal velocity at the exit boundary is forced inside the solution algorithm.

3.3. Validation and Mesh Dependency Study

The numerical procedure introduced and explained in the previous sections has to be validated in order to make certain about its accuracy and efficiency. For this purpose, flow around the rectangular cylinder is solved for a range of low Reynolds numbers for which a steady solution is possible.

No experimental results is available in the literature for 2-dimensional low Reynolds number flow around a square cylinder. Therefore, viscous laminar flow over rectangular cylinder is compared against the numerical results of Breuer et al. [1] using Finite Volume Method (FVM). The blockage ratio for the case of the square cylinder is set to be $B = 1/8$. The critical Reynolds number for the rectangular cylinder in which the vortex shedding will occur is around 60 as stated in different studies presented in the literature. As a result, flow around the rectangular cylinder is simulated in a range of low Reynolds numbers from 5 to a little less than 60 (i.e. 58) with the step of 5. The recirculation length as the length of the closed near-wake is measured for different test cases and the results are shown in Figure 3. The results are in very good agreements with the results of Breuer et al. [1] using a FVM method. The non-dimensional recirculation lengths based on the square hydraulic diameter (D) at different Reynolds numbers are plotted and a curve fit on the results give a linear relationship between the recirculation length and the Reynolds number which is very close to the results of previous studies.

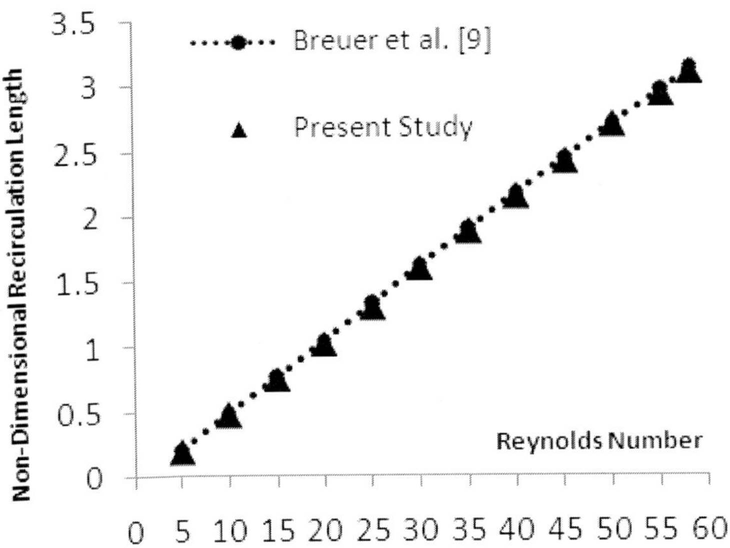

Figure 3. Non-dimensional recirculation length vs. Reynolds numbers (using finee grid [Grid 3])

In order to achieve mesh independent results, a mesh dependency study is done for the current case. This mesh convergence test is based on the diamond-shaped obstacle placed in the middle of the flow inside the confined channel. Four different grids are generated from the coarsest grid which consists of 40,000 cells to the finest grid with 2,560,000 cells. Different grids that were used for the mesh dependency test are listed in Table 2. This test is performed with respect to the validation test which was described earlier. The respective errors for the recirculation length in comparison with the results of Breuer et al. [1] are mentioned in the last column of Table 2.

Table 2. Grids used for the mesh convergence test

Grid No.	Grid Dimension	Total Number of Cells	Grid size	Error
Grid 1	500×80	40,000	0.05	1.23 %
Grid 2	1000×160	160,000	0.01	0.85 %
Grid 3	2000×320	640,000	0.005	0.23 %
Grid 4	4000×640	2,560,000	0.001	0.22 %

Table 2 shows that, by further grid refinement (Grid 4), no major improvement is achieved in the numerical results and therefore, the fine grid (Grid 3) with the grid size of 0.005m (5mm) will be sufficient for the accurate and efficient simulation of the flow around the diamond-shaped obstacle. Another result was that by using the coarsest grid, the calculated recirculation length is smaller than that of the medium, fine and finest grids.

4. RESULTS AND DISCUSSIONS

Flow around the diamond-shaped obstacle has been simulated at a Reynolds number range $1 \le Re < 200$ using the in-house finite difference Navier-Stokes solver code, 2D_NAVISTO. The laminar viscous flow is solved for different geometrical representations of a diamond-shaped obstacle. The Reynolds number is adjusted by altering the fluid's dynamic viscosity and keeping the obstacle's hydraulic diameter and the maximum value of the parabolic inlet velocity profile u_{max} fixed. In the following sections, first the steady flow for Reynolds numbers lower than the critical value is presented. An extensive study is done on a wide range of Reynolds numbers with small increments in order to find the critical Reynolds number. This study shows that the critical value of Re is varied from one geometry to another. It has been seen that the critical Reynolds number, Re_{cr} is increased with the increase of aspect ratio. This is in agreement with our initial perception that the critical Reynolds number would be lower for blunt bodies (with low aspect ratios) than for the streamlined bodies. It is known that, when the drag is dominated by viscous drag, we say the body is streamlined, and when it is dominated by pressure drag, we say the body is bluff. Therefore, flow around streamlined bodies will experience separation and unsteady boundary layer at higher Reynolds numbers. Critical Reynolds numbers for different geometries based on their aspect ratios are shown in Figure 4. As the Reynolds number increases from this critical value, vortex shedding will occur and flow will enter into an unsteady phase.

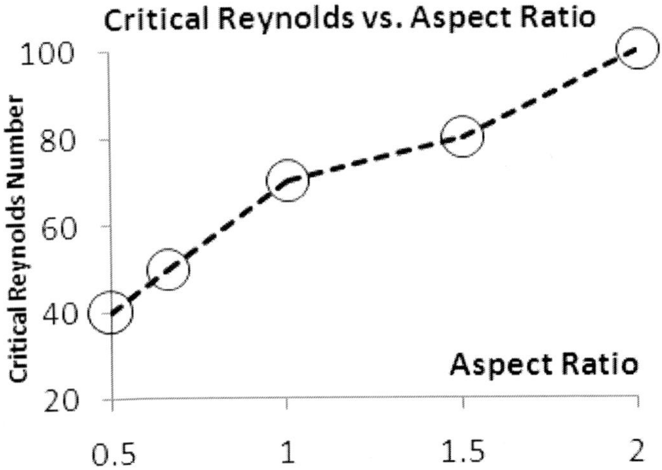

Figure 4. Critical Reynolds number vs. Aspect Ratio (AR) of the obstacle (for different geometries from left to right: H2, H1, ST, L1 and L2)

Discussion on the flow patterns for the unsteady regime is presented in section 4.2 and the flow characteristics are discussed in depth. Finally, the vortex shedding phenomena and a study of the Strouhal number are presented in section 4.2.1.

4.1. Steady Flow: $1 \leq Re < Re_{cr}$

The streamlines for the low Reynolds number flow around the diamond-shaped obstacle with different geometries are shown in Figure 5 through Figure 9. The highly viscous flow at low Reynolds number, $Re \leq 1$, is known as the creeping flow. This steady flow will persist without separation while the viscous forces are largely dominant. Here, flow around different geometries is shown for the creeping flow at *Re* = 1 and for a Reynolds number of about half of the critical value for that particular section.

As the Reynolds number increases, viscous forces will decrease until the laminar boundary layers will experience separation. In this region, a steady recirculation region will appear with two vortices that are placed symmetrically on each side of the wake behind the obstacle. Due to the sharp corners of this diamond-shaped obstacle, points of separation are fixed at these edges on top and bottom of the obstacle.

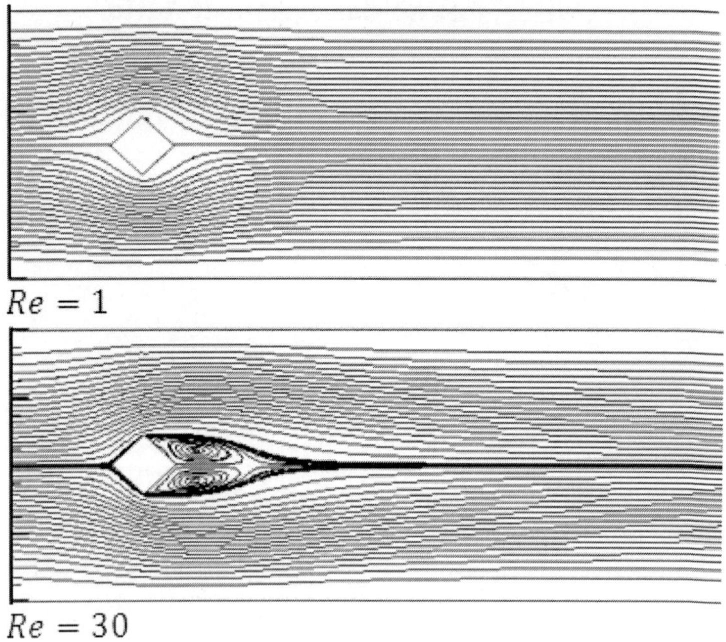

Figure 5. Streamlines for the creeping (up) and steady flow (down) around the diamond-shaped obstacle (ST)

Reynolds number for the steady solution of vortices behind the obstacle is set at a value about half of the critical limit. As a result, the streamlines are shown at different Reynolds numbers based on different critical values, $Re_{cr}$for five geometries.

Final Reynolds number at which the flow has been simulated is set to be *Re* = 200. The reason for this limit can be sought in the change of principal behaviors in the flow pattern with further increase of the Reynolds number from this particular limit.

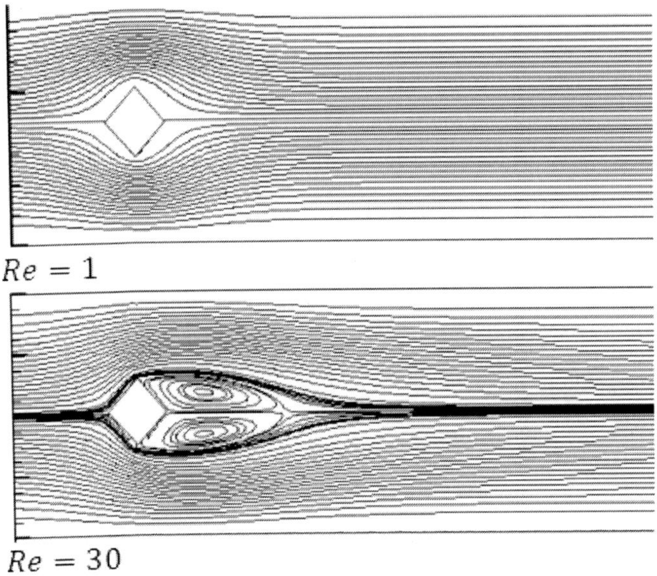

Figure 6. Streamlines for the creeping (up) and steady flow (down) around the diamond-shaped obstacle (H1)

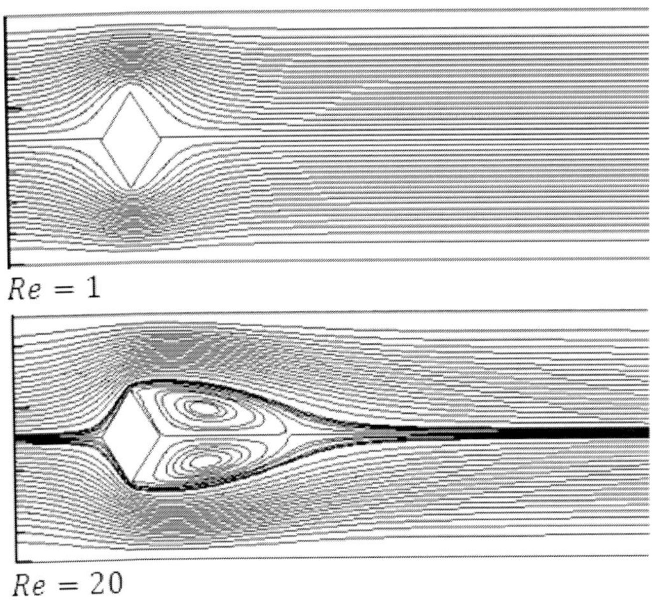

Figure 7. Streamlines for the creeping (up) and steady flow (down) around the diamond-shaped obstacle (H2)

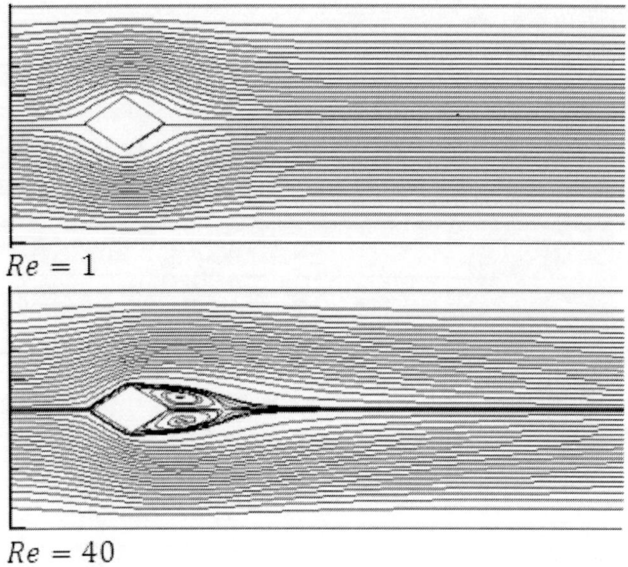

Figure 8. Streamlines for the creeping (up) and steady flow (down) around the diamond-shaped obstacle (L1)

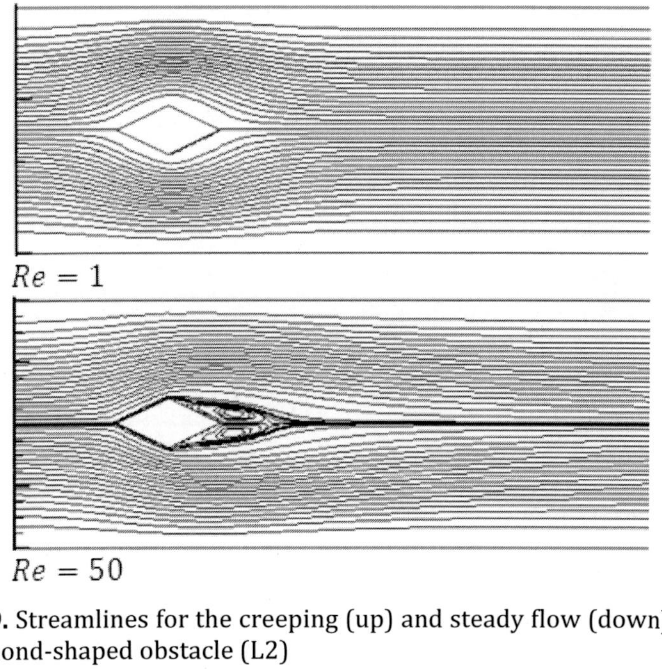

Figure 9. Streamlines for the creeping (up) and steady flow (down) around the diamond-shaped obstacle (L2)

4.2. Unsteady Flow: $Re_{cr} \leq Re < 200$

With the increase of Reynolds number from the critical limit, set of vortices elongate and become unstable. Therefore, at the end of this wake, transverse oscillations will appear and this will cause the generation of waves along the trail. Source of this instability has been studied for the case of circular cylinder and it was said to be originated in the Hopf bifurcation phenomenon as stated by Williamson [19].

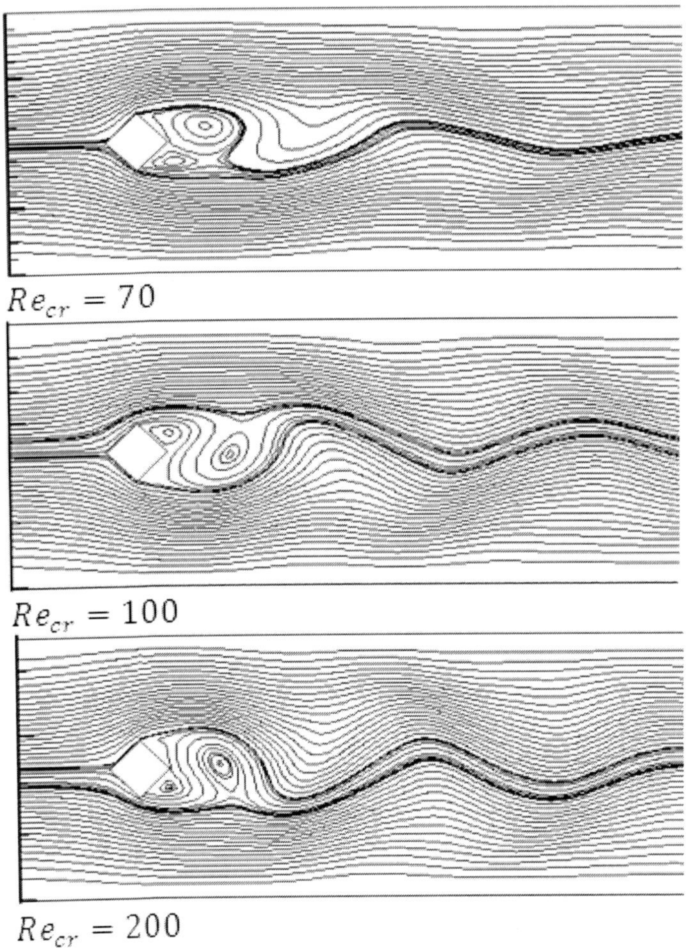

Figure 10. Streamlines for flow around the diamond-shaped obstacle (ST) at Re_{cr} and at $Re = 100, 200$

For the case of rectangular cylinder, less argument is available in the literature while a critical value of about $Re \approx 60 \pm 5$ has been determined by Klekar and Patankar [4] and also in Breuer et al. [1]. It has been found that this limit depends on flow parameters such as blockage ratio and therefore in the present study, different critical values have been found for various geometries that were examined. The well-known phenomenon of Von Kármán vortex street or in general vortex shedding will appear with further increase of *Re* when the formation of eddies will occur. Wavelength of vortex shedding will decrease as the Reynolds number rises. In Figure 10 through 14, the streamlines for the flow around diamond-shaped obstacle are presented at the critical Reynolds number for the particular geometry and at $Re = 100$, 200 for all different cases, except for L2 case at $Re = 150$, 200.

The frequency of vortex shedding will rise and therefore the laminar shedding will no longer remain 2-dimensional. Further studies in 3-dimensional case has proved this statement that the flow at Reynolds numbers above $Re = 200$ will experience 3-dimensional instabilities and vortices. Also, in 2-D simulations, deviations from fully periodic structures have been recorded which will restate the above theory. It should also be noted that Reynolds number based on the height of the inlet boundary has already reached the value of $Re = 1000$ and therefore further increase of Reynolds number would need special treatment for the turbulence flow parameters.

In the case of fully oscillating flow in 2-D simulations, vorticity contours will give a better representation of the Von Kármán vortex street. Therefore, these contour lines are presented in Figure 15 for all different geometrical cases at $Re = 200$.

4.2.1. Vortex Shedding and Strouhal Number

In the case of unsteady flow around the diamond-shaped obstacle, an important quantity has to be taken into account and that is the Strouhal number (*St*) which is computed based on the hydraulic diameter of the cylinder, the frequency of the vortex shedding f and the inlet velocity

$$St = \frac{fD}{u_{inlet}} \qquad (10)$$

In order to measure the frequency of vortex shedding f, a spectral analysis based on fast Fourier transformation, FFT, has been performed on the time series of oscillating lift forces and the characteristic value of f has been calculated. It was seen that the Strouhal number will increase with rising *Re* in the range of $Re_{cr} \leq Re \leq 200$ which is in good agreement with the literature. This rise in Strouhal number will have a maximum in a particular value which varies from one geometry to another based on different aspect ratios and Re_{cr} they have and will decrease again for higher Re until the limit of 2-D simulation, i.e. $Re = 200$.

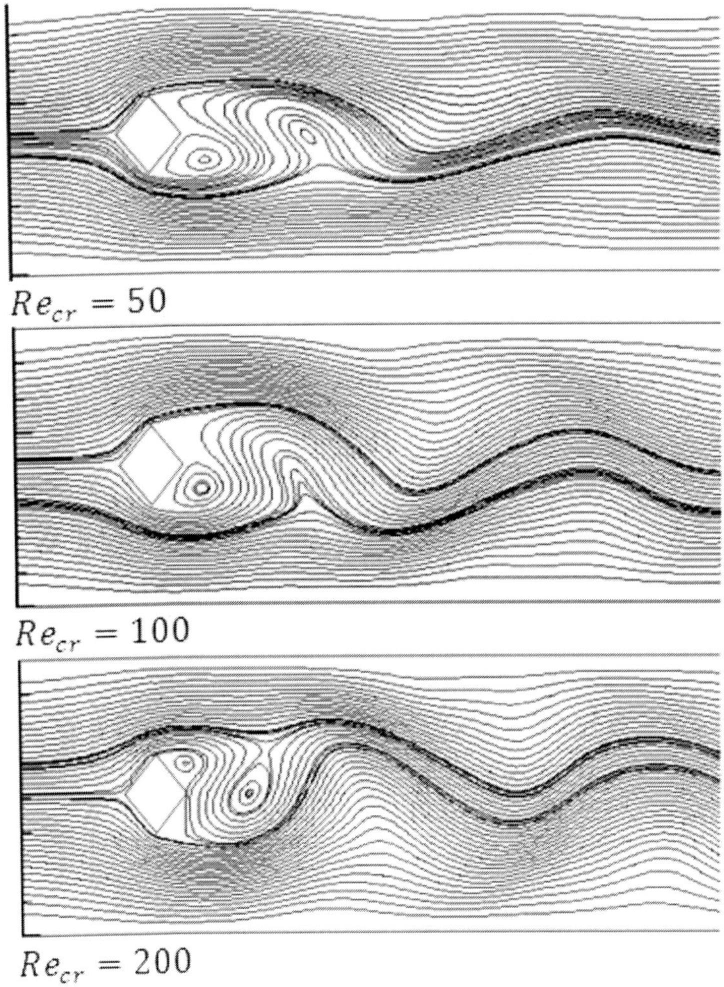

Figure 11. Streamlines for flow around the diamond-shaped obstacle (H1) at Re_{cr} and at $Re = 100, 200$.

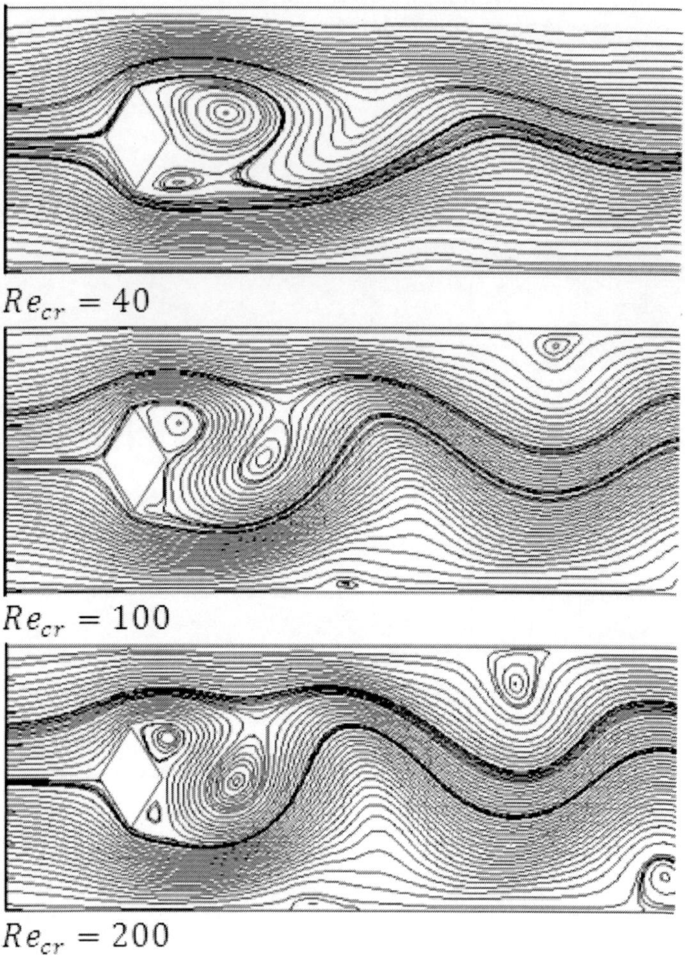

Figure 12. Streamlines for flow around the diamond-shaped obstacle (H2) at Re_{cr} and at Re = 100, 200

Strouhal numbers for unsteady flows in the range of $Re_{cr} \leq \mathrm{Re} \leq 200$ for different geometries are plotted in Figure 16. The behavior of this non-dimensional number based on the previous statements is in good agreement with the results of previous investigations. Although there is a lack of similar 2-D laminar studies in the case of flow around diamond-shaped obstacles, but it can be noticed that there would be a rise of Strouhal number with an increase in the blockage ratio as presented in the literature, [1, 5]. Also, a local maximum in the *St* has been reported by various works in the case of flow around rectangular cylinders.

The same behavior was recorded for the present case of flow around diamond-shaped obstacles but at different *Re* based on different blockage ratios for each case.

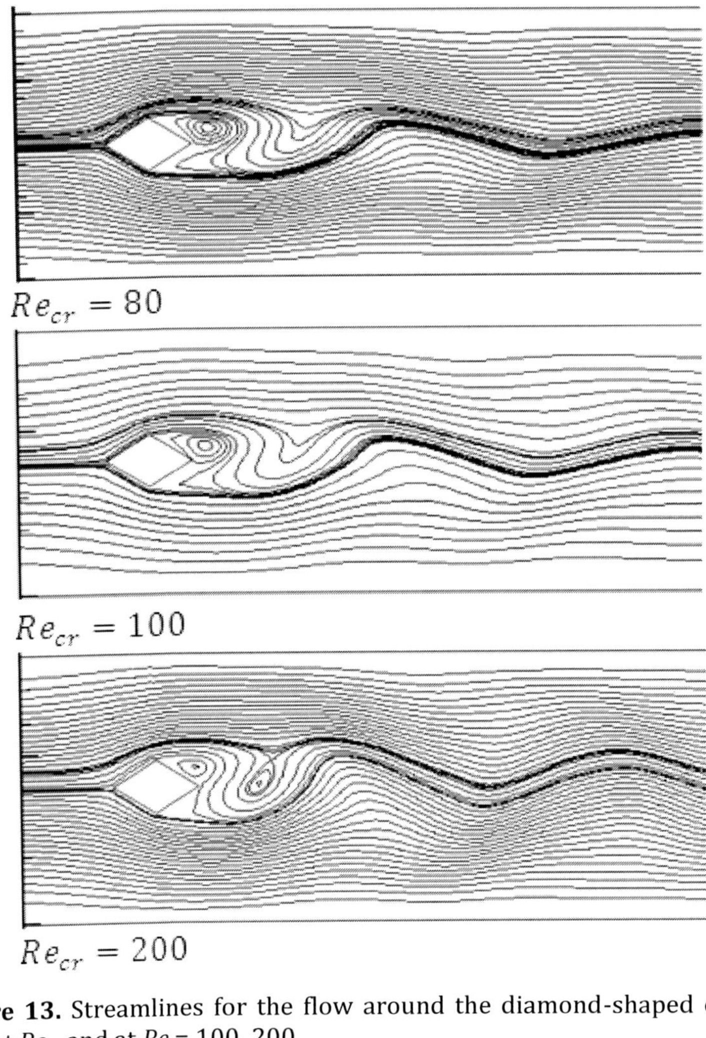

Figure 13. Streamlines for the flow around the diamond-shaped obstacle (L1) at Re_{cr} and at $Re = 100, 200$

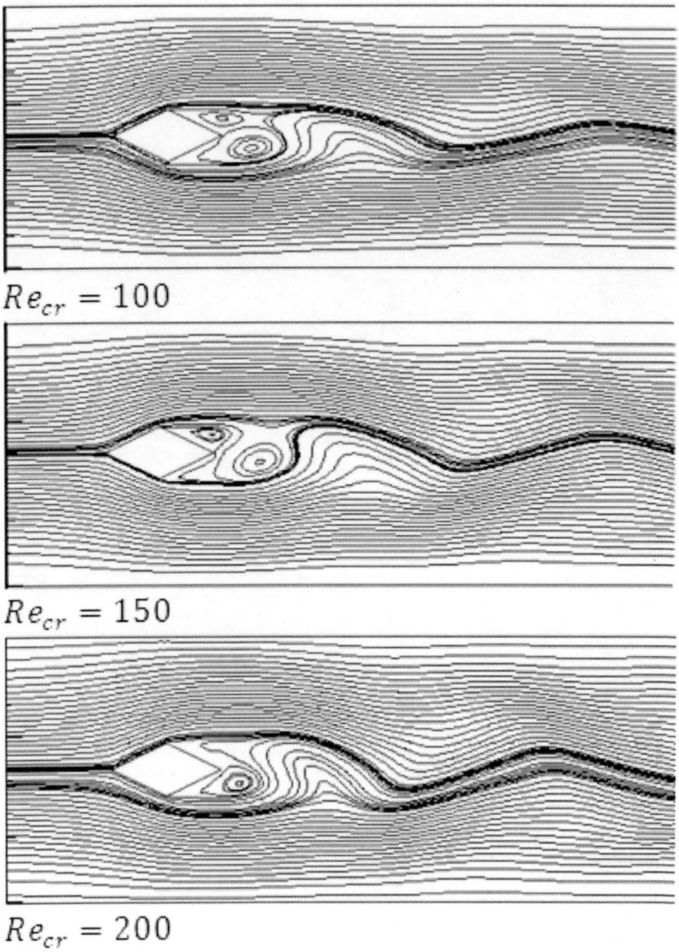

Figure 14. Streamlines for flow around the diamond-shaped obstacle (L2) at Re_{cr} and at Re = 150, 200

5. CONCLUSIONS

In the present study, the laminar 2-D flow around a diamond-shaped obstacle was investigated using a finite difference Navier-Stokes solver. Due to the lack of accurate and detailed data in the literature for the present case, the flow was simulated in a range of Reynolds numbers and with five different geometries creating a range of aspect ratios and blockage coefficients. A diamond-shaped obstacle was placed in a confined channel with sufficient distance from both upstream and downstream boundaries. A finite difference Navier-Stokes solver, 2D_NAVISTO, using staggered grid arrangement and Chorin's projection method was developed as an in-house computer code in order to simulate fluid flow over the obstacles in the confined channel. A mesh dependency test was carried out in order to reach a minimum grid resolution for the required accuracy of the solution.

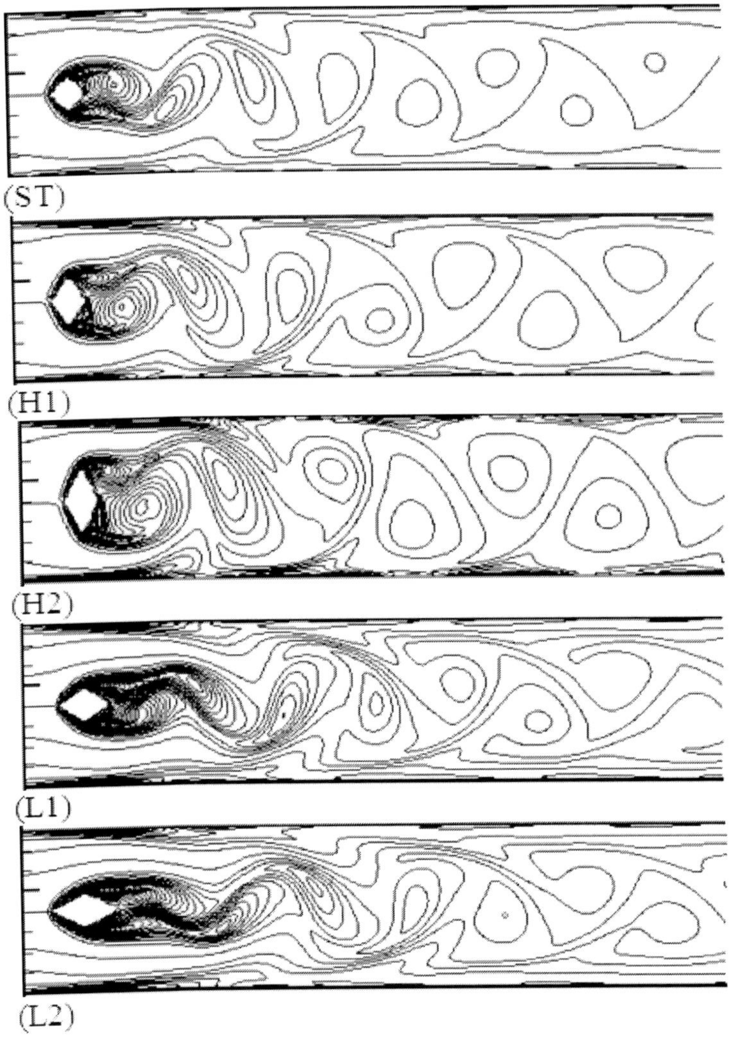

Figure 15. Vorticity contours for flow around the diamond-shaped obstacles at $Re = 200$

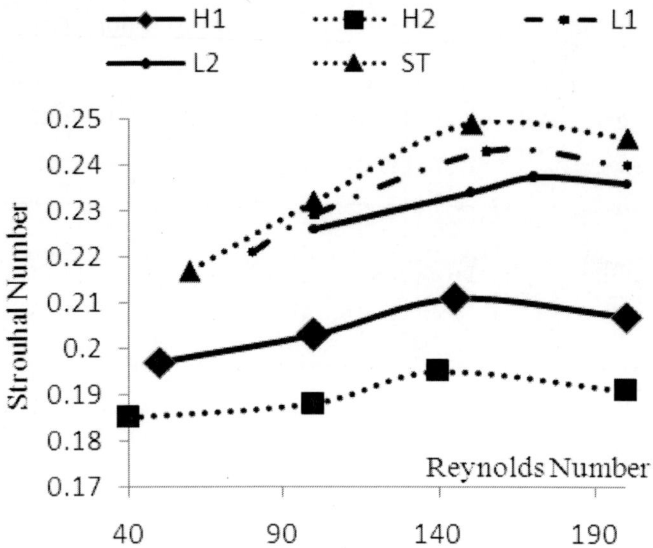

Figure 16. Computed Strouhal numbers vs. Reynolds number for different obstacles (ST, H1, H2, L1 and L2)

Based on the lack of detailed and similar data for the case of diamond-shaped obstacle, the numerical solver was validated using the results of flow over the rectangular cylinder which has almost similar characteristics based on its sharp corners.

The flow simulation was divided into two different groups. First the steady flow over the obstacles was investigated with the Reynolds numbers range from $1 \leq Re \leq Re_{cr}$. The fully viscous or the creeping flow at $Re = 1$ shows no sign of separation while the flow at higher Reynolds number and below the critical limit, presents two symmetrically placed vortices in the wake of the obstacle. Next, the unsteady flow is simulated with the range of $Re_{cr} \leq \mathrm{Re} \leq 200$. At the critical Reynolds number Re_{cr}, the vortices will become instable and at higher Re, the well-known Von Kármán vortex street will occur. The limit $Re = 200$ is suggested in the literature for the maximum value at which a 2-D simulation is accurate. Preliminary 3-D studies have shown that at higher *Re* the vortices will have oscillations in 3-dimension which makes it impossible to achieve accurate results in 2-D simulations.

Diamond-shaped obstacles are altered in order to obtain different aspect and blockage ratios which will have significant effect on the critical Reynolds number Re_{cr}. Numerical results prove that bluff bodies will have a lower critical value of Reynolds number than the streamlined bodies. Another study is presented based on the Strouhal number for each case at different Reynolds numbers. It has been found that there is an increase in *St* with the rise of *Re* until a maximum value that is found to be different for each geometry.

REFERENCES

1. Breuer M, Bernsdorf J, Zeiser T, Durst F. Accurate computations of the laminar flow past a square cylinder based on two different methods: lattice-Boltzmann and finite-volume. Int. J. Heat Fluid Flow, 21: 186-196, 2000.
2. Franke, R., Numerische Berechnung der instationnaren Wirbelablosung hinter zylindrischen Korpern. Ph.D. Thesis. University of Karlsruhe, 1991.
3. Okajima A. Strouhal numbers of rectangular cylinders. J Fluid Mech; 123: 379-98, 1982.
4. Klekar, K.M., Patankar, S.V. Numerical prediction of vortex shedding behind square cylinders. Int. J. Numer. Meth. Fluids., 14: 327-341, 1992.
5. Davis RW, Moore EF, Purtell LP. A numerical-experimental study of confined flow around rectangular cylinders. Phys. Fluids, 1984; 27(1): 46-59, 1984.
6. Mukhopadhyay A, Biswas G, Sundararajan T. Numerical investigation of confined wakes behind a square cylinder in a channel. Int. J. Numer. Meth. Fluids, 14: 1473-1484, 1992.
7. Suzuki H, Inoue Y, Nishimura T, Fukutani F, Suzuki K. Unsteady flow in a channel obstructed by a square cylinder rod (crisscross motion of vortex). Int. J. Heat Fluid Flow, 14(1): 2-9,1993.
8. Sohankar A, Norberg C, Davidson L. A numerical study of unsteady twodimensional flow around rectangular cylinders at incidence. Internal Report No. 96/25, Department of Thermo and Fluid Dynamics, Calmers University of Technology, Gothenburg, Sweden; 1996.
9. Nakamura Y, Ohya Y, Ozono S, Nakayama R. Experimental and numerical analysis of vortex shedding from elongated rectangular cylinders at low Reynolds numbers 200-103. J Wind Eng. Ind. Aerodyn. 65: 301-8, 1996.
10. Dutta SH., Panigrahi P.K. Muralidhar K, Experimental Investigation of Flow Past a Square Cylinder at an Angle of Incidence, J Engineering Mechanics, 134(9): 788-803, 2008.
11. Berrone S., Garbero V., Marro M. Numerical simulation of low-Reynolds number flows past rectangular cylinders based on adaptive finite element and finite volume methods. J. Computers & Fluids, 40: 92-112, 2011.
12. Bruno L, Fransos D, Coste N, Bosco A. 3D flow around a rectangular cylinder: a computational study. In: BBAA VI international colloquium on bluff bodies aerodynamics & applications, Milano, Italy; 20-24, 2008.
13. John V, Rang J. Adaptive time step control for the incompressible Navier- Stokes equations. Comput Methods Appl Mech Eng., 199: 514-24, 2010.
14. Xin liang Tian, Muk Chen Ong , Jianmin Yang , Dag Myrhaug. Unsteady RANS simulations of flow around rectangular cylinders with different aspect ratios. J Ocean Engineering 2012; 58 208-216.
15. Kogaki T., Kobayashi T., Taniguchi N., Large eddy simulation of flow around a rectangular cylinder. Fluid Dynamics Research 20(1-6): 11-24, 1997.
16. Regulski W., Szumbarski J., Numerical simulation of confined flows past obstacles -the comparative study of Lattice Boltzmannand Spectral Element Methods, J Arch. Mech., 64: 423-456, 2012.
17. Peric, M., Kessler, R. and Scheuerer, G. Comparison of finite-volume numerical methods with staggered and collocated grids, Computers & Fluids, 16(4): 389-403, 1988.
18. Chorin, A. J., Numerical Solution of the Navier-Stokes Equations, Math. Comp. 22: 745-762, 1968.
19. Williamson, C.H.K., Vortex dynamics in the cylinder wake. Annu. Rev. Fluid Mech. 28: 477-539, 1996.

CHAPTER 2

Time-Dependent Contaminant Transport in Ventilating Air from a Moving Source

George L. Danko, William Asante

University of Nevada, Reno, USA

ABSTRACT

A new method which combines the Eulerian, fixed control volume with a moving, Lagrangean flow channel is described for the solution of the conjugate, advection-diffusion problem for modeling transport processes of contaminant species. The transport model is presented as a conservative mass balance equation in a state-flux, species transport form in the space-time domain. A fully-implicit, general solution scheme is formulated with matrix operators in the space-time domain. The particular solutions for specific initial and boundary conditions and source term are constructed with the help of a single, inverse matrix operator, A^{-1}, which has to be calculated only once for all possible particular problems. Although A^{-1} involves a large number constants, all are independent from the initial, boundary, and source term input vectors. The multi-level, state-flux, space-time (SFST) scheme brings a significant computational acceleration since A^{-1} has to be calculated only once, such as in mine ventilation cases involving long drifts with constant air flow velocities. Such application is shown in an example for analyzing the transport and concentration distributions of diesel particulate matter (DPM) in the ventilation air at the working area with the interactions between ventilation and a moving diesel loading machine. Comparison between simulation and in situ DPM monitoring results suggests that reliable evaluation of average exposure of DPM to mine workers may be accomplished directly from tailpipe DPM emission data, ventilation air velocity, and mine geometry with the use of the SFST model even in a highly dynamic working area, potentially reducing the need for real-time DPM monitoring.

Keywords: DPM Concentration, Species Transport, Numerical Modeling

1. INTRODUCTION

The transport of contaminant species by the moving air is of great importance in the natural atmosphere as well as in the built environment. Concentration of harmful chemical components or dust particles negatively affects health and safety of humans as well as the biosphere and must be controlled. Specific interest is of concentration distribution spread in space and with time originating from contaminant sources. Analysis, numerical model simulation and measurement are often needed to understand the transport of contaminant species for checking the concentration against regulatory compliance.

The transport mechanism of contaminants is complex involving advection, diffusion, dispersion, convection and accumulation in the moving air. Further complexity is added if the contaminant source is moving relative to the flow of air. Such problems are encountered in transportation tunnels, congested cities, and underground mines. Fick's second law may serve as the fundamental component of the governing equations [1] [2] , but the formulation is based on the Eulerian, fixed control volume approach which requires fine spatial discretization. In the case of advection and diffusion, the Courant number, Cu, must be kept at Cu = 1 for accurate prediction from time-dependent simulation [3] [4] [5] . In case of moving source and independently moving air, the grid selection must be further constrained. The resulting solution in the form of a CFD (Com- putational Fluid Dynamics) model [6] [7] often require millions of spatial grid elements in sub-meter, or even micrometer size and of corresponding time steps in seconds, or down to even shorter temporal divisions.

A new method for modeling transport processes has been introduced that combines the Eulerian, fixed control volume with a moving, Lagrangean flow channel for a solution scheme for advection-diffusion problems [5] . The method promises a reduced grid sensitivity due to the intrinsic Cu = 1 condition built into the solution. The method, which can be used for the simulation of macroscopic flow, heat, and momentum transport, is briefly described here for contaminant transport studies. An example is provided for studying diesel emission variation in a ventilated underground working area with a moving loader machine. The numerical simulation results are compared with measurement data collected with stationary as well as moving sensors. The results are discussed and conclusions are drawn from the exercise.

2. EULERIAN BALANCE EQUATION WITH LAGRANGEAN INTERNAL TRANSPORT

The general balance equation for advection, diffusion, dispersion, convection and accumulation of species e in the moving air is described in [5] in a stationary Eulerian volume V, swept with a Lagrangean flow channel of advection volume V_a. The notations are shown in Figure 1 with a stagnant volume space as $V-V_a$, filled by eddies, but with no net advection transport. The advection flux density component, q_a flows as a Lagrangean wave front traveling at velocity v from A_{in}through A_{out}.

The transport balance equation for species e in the advection volume is [5] :

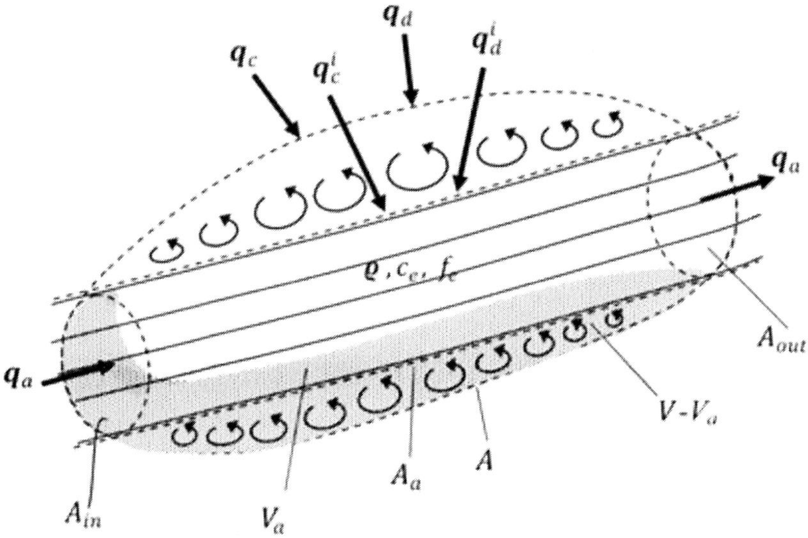

Figure 1. Control volume and surface for advection, convection, diffusion and accumulation (Reprinted from [5] with permission from Springer).

$$\int_{A_{in}} q_a\left[t-\Delta t(A)\right]\mathrm{d}A+\int_{A_{out}} q_a(t)\mathrm{d}A+\int_A q_{d+c}(t)\mathrm{d}A$$
$$+\int_{V-V_a}\frac{\partial\rho_e(t)}{\partial t}\mathrm{d}V=\int_V \rho f_e\mathrm{d}V \tag{1}$$

In (1), q_a is advective mass flux density of species e; $\Delta t(A)$ is the average travel time of the advective fluidoriginating from A_{in}; q_{d+c} is the sum of diffusion, dispersion and convection of e across area, A; ρ_e and ρ are the densities of e and the bulk flow, respectively, and f_e is the source term for species e.

Integral Equation (1) may be written in a finite difference form using finite volume and surface elements. With the introduction of the mass fraction variable, $\omega_e=\rho_e/\rho$ for species e, a State-Flux (SF) model is formulated in which the state, ω_e is the driving force for the transported mass flux density of species e [5] . The SF model is derived directly from (1), substituting volume $V=\Delta x\Delta y\Delta z$; stagnant volume $V_s=S\Delta x\Delta y\Delta z$; and advection channel cross section $A_a=(1-S)\Delta y\Delta z$ surface. Note that the advection travel time, the spatial division, and flow velocity must obey the relationship $\Delta t=\Delta x/v$, that is,

$$Cu=\frac{v\Delta t}{\Delta x}=1,$$

for the validity of (1), where Cu is the Courant number [5] .

The sum of the first and second flux terms in (1) gives the advective flux driven by mass fraction difference, $\omega_{ei}^{n}-\omega_{ei-1}^{n-1}$, where i and i + 1 denotes the input and the output points, respectively [5] :

$$\int_{A_{in}} q_a\left[t-\Delta t(A)\right]\mathrm{d}A+\int_{A_{out}} q_a(t)\mathrm{d}A=(1-S)\rho v\Delta y\Delta z\left(\omega_{ei}^{n}-\omega_{ei-1}^{n-1}\right) \tag{2}$$

The third term in (1) gives the difference of the flux by diffusion and convection (if applicable) driven by mass fraction differences,

$$\left(\omega_{ei+1}^{n}-\omega_{ei}^{n}\right) \text{ and } \left(\omega_{ei}^{n}-\omega_{ei-1}^{n}\right):$$

$$\int_A q_{d+c}(t)\,\mathrm{d}A\, iR^D = -\rho D_e \Delta y \Delta z/\Delta x\left[\left(\omega_{ei+1}^n - \omega_{ei}^n\right) - \left(\omega_{ei}^n - \omega_{ei-1}^n\right)\right] \tag{3}$$

The fourth term in (1) is the accumulation of the substance flux, proportional to the mass fraction change $\left(\omega_{ei}^n - \omega_{ei}^{n-1}\right)$ over the Δt time period:

$$\int_{V-V_a} \frac{\partial \rho_e(t)}{\partial t}\mathrm{d}V = S\rho\Delta x\Delta y\Delta z/\Delta t\left(\omega_{ei}^n - \omega_{ei}^{n-1}\right) \tag{4}$$

Note that the accumulation term is reduced to the stagnant volume only, $V - V_a$

The right side of (1) expresses the source term of species e in control volume V:

$$\int_V \rho f_e \mathrm{d}V = \rho f_{e_i} \Delta x \Delta y \Delta z \tag{5}$$

Choosing a unit advective admittance, $iR^A = (1-S)\rho v \Delta y \Delta z = 1$, the three transport admittances and the source term in (2)-(5) can be reduced to two, non-dimensional parameters, iR^D and iR^S, and a cell source term, Fc_i, following [5] :

$$\left.\begin{aligned}
iR^A &= (1-S)\rho v \Delta y \Delta z = 1\\
iR^D &= \frac{D_e}{v\Delta x(1-S)} = \frac{1}{Re(1-S)Sc}\\
iR^S &= S/(1-S)\\
Fc_i &= F_i/iRA = f_{e_i}\Delta t/(1-S)
\end{aligned}\right\} \tag{6}$$

The reciprocal of the $D_e/(v\Delta x)$ term in Equations (6) is recognized as the multiple of the Reynolds, Re, and Schmidt, Sc, numbers, two basic, non-dimen- sional parameters of transport processes [1] :

$$\left.\begin{aligned}
Re &= v\Delta x/\nu\\
Sc &= \nu/D_e\\
ReSc &= (v\Delta x/\nu)(\nu/D_e) = v\Delta x/D_e
\end{aligned}\right\} \tag{7}$$

The normalized, finite difference form of (2) with the use of (6) and (7) constitutes the SF network equation for a network branch between nodes i – 1 and i, however, with the connection to node i + 1 also due to diffusion, dispersion, and convection [5] :

$$\left(\omega_{ei}^{n}-\omega_{ei-1}^{n-1}\right)-iR^{D}\left[\left(\omega_{ei+1}^{n}-\omega_{ei}^{n}\right)-\left(\omega_{ei}^{n}-\omega_{ei-1}^{n}\right)\right]+iR^{S}\left(\omega_{ei}^{n}-\omega_{ei}^{n-1}\right)=Fc_{i} \quad (8)$$

Note that the condition of validity of (8) is the unity of the Courant number, $Cu = v\Delta t/\Delta x = 1$.

Assuming a homogeneous flow and transport field with constant material properties and transport coefficients, the SF network Equation (8) can be applied to a series of finite volume cells connected together. A fully implicit in space, time-marching in time solution scheme may be used for solving a set of network equations [5] .

The time-marching step must be selected as $\Delta t = \Delta x/v$. A multiple time step, fully implicit, State-flux, Space-Time (SFST) solution scheme can also be constructed offering a closed-form, operator solution for the conjugate advection, convection and diffusion-dispersion problem [5] .

3. A FULLY IMPLICIT, SFST NUMERICAL SOLUTION

A fully-implicit transport network solution is given in [5] using Equation (8) at all the nodes and Equations (6) everywhere at the branches of a high-density internal grid. The transport network for such a multiple-level implicit scheme with inter-connected spatial and temporal grids is shown in Figure 2.

The boundary nodes with known driving potentials at $n = 1, \cdots, M$ time steps are separated from the active nodes of unknown potentials along $i = 1, \cdots, N$ spatial divisions.

The advective transport connections between consecutive time levels are separated by potential followers, in order to provide the correct initial state values from time n – 1 to time n, affecting the substance flux density balance only at time n, but not at time n – 1. The initial potentials are established from time-step to time-step with no feedback flux effect from future time. The size of the network model to be solved simultaneously is (N ´ M)².

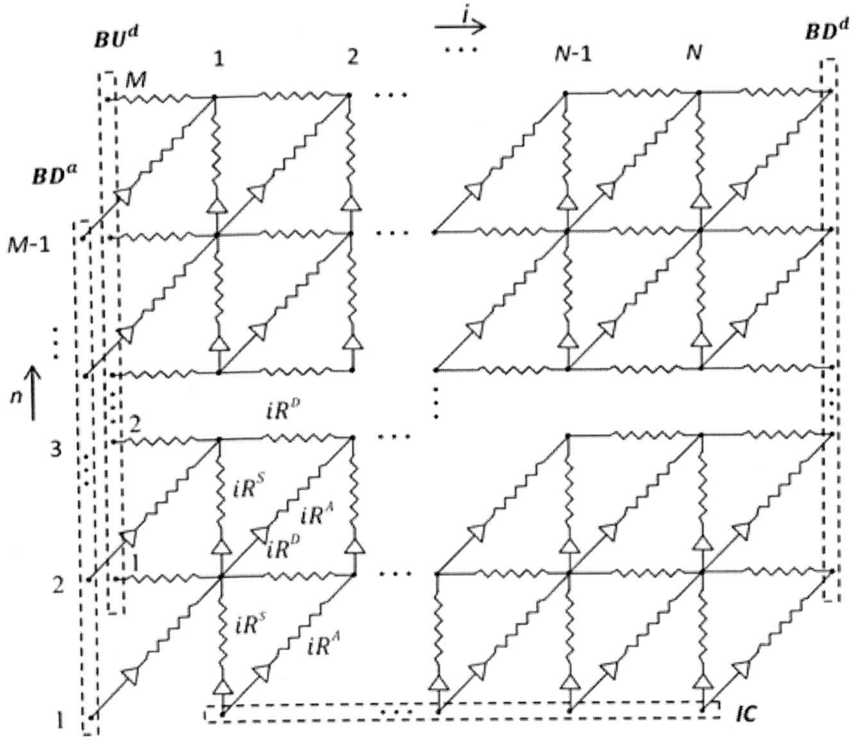

Figure 2. Substance transport network with explicit spatial and temporal grids (v > 0). (Reprinted from [5] with permission from Springer).

It is shown that the solution can be expressed in a matrix-vector form with a five-diagonal admittance matrix, A, [5] . Out of the five diagonals, there is a triple-diagonal strip matrix symmetric around the main diagonal, stretching to the size of (N ′ M)^2. In addition, there are two off-diagonal lines to include transport connections from the previous time interval. One off-diagonal line models the advection connections for each time step with the time-shifted potentials in the new A matrix, with $A(i,i-N-1)=-1, i=N+2,\cdots,N\times M$. The other off-diagonal line includes the accumulation connections for the stagnant volume, $A(i,i-N)=-iRS, i=N+1,\cdots,N\times M$ for each internal time step. The five-diagonal matrix, $A(i,j), A\in\mathcal{R}^2, i=1,\cdots,N\times M, j=1,\cdots,N\times M$ for a positive, left-to-right, v > 0 velocity is [5] :

$$A(i,j)=\begin{cases}-iR^D \text{ if } j=i+1, i=1,\cdots,N\times M-1\\ 1+2iR^D+iR^S \text{ if } j=i, i=1,\cdots,N\times M\\ -iR^D \text{ if } j=i-1, i=2,\cdots,N\times M\\ -1 \text{ if } j=i-N-1, i=N+2,\cdots,N\times M\\ -iR^S \text{ if } j=i-N, i=N+1,\cdots,N\times M\\ 0 \quad \text{otherwise}\end{cases} \tag{9}$$

For a right-to-left, $v_a < 0$ velocity, the advection connections are transposed:

$$A(i,j)=\begin{cases}-iR^D \text{ if } j=i+1, i=1,\cdots,N\times M-1\\ 1+2iR^D+iR^S \text{ if } j=i, i=1,\cdots,N\times M\\ -iR^D \text{ if } j=i-1, i=2,\cdots,N\times M\\ -1 \text{ if } j=i-N-1, i=N+2,\cdots,N\times M\\ -iR^S \text{ if } j=i-N, i=N+1,\cdots,N\times M\\ 0 \quad \text{otherwise}\end{cases} \tag{10}$$

Matrix A may be viewed as a composite array of M ´ M sub-matrices, $a_{i,j}$, i.e., $A(i,j)=\{a_{i,j}\}, i=1,\cdots,M, j=1,\cdots,M$:

$$A(i,j)=\begin{bmatrix} a_{1,1} & \cdots & a_{1,M}\\ \vdots & \ddots & \vdots\\ a_{M,1} & \cdots & a_{M,M}\end{bmatrix} \tag{11}$$

In (11), the diagonal sub-matrices, $a_{i,j}(k,l)$ are N × N in size, $a_{i,i}(k,l)\in\mathcal{R}^2, i=1,\cdots,M, k=1,\cdots,N, l=1,\cdots,N$. The $a_{i,i}$ elements constitute triple-diagonal sub-matrices:

$$a_{i,i}(k,l)=\begin{cases}-iR^D & \text{if } l=k+1, k=1,\cdots,N-1\\ 1+2R^D+iR^S & \text{if } l=k, k=1,\cdots,N\\ -R^D & \text{if } l=k-1, k=2,\cdots N\\ 0 & \text{otherwise}\end{cases} \tag{12}$$

The off-diagonal sub-matrices, $a_{i,j}(k,l)$ are mainly zeros. For left-to-right, v > 0 velocity, the non-zero off-diagonal sub-matrices for $i=2,\cdots,M$ are:

$$a_{i,i-1}(k,l)=\begin{cases}-iR^{S} & \text{if } l=k,\ k=2,\cdots,N\\ -1 & \text{if } l=k-1,\ k=2,\cdots,N\\ 0 & \text{otherwise}\end{cases} \quad (13)$$

The advection connections are transposed for the right-to-left, v < 0 velocity in the off-diagonal elements, $a_{i,j}(k,l)$ as follows for $j=2,\cdots,M$:

$$a_{j-1,j}(k,l)=\begin{cases}-iR^{S} & \text{if } l=k,\ k=2,\cdots,N\\ -1 & \text{if } k=l-1,\ l=2,\cdots,N\\ 0 & \text{otherwise}\end{cases} \quad (14)$$

For all other sub-matrixes, not defined by (12) through (14), are null matrices, $a_{i,j}$ = {0}. Consequently, A(i,j) is dominantly a sparse matrix.

Although the structure of the matrix operators are different for v > 0 from v < 0 velocity directions, the notations are simplified and applied for left-to-right, v > 0 velocity, shown in Figure 2.

For simplicity, sub-matrix notations are used for the upward and downward boundary condition vectors, BU^d, BU^a and BD^d, as well as the initial condition vector, IC. The boundary and initial condition sub-vectors in the matrix-vector balance equation have to be all in the length of N × M, filling them with zeros where no connections are defined in the network of Figure 2, and keeping their non-zero value at the active connections. Using the notation of $bu_i^d(j)=BU^d(n)$, $bu_i^a(j)=BU^a(n)$, $bd_i^d(j)=BD^d(n)$ and $ic_i(j)=IC(j)$, respectively, the boundary vector elements are:

$$bu_i^d(j)=\begin{cases}BU^d(i) & \text{if } j=i, i=1,\cdots,M, j=1,\cdots,N\\ 0 & \text{otherwise}\end{cases} \quad (15)$$

$$bd_i^d(j)=\begin{cases}BD^d(i) & \text{if } j-N+1=i, i=1,\cdots,M, j=1,\cdots,N\\ 0 & \text{otherwise}\end{cases} \quad (16)$$

$$bu_i^a(j)=\begin{cases}BU^a(i) & \text{if } j=i, i=1,\cdots,M, j=1,\cdots,N\\ 0 & \text{otherwise}\end{cases} \quad (17)$$

$$ic_i(j)=\begin{cases} IC(j) & \text{if } i=1, j=1,\cdots,N \\ 0 & \text{otherwise} \end{cases} \tag{18}$$

The unknown mass fraction vector, ω_e, is also used in sub-vectors form of $\omega_i=\omega_e[(i-1)M+j]$:

$$\omega_i(j)=\omega_e[(i-1)M+j], i=1,\cdots,M, j=1,\cdots,N \tag{19}$$

Distributed substance source for each node in Figure 2 may be included, varying with space and time. The source term in sub-vector form is:

$$F_i(j)=Fc_i[(i-1)M+j], i=1,\cdots,M, j=1,\cdots,N \tag{20}$$

The balance equation of the SF network of Figure 2 is now written in sub-matrix notation:

$$\begin{bmatrix} a_{1,1} & \cdots & a_{1,M} \\ \vdots & \ddots & \vdots \\ a_{M,1} & \cdots & a_{M,M} \end{bmatrix} \begin{bmatrix} \omega_1 \\ \vdots \\ \omega_M \end{bmatrix} = \begin{bmatrix} ic_1 \\ \vdots \\ 0 \end{bmatrix} + iR^D \begin{bmatrix} bu_1^d \\ \vdots \\ bu_M^d \end{bmatrix} + \begin{bmatrix} bu_1^a \\ \vdots \\ bu_M^a \end{bmatrix} + iR^D \begin{bmatrix} bd_1^d \\ \vdots \\ bd_M^d \end{bmatrix} + \begin{bmatrix} F_1 \\ \vdots \\ F_M \end{bmatrix} \tag{21}$$

The simultaneous solution for the entire mass fraction field with space and time is:

$$\begin{bmatrix} \omega_1 \\ \vdots \\ \omega_M \end{bmatrix} = A^{-1} \begin{bmatrix} ic_1 \\ \vdots \\ 0 \end{bmatrix} + A^{-1} iR^D \begin{bmatrix} bu_1^d \\ \vdots \\ bu_M^d \end{bmatrix} + A^{-1} \begin{bmatrix} bu_1^a \\ \vdots \\ bu_M^a \end{bmatrix} + A^{-1} iR^D \begin{bmatrix} bd_1^d \\ \vdots \\ bd_M^d \end{bmatrix} + A^{-1} \begin{bmatrix} F_1 \\ \vdots \\ F_M \end{bmatrix} \tag{22}$$

All sub-vectors except for the distributed source term vector are substantially sparse in (22). It is possible to eliminate the zero elements from the terms on the left side of (22) and to return to the full initial and boundary condition vectors [5] . The source term is also reduced to an M-element F^svector by either accepting the average of the nodal sources along each $i=1,\cdots,N$ line, or sampling the values of the $F_i(j)$ vector along a desired trajectory, e.g., along a moving source, in the x-t space within the cell domain. With these simplifications by algebra, a new matrix-vector equation is obtained with matrices of MN rows and M columns, with still full expression for all MN mass fraction values, but in need of M-vectors only on the right side [5] :

$$\begin{bmatrix} \omega_1 \\ \vdots \\ \omega_M \end{bmatrix} = iA^{IC}IC + iA^{BUd}BU^d + iA^{BUa}BU^a + iA^{BDd}BD^d + iA^F F^s \tag{23}$$

In (23), five different coefficient matrices emerged with the definitions as follow:

$$\left.\begin{array}{ll} iA^{IC}(i,j) = A^{-1}(i,j) & \text{for } i = 1,\cdots, M\cdot N,\ j = 1,\cdots N \\ iA^{BUd}(i,j) = A^{-1}(i,j)iR^D & \text{for } i = 1,\cdots, M\cdot N,\ j = 1, N+1, 2N+1,\cdots,(M-1)N+1 \\ iA^{BUa}(i,j) = A^{-1}(i,j) & \text{for } i = 1,\cdots, M\cdot N,\ j = 1, N+1, 2N+1,\cdots,(M-1)N+1 \\ iA^{BDd}(i,j) = A^{-1}(i,j)iR^D & \text{for } i = 1,\cdots, M\cdot N,\ j = N, 2N,\cdots, M\cdot N \\ iA^F(i,j) = A^{-1}(i,j)A^s(i,k) & \text{for } i = 1,\cdots, M\cdot N,\ k = 1,\cdots, M;\ A^s: \text{samples of } F^s \end{array}\right\} \tag{24}$$

The coefficient matrices in (23) and (24) all have $M \cdot N$ rows, but the number of columns is N in iA^{IC} and M in all the other four matrices. Therefore, the full matrices in (23) are different from each other; and all have much smaller size that A^{-1} in (22).

Note that the last row in each matrix in Equation (23) is not sparse and may involve a large number constants, carrying all of information necessary for expressing the solution the conjugate advective and diffusive transport in the ST domain with matrix operators, all independent from the IC, BU, and F^S input vectors. At the same time, the multi-level scheme brings a significant computational acceleration if A^{-1} has to be calculated only once, such as in mine ventilation for each long drift with a constant, advection velocity. Such application is shown in an example as follows.

4. APPLICATION EXAMPLE OF THE SFST MODEL WITH MOVING SOURCE TERM

Based on an underground mine experiment [9] , an example of the method is described for contaminant concentration distribution in the ventilating air in a mine drift. A mine haulage route for a diesel load-haul-dump (LHD) machine in the experiment is shown in Figure 3. Diesel particulate matter (DPM), a contaminant is modeled as it is transported in the ventilating air by advection and dispersion, emitted from the tail pipe as a moving source of the engine. The machine travels back and forth along the specified route, continuously repeating a loading cycle.

In order to capture the tailpipe DPM contaminants from the exhaust fumes directly, two sampling sensors were placed 2.1 m away from the tailpipe. Three pairs of stationary sensors were placed, two at each location at three different points on the wall, marked as m_1, m_2, and m_3 at 3.1 m, 41 m, and 78 m from the loading area to measure the DPM concentration in the ventilating air.

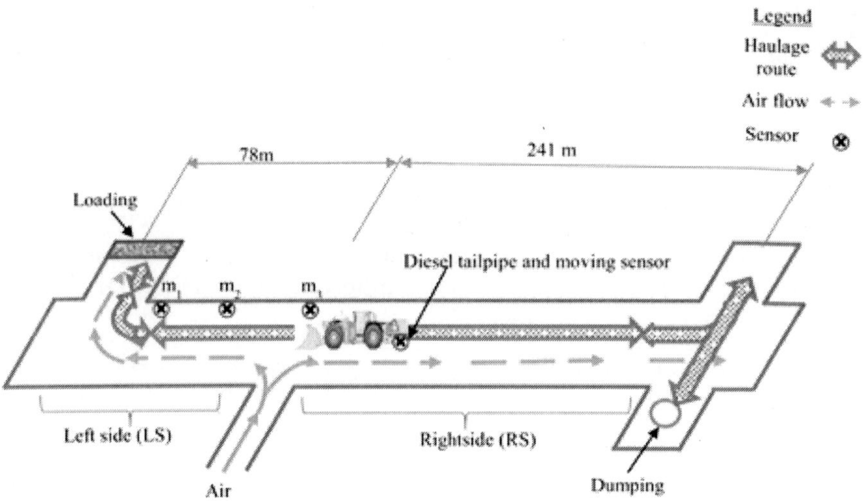

Figure 3. Plan view of mine drift.

The ventilating air enters the mine drift at the mid-section and splits into two directions. The left side (LS) drift segment, 78 m long, leads to the loading zone. The right side (RS) segment, 241 m long, stretches to the dumpling zone. The two drift segments require two separate DPM transport models. The connection between the two model sections is the loading machine crossing from one to the other in a cyclic manner. The diesel exhaust source of the loading machine moves with a travel velocity of $v_T = 2.18\ \text{m/s}$ in either the LS or RS drift section. The airflow velocity of $v_A = 1.3\ \text{m/s}$ is also the same in both the LS and the RS segments.

Figure 4 depicts the cycle time diagram for the loading machine and the airflow directions. The moving sensor measuring concentration is placed

2.1 m behind the moving point source and modeled independently in the two separate sections. The point where the air flow splits into two opposite directions to the left and right sections is used as a reference point for background concentration. The DPM concentration of 90 μg/m³, measured as an incoming value at the intake air from the other areas of the mine, is used in the example as a constant additive to the zero concentration level in the numerical model. The concentration from the moving sensors on the loading machine is modeled by sampling the spatial concentration from the SFST solution at an offset of 2.1 m distance from the exhaust point of the tailpipe.

The airflow directions and the spread of the tailpipe exhaust fumes are illustrated in Figure 5 as the machine travels in the haulage drift in and out of the LS and the RS drift segments. The efficiency of catching the DPM from the exhaust plumes by the moving sensor depends on the direction of the movement of the machine as well as the air flow direction relative to the machine.

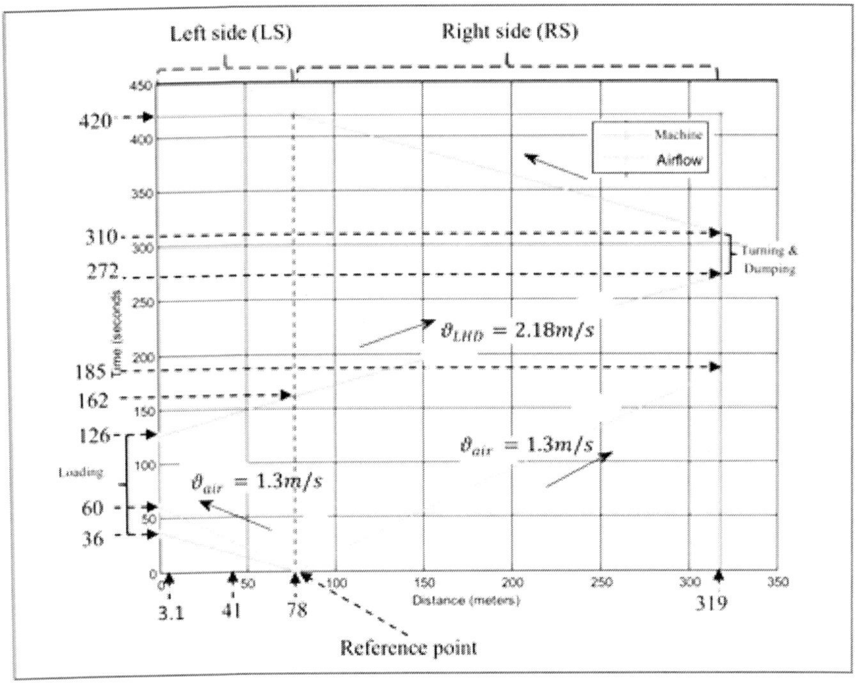

Figure 4. Cycle time diagram for the loading machine and the airflow directions.

A fine spacial mesh of 1.3 m and 1 second of temporal discretization is used in both the LS and RS for the numerical models, satisfying the condition for the Courant number of Cu = 1.3/1.3 = 1. The model is setup using the geometry of the mine experiment. Accordingly, the N and M divisions in the spatial and temporal meshes are N = 60, M = 162 for the LS and N = 186, M = 258 for the RS model segments.

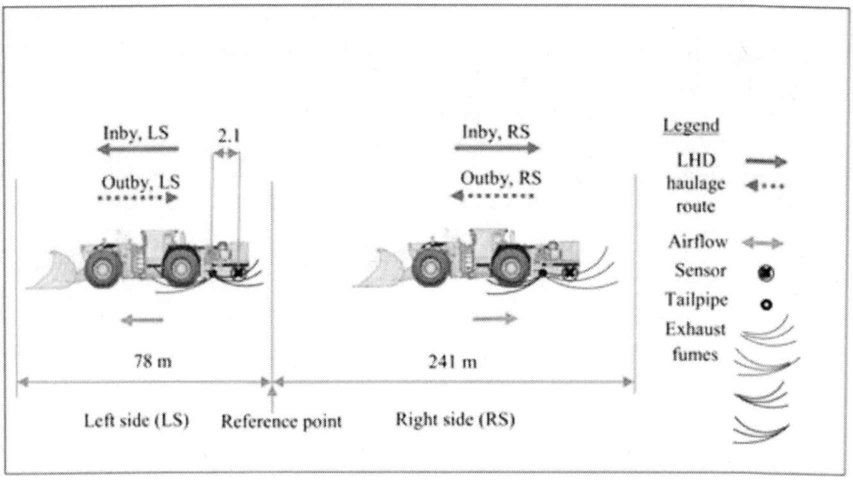

Figure 5. Airflow directions and spread of tailpipe exhaust fumes as a result of the LHD machine movment.

Figure 6 and Figure 7 show the machine velocity versus distance diagram per cycle and machine velocity versus time diagram, respectively. The source term is determined from measured diesel fuel use for the haulage and the tailpie emis- sion data [9] . The DPM concentration source term entering the air flow from the tailpipein the model is 1000 µg/kg/s, determined based on LHD machine data. Note that the moving point source is a cumulative transport term that cannot be converted directly into a mass fraction (or volumetric concentration) in the moving air stream without considering the advetive, dispersive, and accumulation components. Assuming no stagnant volume in the drift cross section (S = 0), no dispersion, (D = 0), and near-zero advective air flow velocity, (such as that $\lim_{v \to 0}\left(\frac{D}{v}\right)=0$), the source term would continuously increase the contaminant mass fraction according to $\omega_i^n - \omega_{i-1}^{n-1} = \Delta\omega = F_c\Delta t$ as seen from Equation (8).

Therefore, the mass fraction (and the corresponding volumetric concentration) would both increase linearly with time to infinity. The transient mass concentration distributions are of interest at various moving and fixed points: at the moving source i.e., at the exhaust tailpipe; at a sensor fastened on the machine with an offset of 2.1 m behind the tailpipe and at three fixed locations in the airway. The fixed locations are: at x = L/25, close to the entrance to the tunnel; at x = L/2; and at x = L.

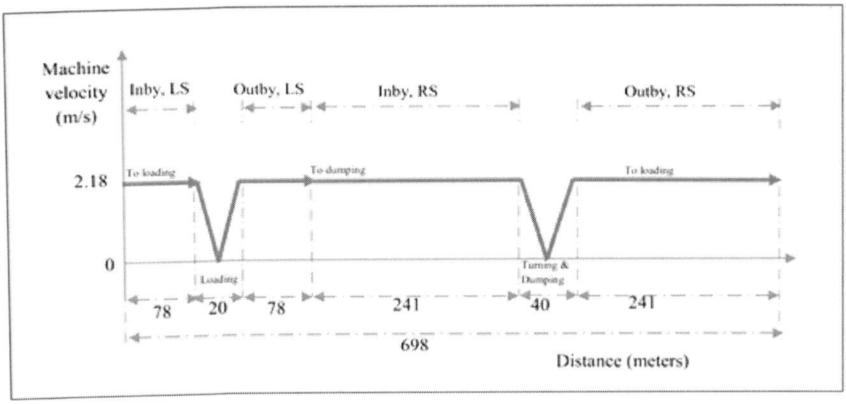

Figure 6. Machine velocity versus distance diagram for a full haulage cycle.

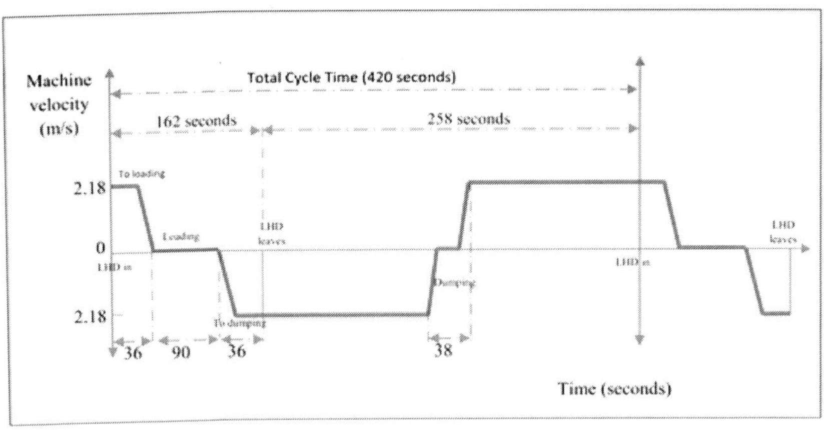

Figure 7. Machine velocity versus time diagram.

Matrix A in (11) and matrices in (23) are calculated with N = 60 and M = 162 for the LS, and N = 186 and M = 258 for the RS drift segments, respectively. The size of A (and A^{-1}) in the operator solution are 9720 × 9720 in the LS and 47,988 × 47,988 in the RS models, both calculated only once in the examples for each dispersion coefficient. The dispersion coefficients is calculated as D = 2.5 m^2/s for turbulent pipe flow in reference to Taylor's results [8] , but two lower values of 0.05 m^2/s and 0.5 m^2/s are also used for sensitivity checks in the example. Zero initial and boundary values, i.e., IC = BU = BD = 0, are assumed in the Eulerian domain. The source points are defined along the traveling trajectory of the machine over the space-time plane intersecting or passing by minimum distance from M grid points which now represent the Lagrangean model domain. The mass concentration values are calculated from the full ST solution of (23), applied separately to the LS and RS drift segments. The spatial and temporal concentration values for the stationary and moving sensors; as well as the tailpipe of the LHD are determined from the full ST solutions by sampling the data sets obtained for the LS and RS drift segments.

The full ST solution is shown in Figure 8(a) and Figure 8(b). Both figures depict the mass concentration field with D = 2.5 m^2/s. For graphical purposes, the large model size is reduced by sampling only every 10th of the DPM mass fraction values, converted to the more convenient concentration units.

The desired solution for concentration both at fixed as well as at the moving points is expressed from the full solution for the entire space-time plane at the space-time grid points. Since both the result and source vectors have M elements, the solution equation is always in the form of:

$$\begin{bmatrix} \omega_1 \\ \vdots \\ \omega_M \end{bmatrix}_R = iA_R^F F^s \tag{25}$$

where iA_R^F is calculated from the common iA matrix operator by multiplication with individual sampling matrices defined in relations (24) for different model resultant solutions, R, according to the selection of the sampled points in the LS and RH segments along the characteristics lines of the traveling machine, shown in Figure 4.

Three different dispersion coefficients, 0.05 m^2/s, 0.5 m^2/s, and 2.5 m^2/s are tested with the model for three independent simulations. The goal is to assess the effect of dispersion on the concentration predicted by the model. The simulated average concentrations are calculated for the moving sensor, as well as the stationary sensors. These values are used to calculate the cycle average concentrations.

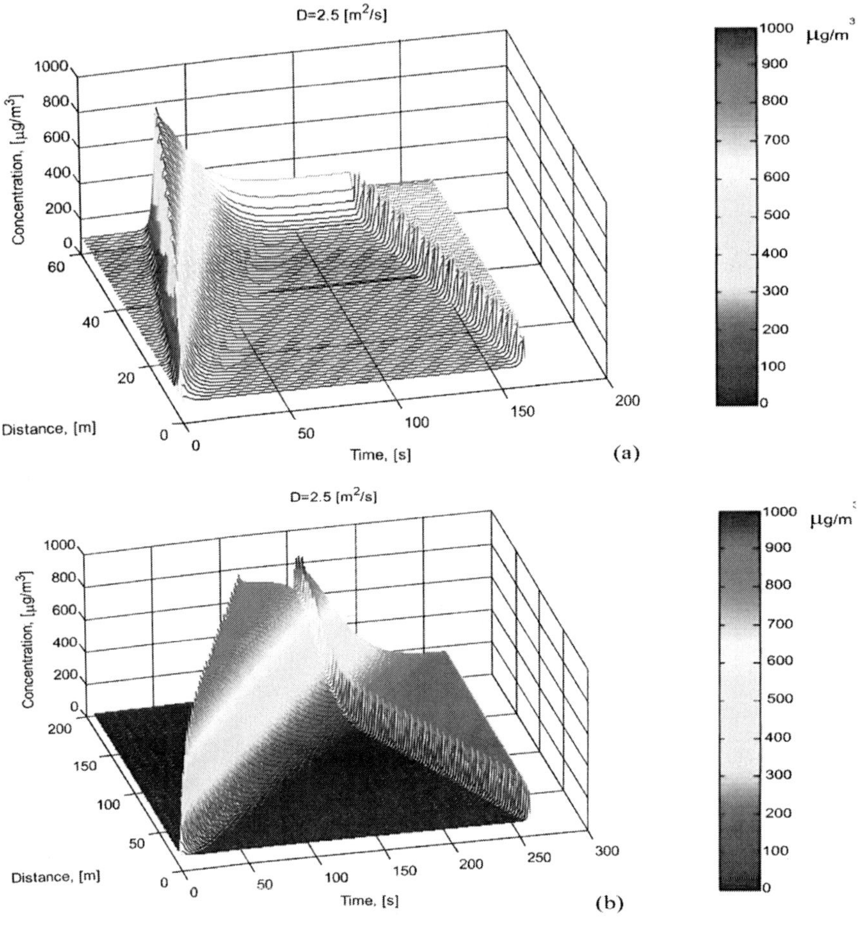

Figure 8. Full SFST solution with D = 2.5 m^2/s; (a) left side, LS; (b) right side, RS.

Figure 9 and Figure 10 show the DPM concentration for the LHD tailpipe and the DPM moving sensors for the LS and RH segments, respectively. Figure 11 depicts the concentrations for the fixed locations sensors with D = 0.05 m^2/s. Two higher dispersion coefficients are used in the simulations for sensitivity examination with D = 0.5 m^2/s and D = 2.5 m^2/s. For D = 0.5 m^2/s, the DPM concentration for the LHD tailpipe and the DPM moving sensors are illustrated in Figure 12 and Figure 13 for the LS and RH segments, respectively. The concentration for the fixed location sensors are shown in Figure 14 for D = 0.5 m^2/s.

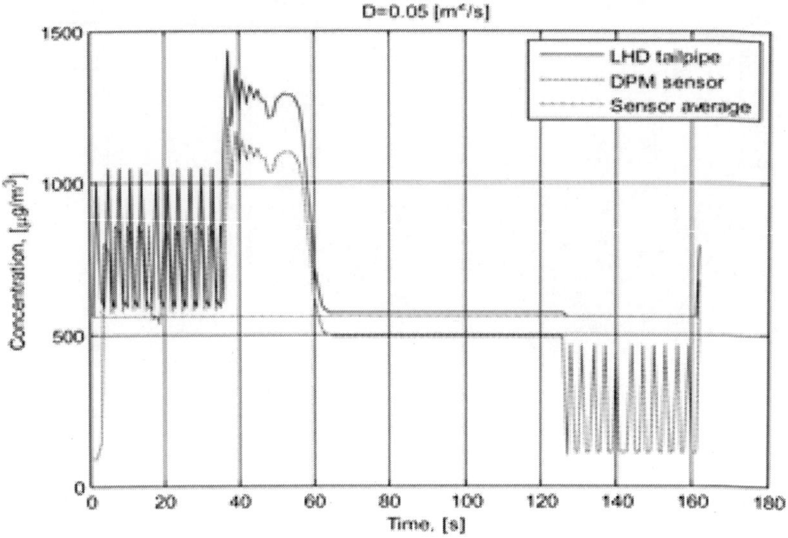

Figure 9. Sampled mass concentrations at moving points on the LS drift segment with D = 0.05 m^2/s.

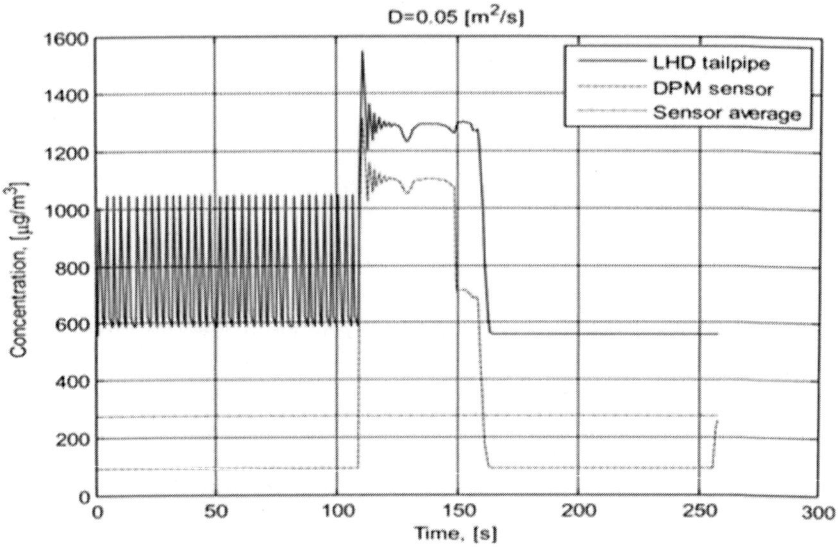

Figure 10. Sampled mass concentrations at moving points on the RS drift segment with D = 0.05 m^2/s.

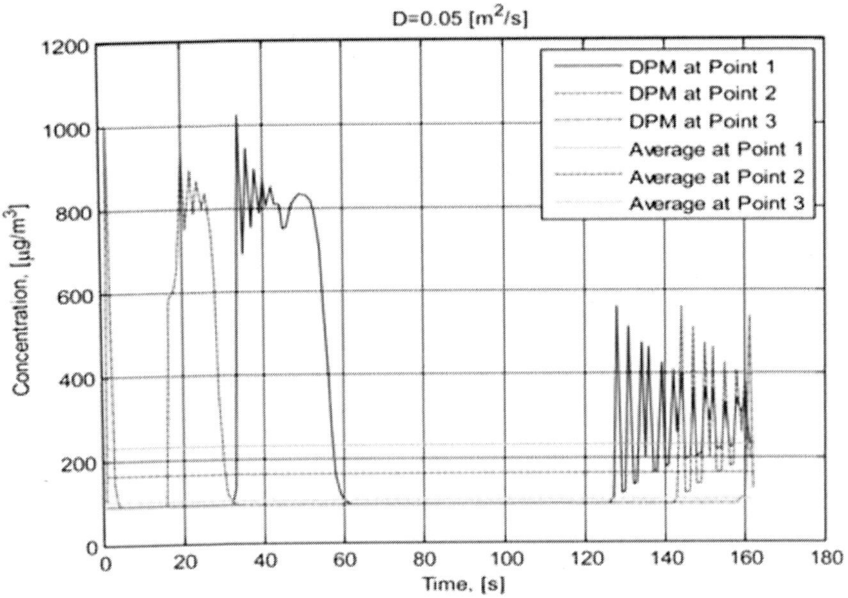

Figure 11. Sampled mass concentrations at fixed points on the LS drift segment with D = 0.05 m^2/s.

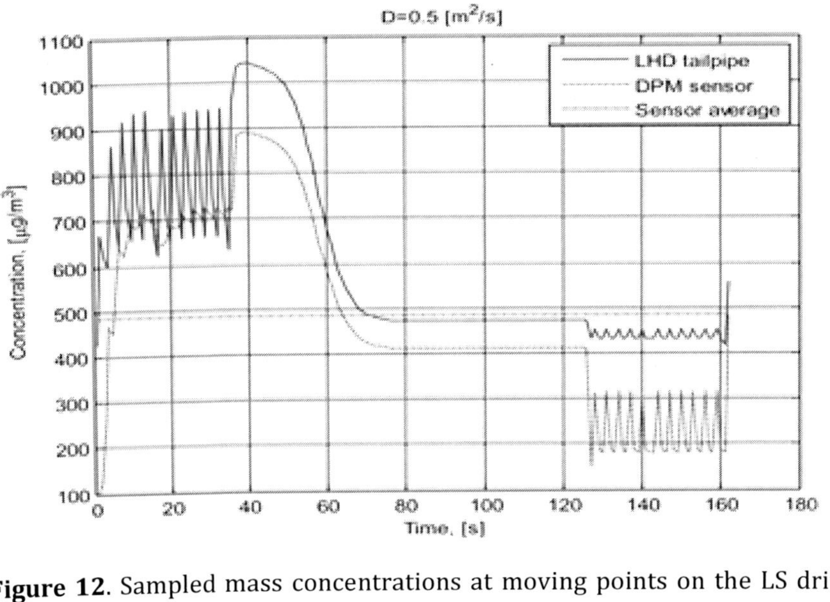

Figure 12. Sampled mass concentrations at moving points on the LS drift segment with D = 0.5 m^2/s.

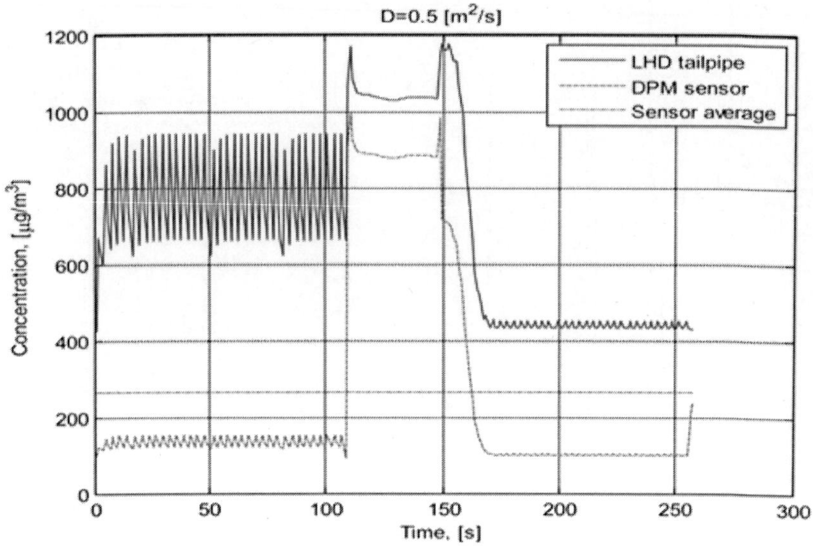

Figure 13. Sampled mass concentrations at moving points on the RS drift segment with D = 0.5 m^2/s.

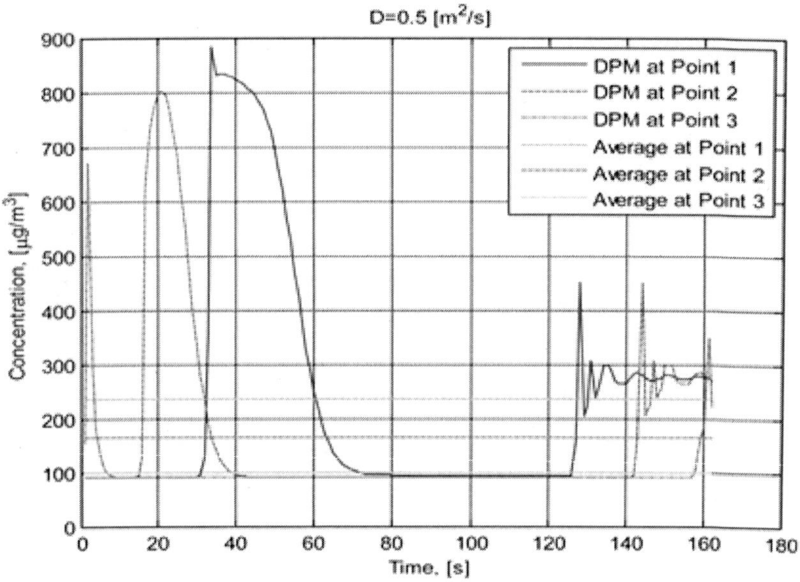

Figure 14. Sampled mass concentrations at fixed points on the LS drift segment with D = 0.5 m^2/s.

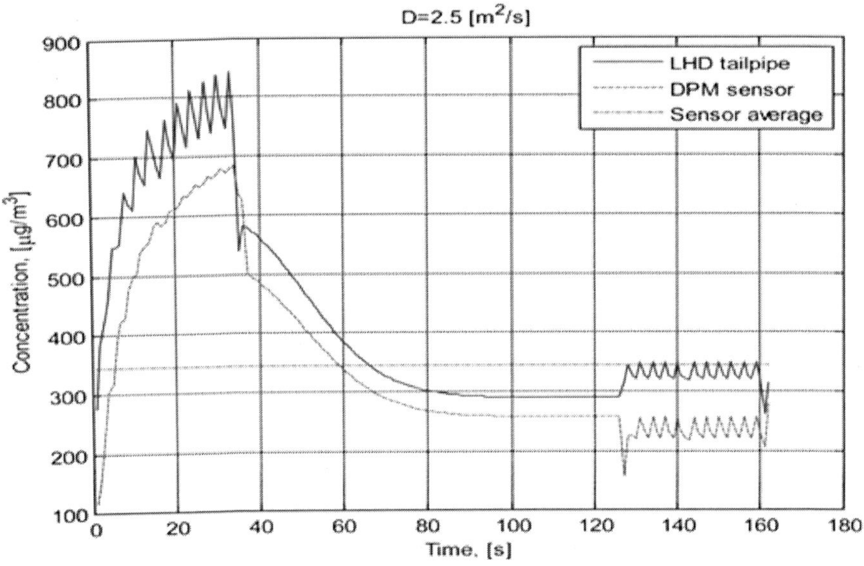

Figure 15. Sampled mass concentrations at moving points on the LS drift segment with D = 2.5 m^2/s.

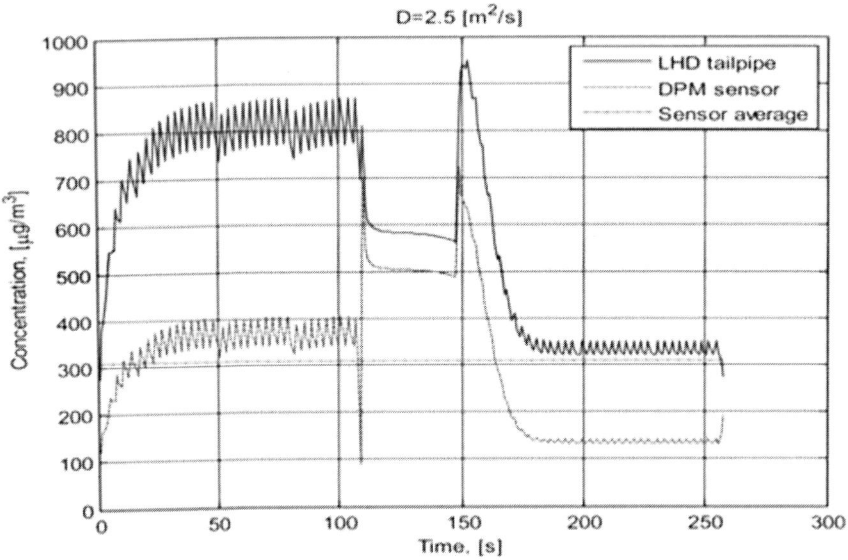

Figure 16. Sampled mass concentrations at moving points on the RS drift segment with D = 2.5 m^2/s.

Figure 15 and Figure 16 depict the DPM concentration for the LHD tailpipe and the DPM moving sensors for the LS and RH segments, respectively, assuming a dispersion coefficient of D = 2.5 m^2/s; the corresponding concentrations for the stationary sensors are shown in Figure 17.

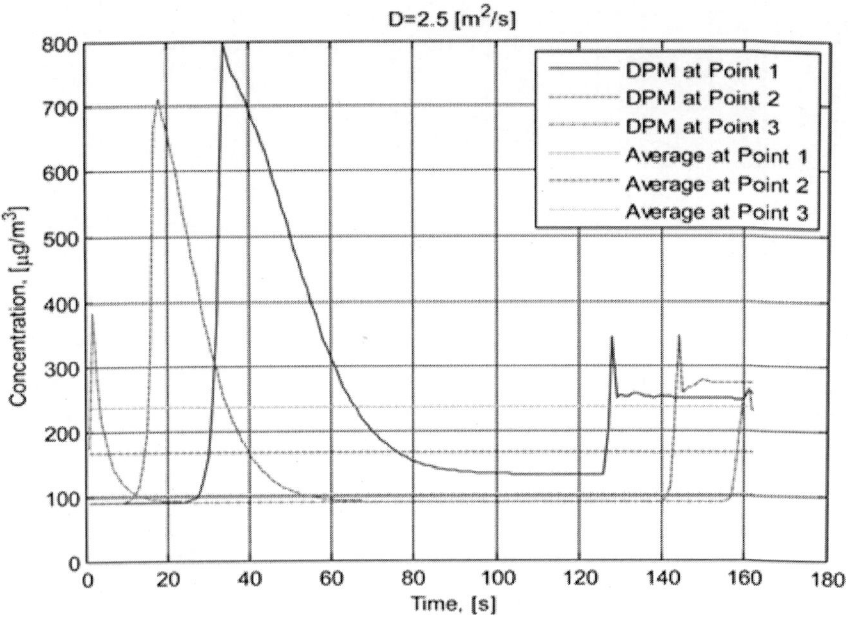

Figure 17. Sampled mass concentrations at fixed points on the LS drift segment with D = 2.5 m^2/s.

5. DISCUSSION OF THE SFST METHOD AND THE RESULTS IN THE EXAMPLE

The SFST solution for a moving substance source is given in a Eulerian domain overlaid with a Lagrangean space. The particular solutions for the initial and boundary conditions and source term are constructed with the help of a matrix operator, A^{-1}, which has to be calculated only once for the LS and the RS drift segment for all particular, initial and boundary problems and even with the moving DPM source. The A^{-1} matrix operators would remain valid for any other movement characteristics of the pollutant source term, however, the sampling matrices in (23) and (24) would have to change.

The method is advantageous to use where the air flow direction and velocity in the underground drifts are known, and the iA = A^{-1} can be conveniently pre-calculated. Such a case is often found in underground mines with constant ventilation and long time periods of continuous loading operations such as in the application example for diesel emission measurement evaluation.

The results for the three independent simulations performed at two separate sections such as LS and RS of the mine drift using the three different dispersion coefficient are delineated as follows.

5.1. LS, LHD Tailpipe and Moving Sensor Concentration Variations

It is interesting to observe that the DPM concentration at the tailpipe exit point over the entire length of the inby section show a saw tooth-shape fluctuation. With low dispersion (D = 0.05 m^2/s, Figure 9), the amplitude of fluctuation does not change due to the fact that the machine always encounters a portion of fresh air at Δx with background concentration at the time of entering a new airway section of which is then charged by the source term resulting in a near-constant $\Delta\omega$ (and corresponding volumetric concentration) change. Since the speed of the LHD machine is higher than that of the air and the machine does not travel perfectly together with the same discrete air volume (however, part of the previous air section is still connected during a Δt time interval), the concentration at the next Δx section starts again at a lower concentration which is interestingly higher than the background concentration due to the near-instantaneous introduction of pollutant source to the control volume upon arrival at Δx at the tailpipe exit point. It is assuring to observe that when the transport connection is spread to a larger volume with increased dispersion (Figure 12, Figure 15), the DPM concentration shifts toward starting from the background concentration, and the amplitude of the saw tooth variation is decreasing.

The DPM concentration is shown to be sensed by the moving sensor very well over the entire length of the inby section in Figure 9, due to the favorable LHD movement direction for the exhaust plume in this section as illustrated in Figure 5, with the DPM sensor in the plume behind the tailpipe. It is assuring to see that the starting concentration of the moving sensor is at the background value and that the amplitude of the saw tooth variation is smaller, due to dispersion over a larger distance of the offset of 2.1 m that is larger than the value of $\Delta x = 1.3$ m. With increased dispersion shown in Figure 12 and Figure 15, the DPM concentration from the moving sensor is getting smoother following very well the concentration of the tailpipe emission curve.

At the loading point, the LHD and the moving sensor are stationary. As shown in Figure 9, the concentration gradually accumulates with time due to the continuous pollutant mass source as well as the arrival of the polluted air that is left behind the faster-moving machine. With increased dispersion shown in Figure 12 and Figure 15, the DPM concentration from the moving sensor at the loading point is getting lower and smoother, but always following very well the concentration of the tailpipe emission.

The DPM concentration at the tailpipe and the moving sensor in the LS drift section during the outby travel of the LHD are very different from that of the inby travel, as seen in Figure 9. The DPM concentration at the tailpipe appears to be smooth due to traveling against the air flow with low dispersion. For the concentration of the moving sensors, three processes compete: 1) the increase of the concentration in the Δx section due to the DPM source which travels against the counter-flow of the air; 2) the instantaneous dispersion along the drift, spreading the concentration from the tailpipe; and 3) the 2.1 m offset of the moving sensor stretching into the next $\Delta x = 1.3$ m section at background concentration. Consequently, a fluctuation is seen showing periodic variations between the background and tailpipe concentrations, the frequency depending on the grid alignment with the traveling characteristics of the sensor during sampling the discretized concentration profile.

5.2. RS, LHD Tailpipe and Moving Sensor Concentration Variations

A similar phenomena as explained in the LS drift section are observed for the DPM concentration at the tailpipe exit point for the inby and outby sections of the RS drift section shown in Figure 10, Figure 13, and Figure 16 with dispersion coefficients of D = 0.05 m^2/s, D = 0.5 m^2/s, and D = 2.5 m^2/s, respectively. However, the difference is that, unlike the LS, the DPM concentration is shown not to be sensed by the moving sensor very well over the entire length of the inby section in Figure 10, Figure 13, and Figure 16. For the outby section, the moving sensor concentration slightly increases at the start of the section due to the LHD meeting the fume that is left behind by the tailpipe source as a result of the short dumping time and the late arrival time of the air flow. However, the concentration of the moving sensor immediately reduces to the background concertation at the arrival of the airflow, diluting the plume. The significant reduction in concentration in Figure 10 can be attributed to the unfavorable LHD movement direction relative to the sensor (with the DPM sensor in the plume being behind the tailpipe) for both inby and outby sections as illustrated in Figure 5.

Similar trends can be seen with increased dispersion coefficients in Figure 13 and Figure 16, albeit the higher dispersions appear to connect the moving sensor to the tailpipe better, elevating concentration levels.

At the dumping point, the LHD and the moving sensor are stationary. As shown in Figure 10, the concentration gradually accumulates with time due to the continuous pollutant mass source as well as the arrival of the polluted air left behind the moving machine. With increased dispersions in Figure 13 and Figure 16, the DPM concentration from the moving sensor at the dumping point is getting lower and smoother, but always following very well the concentration of the tailpipe emission.

5.3. LS, Stationary Drift Location Sensors

The stationary drift sensors at the LS drift section show increasing concentrations in the inby section in Figure 11 as the LHD moves from the reference point to the loading point. This is due to the lower velocity of the air than that of the tailpipe DPM source. The sensors show afluctuation in concentration with a reduced amplitude in the outby section in Figure 11 upon the return of the LHD. This is caused by the counter-direction of the air flow to the LHD and the phenomena of the sampling effects upon the concentration variations with moving air and source explained in the foregoing. The air mixes with the source concentration and dilutes it. During loading, the concentration of the fixed drift sensors are significantly reduced towards the background concentration due to the direct removal of the tailpipe source by the air. The width of the concentration pulses for the sensors at points 3, 2, and 1 located at 78 m, 41 m, and 3.1 m away from the air intake point increase as the distance gets further from the point at background concentration. The reason for this is the different duration of the exposure to the pollutant source. With $D = 0.05\ m^2/s$ in Figure 11, the concentration pulses are nearly rectangular. With increased dispersions shown in Figure 14 and Figure 17, the DPM concentrations from the fixed sensors are taking an asymmetrical shape with sharp increase and a slow descent referred to as "heavy tail" in the transport literature for contaminants.

5.4. Cycle Weighted Averages (CWA) and Comparison of Model Results to Measured Data

Part of the process of evaluating the concentrations at the moving and stationary sensors is to calculate time-averaged values for the haulage cycles in both the LS and RS drift sections. These average values are depicted in Figure 9 through Figure 17. Furthermore, a cycle averaging and weighting method is used to process the average DPM concentration data from the model results shown in Figure 9 through Figure 17. The calculated cycle weighted averages (CWA) using weight factors of 162 seconds for LS and 258 seconds for the RS is as follows:

$$\mathrm{CWA} = \frac{\left(LS_{av}\left(\mu g/m^3\right)\times 162\ \mathrm{s}\right)+\left(RS_{av}\left(\mu g/m^3\right)\times 258\ \mathrm{s}\right)}{420\ \mathrm{s}\left(\text{Total Cycle Time}\right)} \tag{26}$$

where:

LS_{av} = Left side average concentration (μg/m^3),

RS_{av} = Right side average concentration (μg/m^3).

The RS_{av} used for the stationary sensors is 90 μg/m^3, which is the background concentration entered into the model. Table 1 summarizes the CWA obtained from the SFST model for the different dispersion coefficients.

As seen in Table 1, different dispersion coefficients in the model do not have much effect on the average concentration values. This is in spite of the observed changes in the varying shape of concentration spreads in Figure 9 through Figure 17, such as less spread and higher peak concentration values for a low dispersion coefficient due to little mixing of the contaminant with air; and better spread and lower peak concentration values with high dispersion coefficients. However, the time-averaged values are not significantly affected, a favorable outcome for the otherwise uncertain parameter, D, in the mining environment.

The results of the SFST model are compared with in situ DPM in situ measurements performed at an underground mine [9] . The DPM measurement results are summarized in Table 2.

Comparison of DPM measurement data with the SFST model simulation results is shown in Table 3for different dispersion coefficients. As seen, the dispersion coefficient values have a relatively minor effect on the average concentrations.

The concentration values predicted by the SFST model for all different dispersion coefficients are in very good agreement with the measured concentration values from the mine. The model error selecting the D = 0.5 m^2/s, dispersion coefficient relative to the measured concentration is 5.8% for moving sensor and less than 10% for the average of the data from three stationary sensor locations.

Table 1. Summary of CWA from the SFST model for different dispersion coefficients.

$D = 0.05\ m^2/s$	
	CWA (μg/m³)
Moving sensor:	384
Fixed point 1 (m_1):	144
Fixed point 2 (m_2):	118
Fixed point 3 (m_3):	93
$D = 0.5\ m^2/s$	
Moving sensor:	349
Fixed point 1 (m_1):	145
Fixed point 2 (m_2):	119
Fixed point 3 (m_3):	94
$D = 2.5\ m^2/s$	
Moving sensor:	321
Fixed point 1 (m_1):	147
Fixed point 2 (m_2):	119
Fixed point 3 (m_3):	94

Table 2. DPM measurement results.

Location	DPM Concentration (μg/m³)	Average (μg/m³)
Tailpipe sensor #1	380	
Tailpipe sensor #2	280	330
Point 1 sensor # 1	120	
Point 1 sensor #2	160	140
Point 2 sensor #1	91	
Point 2 sensor #2	85	88
Point 3 sensor #1	110	
Point 3 sensor #2	97	103.5
Background (Drift entry point)		90

The numerical simulation and the comparison with in situ measurement results prove that the average DPM concentration in the air at any area in the mine can be predicted from the DPM mass transport model using the SFST simulation. The DPM source term from the tailpipe DPM concentration and their fuel consumption data can be evaluated as an input to the SFST simulation model. The mass balance and transport network modeling method ensures a cost effective and accurate way of predicting average DPM concentrations in underground mines, reducing the need for real-time monitoring for appropriate ventilation design.

Table 3. Comparison of measurement data with model results using different dispersion coefficients.

Sensor Location	Measured μg/m^3	Model with $D = 0.05$ m^2/s (CWA) μg/m^3	Model with $D = 0.5$ m^2/s (CWA) μg/m^3	Model with $D = 2.5$ m^2/s (CWA) μg/m^3	Matching error between measurement and SFST simulation at $D = 0.5$ m^2/s (%)
Moving Sensor	330	384	349	321	+5.8
Point 1	140	144	145	147	+3.6
Point 2	88	118	119	119	+35
Point 3	103	93	94	94	−8.7

6. CONCLUSIONS

- A new, powerful, fully-implicit SFST solution is applied for interpreting mea- surement results for DPM contaminant concentration variations from a moving machine in an underground mine.
- The mathematical model provides a link between the time-averaged and the peak DPM concentration values at the tailpipe.
- Very good match was obtained for all three drift stationary sensors as well as the moving sensor with D = 0.05, 0.5 and 2.5 m^2/s between in situ measurement results and SFST model simulation.
- The DPM concentration variations with location and time in the air of the mine can be predicted from the known tailpipe DPM concentration from machine smog tests and the fuel consumption of the diesel machine.
- Therefore, the mathematical model may be used to evaluate the average concentration exposure value of the DPM for compliance analysis without real-time, complicated DPM measurements, relying basically on tailpipe smog test, fuel consumption and the SFST

contaminant transport model, incorporated in the mine ventilation model.

- With the simulation of total, accumulated DPM concentration at the working area, mining companies will be able to implement the right ventilation strategies to reduce or eliminate harmful DPM exposure to mine workers.

ACKNOWLEDGEMENTS

A research support grant from the National Institute of Occupational Safety and Health, USA (NIOSH) is thankfully acknowledged.

REFERENCES

1. Bird, R.B., Steward, W.E. and Lightfoot, E.N. (1960) Transport Phenomena. John Wiley and Sons, New York.
2. Welty, J.R., Wicks, C.E. and Wilson, R.E. (1984) Fundamentals of Momentum, Heat and Mass Transfer. 3rd Edition, John Wiley and Sons, New York.
3. Le Veque, R.D. (2002) Finite Volume Methods for Hyperbolic Problems. Cambridge University Press, Cambridge.
4. Peacemen, D.W. (1977) Fundamentals of Numerical Reservoir Simulation. Elsevier Scienticic, New York, 65-82.
5. Danko, G. (2016) Model Elements and Network Solutions of Heat, Mass and Momentum Transport Processes. Springer-Verlag GmbH, Berlin.
6. Fluent 5.5, ANSYS (1997) Copyright Fluent Inc., Lebanon.
7. Cradle (2012) Thermofluid Analysis System with Unstructured Mesh Generator SC/Tetra Version 10 User's Guide. Software Cradle Co., Ltd.
8. Taylor, G. (1954) The Dispersion of Matter in Turbulent Flow through a Pipe. Proceedings of the Royal Society of London, 223, 446-468.
9. Asante, W. (2014) Mine-Wide Diesel Particulate Matter (DPM) Monitoring Applications. MS Thesis, University of Nevada, Reno, 1-89.

CHAPTER 3

Multi-Objective Optimization of University Bus Based on Passenger Probability Density Estimation

Jingcheng Zhao, Yang Liu, Yufeng Gui*

College of Science, Wuhan University of Technology, Wuhan, China

ABSTRACT

In this paper, a multi-objective optimization model is presented, based on Parzen Window method which can make a theoretical analysis of the probability density function of passengers and determine the demand over a corresponding period by integration. This model can minimize the total of waiting time of passengers and the running cost of school buses with several constrains including the passengers' numbers and the longest endurance time. Moreover, the data from Wuhan University of Technology, used as an example, verify the feasibility of the model. At last, the whole results are calculated by the Particle Swarm Optimization.

Keywords: School Bus, Parzen Window, Multi-objective Optimization, Particle Swarm Optimization

1. INTRODUCTION

With the development of education, there are more colleges establishing some new campuses which need school buses back and forth to meet students' demand their daily life. However, because of the unreasonable arrangement of buses' sites, timetable and routes, there exists the waste of the spending and the oil resources. Moreover, students are now wasting too much time waiting for the buses. Therefore, solving these problems of school bus in time is very essential and the interest of this work is two-fold.

In the past, early models of school bus problem mainly concerned on the route and sites setting, such as Zhang Fu, Zhu Taiying (2012) [1] , Hao Zhongna (2016) [2] , Zhang Miao (2008) [3] . Recently, there are some researches focusing on the timetable like Zhao Shasha, Xiong Wei (2015) [4] . They analyzed the congestion time and gave a time schedule and the departure number of the school bus. Unfortunately, due to the lack of real-time data, the paper couldn't give the specific departure time of each vehicle. However, with the emerging of travel credit card, it becomes much easier to collect the demanding data. Hence, to cope with the problem, we now use the real-time data to estimate the distribution of passengers by the non-parametric estimation method and then establish a multi-objective optimization model to specify the timetable and departure bus numbers.

Realistically, the advantages of the optimization of the school buses' timetable and departure numbers are the following:

1) On one hand, it can cut down the waste of resource and reduce the school investment when demands vacant by shortening the total operating time. On the other hand, it can shorten the waiting time of students which can bring a great convenience.
2) The optimization model can provide some theoretical and practical guidance to the arrangement of transportation in other field, for example, the city bus.
3) As a branch of vehicle target planning, the research has further developed the theory and practice of vehicle target planning.

2. PRELIMINARY DATA PROCESSING

2.1. Advantages of Prazen Window Estimation

It is tough to get the real-time number of passengers when students take a school bus. In order to easily obtain the real-time data, we utilize the school bus charge card system. Every student can only pay the fee through the machine when they need to take the bus. By recording "1" when someone get on at a moment or recording "0" contrarily, the numbers of passengers can be known. In the actual situation, it is not determined whether there will be passengers show up. A student who took the school bus at 12:35:01 is possible to take the school bus at 12:36:59 the next time. The possibility is influenced by many conditions. Hence, it is required to take all the sample points into consideration by using the Parzen Window estimation method and then figure out the number of passengers.

2.2. Calculate the Window Function and Window Width

Assuming that the data $x_1, x_2, \cdots, x_n$ are taken from a continuous distribution [5] . A kernel density estimate at any point x is defined as:

$$\hat{p}_n(x) = \frac{1}{nh}\sum_{i=1}^{n} K\left(\frac{x - x_i}{h}\right) \tag{1}$$

where n is the sample size, $K(\cdot)$ is defined as Window function, h is the bandwidth. In order to ensure the rationality of the probability density function, it is necessary to ensure its value the non-negative and the total results of integration will add up to 1. And those conditions can be obtained by requiring the kernel function $K(x)$ the distribution density.

Generally, it can choose square window, normal window to be the window function. But considering that the smoothness of the normal function will make the estimation function changes smoothly and changes only with one parameter change, the normal window is decided to be the kernel function.

So the kernel function is:

$$K(x, x_i) = \frac{1}{\sqrt{2\pi}h} \exp\left\{-\frac{(x - x_i)^2}{2h^2}\right\} \tag{2}$$

The window width h is the only parameter needed to be adjusted. If the width h is too long, then the resolution will be very low. But if the width h is too short, the statistic change will be too high. Generally as long as the width h can minimize AMISE (Asymptotic Mean Integrated Squared Error), it is the optimal window width. However, in this way, the bandwidth can only be used as a starting point to explore the optimal estimation. From the practical point of view, Rudemo (1982) and Bowman (1984) propose a recursive method to determine the final bandwidth by cross validation method [5] .

Determine h to minimize the $ISE(h)$:

$$ISE(h) = n^{-2}h^{-1}\sum_{i=1}^{n}\sum_{j=1}^{n} K \cdot K\left(\frac{X_i - X_j}{h}\right) - 2n^{-1}\sum_{i=1}^{n}\hat{p}_{-i}(X_i) \tag{3}$$

where $K \cdot K\left(\frac{X_i - X_j}{h}\right)$ is the convolution, $\hat{p}_{-i}(X_i)$ is the estimation of probability density after removing the point X_i.

2.3. Calculate the Specific Numbers of Passengers over the Corresponding Period

According to the data from the card management center, the record of passenger's at every second is "1"or "0". So with the record in the total time T, the Parzen Window method can figure out the probability density function: $p(x)$, which have the features:

$$\begin{cases} p(x) \geq 0; \\ \int_0^T p(x)\,dx = 1 \end{cases} \tag{4}$$

To visually represent the probability value of having a passengers arriving at the parking site per second, $p(x)$ needs to be disposed as following:

$$p_t = p(x)\big|_{x=t} \cdot N \tag{5}$$

which N indicates the total number of passengers in the whole day at the school bus-stops. Then integrate p_t which in the section of $[T_1, T_2]$, and the number of passengers during the time slot $[T_1, T_2]$ which is remarked as $N_{T_1}^{T_2}$ can be calculated:

$$N_{T_1}^{T_2} = \int_{T_1}^{T_2} p_t \mathrm{d}t = \int_{T_1}^{T_2} p(x)\Big|_{x=t} \cdot N\mathrm{d}t = N \cdot \int_{T_1}^{T_2} p(x)\mathrm{d}x \tag{6}$$

3. MODEL ANALYSIS AND ASSUMPTIONS

3.1. Model Analysis

This paper intends to draw a conclusion about the reasonable timetable and number of the departure school bus Firstly, based on the real-time data, Parzen Window method can figure out the demanding number of passengers at any different time. Then, establish multi-objective optimization model and calculate the rational allocation of the school bus.

Considering the objective function, firstly from the perspective of students, the aim is to minimize the waiting and travelling time; secondly from the perspective of school, the aim is to minimize the total travelling cost and the loss due to the empty seat during the travelling.

Moreover, considering the constraints, one constraint is the capacity of the school buses. Their real carrying number can't beyond the maximum capacity. Besides, another constraint is the waiting time, each student staying in the parking site or staying in the bus has a maximum endurable waiting time, so every school bus no matter in which station should have a maximum time to stop and cannot overstay.

3.2. Model Assumptions

In order to facilitate the establishment and solution of the model, this paper makes the corresponding assumptions based on the actual situation:

1) Assuming that the school bus is always at a constant speed during travelling.
2) Assuming that there is no overload during travelling.
3) Assuming that all the school buses are operating properly without any maintenance and emergency situations...

4. THE MATHEMATICAL MODEL

4.1. Establishment of Objective Function I

The shorter the total waiting time of students spend, the better. The student arrives at the stop at the time t_i in the probability of P_i and takes the bus at the time t_j. The total time for all students to take the school bus notes as W_T. The time section that the school bus running from the starting time T_1 to the ending time T_2 is divided into seconds from second 0 ~ second N_T.

So the objective function is as following [6] :

$$\min W_T = \sum_{i=0}^{N_T} P_i \left(t_j - t_i\right) \quad j \ge i \tag{7}$$

4.2. Establishment of Objective Function II

The less the total cost of the school bus is, the better. With reasonable scheduling as the main target, the model only considers the cost and loss of the school bus seat rates, regardless of the bus purchase cost and school staff salaries and other expenses. Assuming that the school bus only consider the distance cost when it reaches its maximum capacity.

Assuming that the university has N_C school buses and N_S school bus-stop sites, then the other objective function is represented as shown below:

$$\min C_T = \sum_{k=1}^{N_C} \sum_{i=1}^{N_S} \sum_{j=1}^{N_S} \left(D_{kij} + E_{kij}\right) \tag{8}$$

$$D_{kij} = N_{kij} \times d_{kij} \tag{9}$$

where C_T is total costs of the school bus operation, D_{kij} is travel costs that school bus k travel from the site i to the site j, E_{kij} is costs that school bus k travel from site i to site j due to empty seats, N_{kij} is driving times that school bus k travel from site i to site j, d_{kij} is the cost of a single trip that school bus k travel from site i to site j.

4.3. Establishment of Constraint I: School Bus Capacity Constraint

According to the actual situation, we take school bus safety and operating costs into consideration, that is to say the number of students must reach a certain number and cannot exceed its capacity whenever the school bus leaves the station. This constraint can be represented as follows:

$$0.5Q_k \le L_{ik} \le Q_k \tag{10}$$

where Q_k is the maximum capacity of the school bus k while L_{ik} is the number of students that school bus k have accumulated at the site i.

4.4. Establishment of Constraint II: Waiting Time Constraint

Due to the limited time of students after class and normal endurable waiting, the sum of the time T_{kij} required for school bus k to arrive at site j from the other station i and the residence time B_{kj} of the school bus k at the i site cannot exceed the waiting limit T_c, which is determined by the questionnaire inferred. We set the speed of each school bus is equal and the number of station is P. The constraint is listed below [7] :

$$\sum_{(j=1,2,\cdots,P)} T_{kij} + B_{kj} \ (i \ne j) \le T_c \tag{11}$$

$$0 \le T_{kij} \le T_1 \tag{12}$$

$$0 \le B_{kj} \le T_2 \tag{13}$$

In summary, the multi - objective optimization model is as follows:

Objective functions:

$$\begin{cases} \min W_T = \sum_{i=0}^{N_T} P_i \left(t_j - t_i \right) \quad j \geq i \\ \min C_T = \sum_{k=1}^{N_C} \sum_{i=1}^{N_S} \sum_{j=1}^{N_S} \left(D_{kij} + E_{kij} \right) \end{cases} \tag{14}$$

Constraints:

$$\begin{cases} 0.5Q_k \leq L_{ik} \leq Q_k \\ \sum_{(j=1,2,3,4)} T_{kij} + B_{kj} \; (i \neq j) \leq T_c \\ 0 \leq T_{kij} \leq T_1 \\ 0 \leq B_{kj} \leq T_2 \end{cases} \tag{15}$$

5. MODEL SOLUTION AND EXAMPLE

5.1. Model Solution

The traditional multi-objective optimization method solves the multi-objective problem by weighted all the targets and summed them into single objective problems. Since this method requires a priori knowledge of the problem, so in the realistic issues, it is really complicated to determine the weight of each object. It could be more appropriate to use a new way called evolving calculation to solve the multi-objective problem. Particle swarm optimization algorithm (particle swarm optimization, PSO) is a kind of optimization algorithm proposed by Kennedy and Eberhart in 1995, because of its easiness to understand and implement, has been successfully applied in many optimization problems.

The minimum of student's waiting time and the running costs is in conflicted with each other. Practically, if we enable students to get on the bus as soon as possible, the bus needs to launch frequently. However, by this way, the driving time and distance of the bus would be boosted, so the cost will increased relevantly. This particularity makes the problem difficult to be solved by the traditional weighted sum into a single objective, so we use the particle swarm optimization algorithm to solve the problem.

The flow chart of multi-objective search algorithm [8] [9] which based on the particle swarm optimization is shown in Figure 1.

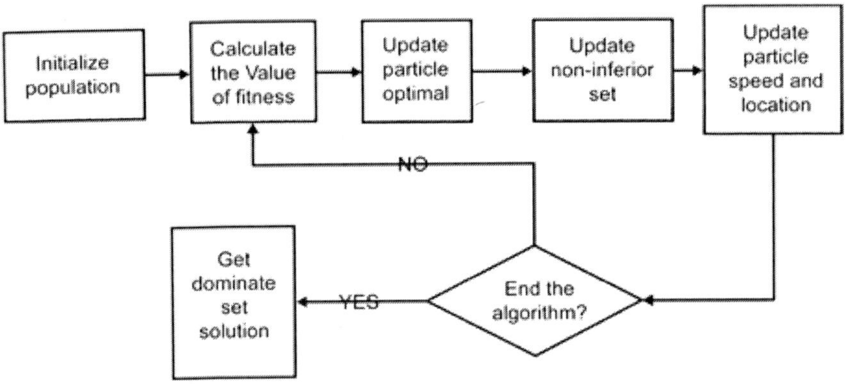

Figure 1. The flow chart of multi-objective search algorithm.

Where the module of population initialization will initialize the position and velocity of the particle randomly; when calculating the value of fitness, each individual has two fitness value including the student's waiting time and the school buses' operation cost, besides, each individual are required to meet all the constraints; the particle update optimization means to update the individual optimal particle according to the new particle position; updating the non-inferior solution set is to filter the non-inferior solution according to the now particle dominance relation.

The particle velocity and position update formula is as follow:

$$V^{k+1} = wV^k + c_1 r_1 \left(P_{id}^k - X^k\right) + c_2 r_2 \left(P_{gd}^k - X^k\right)$$
$$X^{k+1} = X^k + V^{k+1} \tag{16}$$

where w is the inertia weight, r and r_2 is the random numbers between $[0,1]$, k is times of current iteration, P_{id}^k is the best location of individual particle, P_{gd}^k is the best global optimal particle position, c_1 and c_2 are the constant, V is the particle speed, X is the particle position. When the particle adjusts its position depending on the speed, it will be limited to the maximum speed

V_{max}. When V_i exceeds V_{max}, V_i are still be valued as V_{max}. In other way, V_{max} is the width of the search space [8] .

Among all the parameters, w, c_1 and c_2 account greatly on the performance of the algorithm. There are many scholars have studied their setting and adjustment. In terms of choosing parameters, the literature review [10] , through the simulation of different parameters on the particle exploration ability, algorithm and the successful rate of the algorithm performance, recommend a fixed set of parameters: $w = 0.7298$, $c_1 = 2.0434$, $c_2 = 0.9487$.

5.2. Case Study

Since the founding of Wuhan University of Technology, there exists many problems sin the school bus operating because of the much too distributed campus. Those issues have already affect students' daily life. Hence, we take Wuhan University of Technology as the research object. To understand students' demand in different period for different school buses, we did a survey on students through questionnaire both online and offline and collect the real-time data from the center of bus data.

5.2.1. Data Collection

In order to ensure the rationality and practicability of the scheme, the data source must be true and reliable, we collected sufficient raw data through questionnaire and investigated the data from center of school bus. We know that the total number of school bus N_C now is 15. There are three types of the school bus including large bus (L), medium-sized bus (M), small bus (S) whose maximum capacity is 60, 35, 15 respectively. Besides, the total number of the school bus-stop $N_S$is 4. And the longest residence time of the school bus T_1 is 10 minutes and the longest running time between two stops T_2 is 15 minutes. And the longest endurable waiting time of a student T_c is 10 minutes.

5.2.2. Data Processing

The probability distribution of the student passengers in South Lake campus on Monday which is estimated by the Parzen Window method is shown in Figure 2.

Where the plot represents the time whose unit is 10^4 seconds, and the ordinate represent the probability whose unit is 10^{-3}. According to the distribution, we can figure out that the peak probability usually distribute at the moment about 7:30, 9:40, 11:50, 13:30, 15:40, 17:40 which are the time of just finishing classes. So we can draw the conclusion that the result of the Parzen window is credible because the probability distribution is consistent with students' actual travel rules.

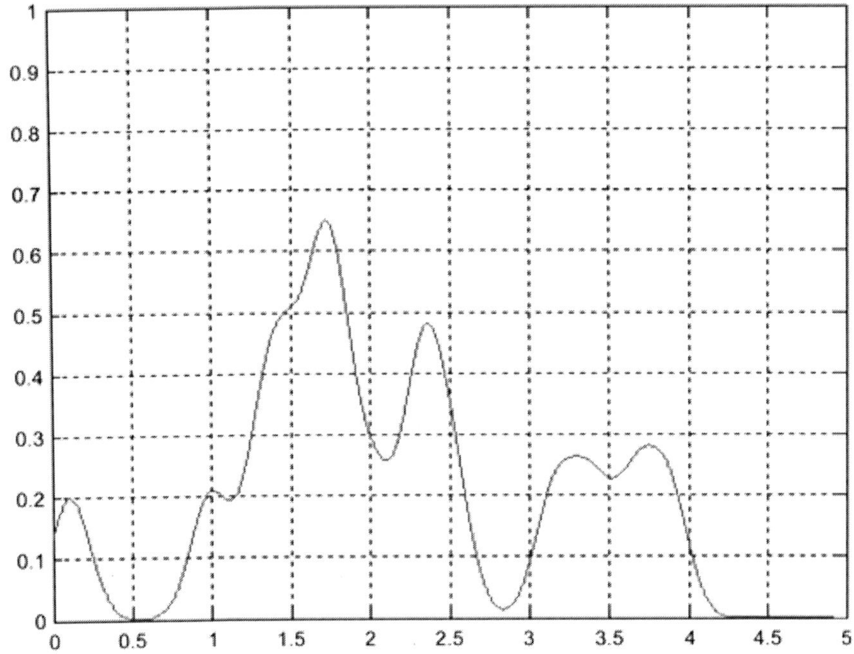

Figure 2. The probability density estimation.

5.2.3. PSO Parameter Setting

In the last part of 5.1 Model Solution, the PSO parameter setting is: $w = 0.7298$, $c_1 = 2.0434$, $c_2 = 0.9487$, the size of initial population $N = 50$, $V_{max} = 4.9 \times 10^4$.

5.2.4. Result

Through the MATLAB programming solution, we can get the best scheduling of the school bus for every campus within a week. Because of the limited space, we now only show the Monday schedule of each campus as follow:

Timetable \ Bus-stops	Nanhu	Dongyuan	Shengsheng	Jianhu
8:30 - 9:00	8:35(M)	8:50(S)		
9:00 - 9:30	9:03(M)	9:22(M)	9:17(M)	9:08(M)
9:30 - 10:00	9:41(M) 9:45(L)	9:35(M) 9:43(L)	9:31(M)	
10:00 - 10:30	10:11(M)	10:20(S)		10;17(M)
10:30 - 11:00	10:46(M)	10:33(M) 10:52(L)	10:53(M)	10:51(M)
11:00 - 11:30	11:18(M)	11:22(M)	11:25(M)	11:07(M)
11:30 - 12:00	11:45(M) 11:53(L) 11:58(L)	11:53(M) 11:58(L)	11:35(S)	11:52(M) 11:57(L)
12:00 - 12:30	12:05(M)	12:10(S)		12:12(M)

6. CONCLUSION

In this paper, we use the non-parametric estimation called the Parzen Window Method instead of a simple estimate of the probability distribution to estimate the probability distribution of students' taking school bus every day. This method takes all the sample points' influences into consideration rather than a part of them, so the estimated probability is more realistic. Next, according to the demand of student passengers, a multi-objective programming model is established, which can be adjusted and allocated to different time periods, different school districts, different types of school bus and the number of buses they own. Through the modeling of the Wuhan University of Technology, the model can get the detailed timetable. However, because we did not consider that the congestion period will have influences on the time of school bus travelling between two bus-stop stations, the final result is a little bit ideal. If the following research can fix this drawback, the model can be more consummative.

REFERENCES

1. Zhang, F. and Zhu, T.Y. (2012) Optimization of School Bus Station and Line. Mathematics in Practice and Cognition, 4, 141-146.
2. Hao, Z.N. (2016) Study on Optimization Model and Algorithm of School Bus Route. Journal of Chongqing Jiaotong University (Natural Science Edition), 2, 126-130, 162.
3. Zhang, M. (2008) Research on Multi-Objective Routing Problem Optimization Based on Bi-level Programming. Southwest Jiaotong University.
4. Zhao, S.S., Xiong, W., Yu, W.Y., Wang, Z. and Lu, Q. (2015) Optimal Dispatching and Arrangement of University Buses—Taking Three Gorges University as an Example. Science and Technology Innovation and Application, 32, 7-9.
5. Wang, X. (2005) Nonparametric Statistics. Renmin University of China Press, Beijing, 213-218.
6. Zhou, Q., Zheng, Z. and Su, R.S. (2008) Research on Optimization of School Bus Dispatching. Journal of Hebei University of Technology (Social Science Edition), 23, 32-35.
7. Liu, W. (2013) Research on Optimal Scheduling Algorithm and Model of School Bus. Journal of Tsinghua University, Natural Science Edition, 2, 247-251.
8. Zhang, L.B., Zhou, C.G. and Ma, M. (2004) Solving Multi-Objective Optimization Problem Based on Particle Swarm Optimization. Journal of computer research and development, 41, 1286-1291.
9. Wen, Y., Liao, W.Z. and Be, I.Z. (2010) An Adaptive Particle Swarm Optimization Algorithm for Solving Multi-Objective Optimization Problems. Computer engineering and application, 46, 38-40.
10. Wang, D.F. and Meng, L. (2016) Performance Analysis and Parameter Selection of Particle Swarm Optimization. Automation, 10, 1552-1561.

CHAPTER 4

Understanding the Assembly Mechanism of Proteins from Monte Carlo Simulations

Stéphane Cuenot[1*], Agata Zykwinska[2], Sadia Radji[3], Jean-Philippe Bouchara[4,5]

[1]*Institut des Matériaux Jean Rouxel (IMN), Université de Nantes, Nantes, France*

[2]*IFREMER, Laboratoire Ecosystèmes Microbiens et Molécules Marines pour les Biotechnologies, Nantes, France*

[3]*IPREM Equipe de Physique et Chimie des Polymères, Université de Pau et des Pays de l'Adour, Pau, France*

[4]*UNIV Bretagne-Loire, Université d'Angers, Université de Brest, Groupe d'Etude des Interactions Hôte-Pathogène, Angers, France*

[5]*Laboratoire de Parasitologie-Mycologie, Centre Hospitalier Universitaire, Institut de Biologie en Santé, Angers, France*

ABSTRACT

Understanding the molecular mechanism of the protein assembly still remains a challenge in the case of many biological systems. In this frame, the mechanism which drives RodA hydrophobins to self-assemble onto the surface of the conidia of the human fungal pathogen Aspergillus fumigatus into highly ordered nanorods known as rodlets, is still unresolved. Here, AFM investigations were combined with Monte Carlo simulations to elucidate how these small amphiphilic proteins self-assemble into tightly packed rodlets and how they are further organized in nanodomains. It becomes that the assembly of RodA hydrophobins into rodlets and their parallel alignment within nanodomains result from their anisotropic properties. Monte Carlo simulations allowed us to confirm that anisotropic interactions between macromolecules are sufficient to drive them to assembly into rodlets prior to nanodomains formation. Better knowledge of the mechanism of hydrophobins assembly into rodlets offers new prospects for the development of novel strategies leading to inhibition of rodlet formation, which should allow more rapid detection of the conidia by the immune system.

Keywords: Protein Assembly, Aspergillus fumigatus, Anisotropic Interactions, Monte Carlo Simulations

1. INTRODUCTION

The self-assembly of proteins into small-scale complexes is a ubiquitous phe- nomenon that is of great importance to biology. For instance, protein aggregation occurs in several diseases such as Alzheimer's [1] and Parkinson's [2] diseases, as well as in type II diabetes [3] . Understanding the mechanism of protein assembly to develop strategies to prevent this phenomenon becomes therefore fundamental. Besides experiments, modeling and numerical simulation methods offer an interesting alternative approach to decipher the pertinent parameters involved into these complex mechanisms of assembly. In this way, Monte Carlo simulations have been applied to get further insight into assembly process of various proteins by using their anisotropic character both in their shape and in their interactions [4] . One of the advantages of Monte Carlo methods concerns the great flexibility to model different kinds of interactions between particles and to easily check their effect on the resulting assembly. Recently, several simulation works have shown that proteins interacting via anisotropic or "patchy" attractions can spontaneously assemble into anisotropic structures [5] . The use of particles with anisotropic interactions (patchy particles) in Monte Carlo simulations represents a reasonable compromise between the complexity of interactions involved in assembly processes and the relative simplicity required for modelling [5] [6] . By playing with several parameters, such as the amplitude of interactions and the number of inter-particle contacts, various self-assembly processes of polymers, colloids and proteins were simulated [5] [6] . In particular, assemblies of peptides and proteins into amyloid fibrils were also obtained by considering strongly anisotropic and highly directional interactions which mimic the cross-β spine structure [4] [7] . Such structure favours the assembly of proteins via β- sheets which confer an important anisotropic character to proteins.

In the present study, Atomic Force Microscopy (AFM) experiments and Monte Carlo simulations were combined to address a biologically pertinent ques- tion relative to the formation mechanism of protein nanodomains on the conidial surface of one of the major human pathogens, Aspergillus fumigatus. Indeed, this filamentous fungus represents a huge threat for immunocompro- mised patients by causing severe and often fatal disseminated infections [8] [9] . This is because of difficulties establishing a prompt diagnosis and of the low efficiency of current systemic antifungals [8] [9] . The initial step in the majority of infectious diseases is the adhesion of pathogenic organisms to host cells and tissues [10] . Being responsible for the attachment to host tissues, the surface of A. fumigatus conidia plays a crucial role in the infection and its investigation repre- sents an important concern for the development of new therapeutic strategies. Indeed, like few other filamentous fungi, A. fumigatus produces conidia, which are covered by a characteristic layer of parallel and regularly spaced nanorods, called rodlets. Rodlets are composed of amphiphilic proteins, so called hydrophobins and RodA hydrophobins were demonstrated to be responsible for the rodlet structuring on the surface of A. fumigatus mature conidia [11] [12] . The rodlet layer serves multiple purposes, including conferring water resistance to conidia for easier dispersion in air currents and enhancing adherence to some hydrophobic surfaces. Aimanianda et al. [13] have recently demonstrated that the rodlet layer plays a crucial role in the pathogenicity of A. fumigatus since it masks the cell wall components from their recognition by the immune system. Therefore, it appears clearly that rodlets can be identified as an important virulence factor and it becomes crucial to elucidate their formation mechanism. Up to now, the mechanism which drives RodA hydrophobins to assemble into highly ordered rodlet structures has not been fully elucidated. In particular, the organization of hydrophobins into parallel nanorods regularly spaced and the formation of nanodomains covering the whole surface of conidia remain to be determined. In this context, Monte Carlo simulations were devoted here to test a possible mechanism of RodA hydrophobins self-assembly into rodlets covering the surface.

2. MATERIAL AND METHODS

2.1. Fungal Strains

Experiments were conducted on A. fumigatus reference strain IHEM 18963 (Af- 293) which was previously used for genome sequencing of A. fumigatus and that produces typical, dark blue green powdery colonies on yeast extract-peptone- dextrose agar (YPDA).

2.2. Media, Growth Conditions and Preparation of Conidial Suspensions

The fungal strain was maintained by weekly passages on YPDA medium (containing in g/L: yeast extract, 5; peptone, 10; glucose, 20; and agar, 20) supplemented with chloramphenicol 0.5 g/L. Conidia from 5-day-old cultures at 37°C were then fixed on glass coverslips previously coated with laminin (Sigma-Al- drich). After incubation for 30 min at 37°C under constant shaking, coverslips were washed in PBS to remove unbound laminin, and then applied directly at the surface of sporulating colonies of the fungus (5-day-old cultures on YPDA). Finally, coverslips were washed in PBS buffer and then dried at room temperature before AFM analysis.

2.3. Atomic Force Microscopy Measurements

The surface of A. fumigatus conidia was imaged using a NanoWizard® atomic force microscope (AFM, JPK, Germany) operating in intermittent contact mode under ambient conditions. A standard rectangular cantilever (Nanosensors NCL- W) was used for imaging, with a free resonance frequency of 165 kHz and a typical spring constant of about 40 N/m. The radius curvature of the tip was ~10 nm. The detailed analysis of AFM images were performed using JPK Data Pro- cessing software (JPK, Germany).

3. RESULTS AND DISCUSSION

3.1. Surface Characterization of A. fumigatus Conidia

Atomic Force Microscopy (AFM) image presented on Figure 1(a) reveals that the outer surface of A. fumigatus conidia is fully covered by nanodomains composed of parallel and regularly spaced rodlets [14] . The size of nanodomains, composed of at least three rodlets, varied from 120 to 220 nm in length and from 60 to 100 nm in width. Although it was previously demonstrated that each rodlet constitutes an assembly of several RodA hydrophobins [11] [12] , the mechanism which drives hydrophobins to assemble into highly ordered rodlets and their further structuring into nanodomains has not been elucidated yet. Recently, the detailed structural work performed on similar hydrophobins, namely EAS [15] [16] allowed to generate a molecular model for their self-assembly into a rodlet monolayer [17] . It was shown that EAS hydrophobins are organized in a cross-β spine structure with β-sheets formed from extended strands of the protein segment and hydrogen bounded up and down to the identical macromolecules. Such an amyloid structure has already been proposed for other amyloid-forming proteins and peptides [18] [19] .

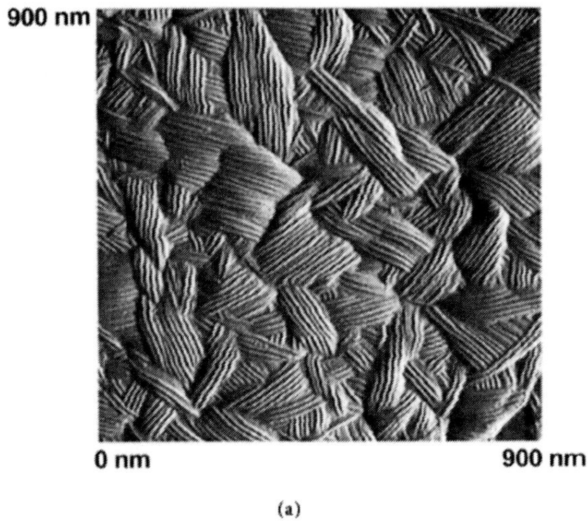

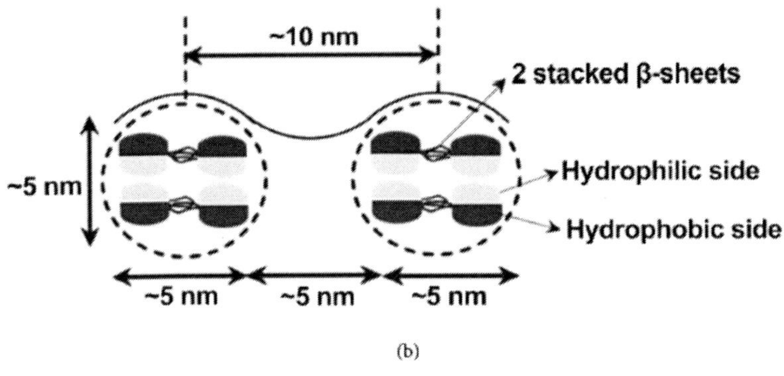

Figure 1. (a) AFM amplitude image of rodlets organized within nanodomains on the surface of a single A. fumigatus conidium; (b) Schematic representation of rodlets cross- section where each rodlet is constituted by a bilayer of two RodAhydrophobin dimers (the geometric dimensions are extracted from the AFM image).

Rodlet forming hydrophobins were shown to be globular proteins of similar diameters, close to 2.5 nm [17] . Therefore, taking into account the A. fumigatus rodlet thickness of ~5 nm, measured from AFM images, and the diameter of a single hydrophobin of ~2.5 nm, it comes out that A. fumigatus rodlets are most likely organized in bilayers. In addition, the distance between the tops of two successive rodlets, i.e. the rodlet inter-distance of ~10 nm measured from AFM images suggests both that each rodlet is spaced out from its neighbouring rodlet by a distance of ~5 nm and that the section of each rodlet is approximately a circle of ~5 nmin diameter. One possible configuration to fulfil such a rodlet dimension is the Rod A. assembly into bilayers composed of superimposed dimers stabilized by β-sheet stacks between anti-parallel molecules (Figure 1(b)). Since the outer surface of A. fumigatus conidia is uniformly hydrophobic, as shown by adhesion measurements performed by AFM [20] , hydrophobic domains of RodA dimers point outward, while their hydrophilic parts point inward [21] .

During conidia formation, hydrophobins are secreted by the fungus into its surrounding environment (e.g. hydrated growth medium). The amphiphilic nature drives hydrophobins to self-assembly into rodlets that pack tightly to form amphipathic layers at different hydrophobic-hydrophilic interfaces (e.g. air- growth medium) [22] . Such interfacial assembly can directly be related to the protein interface concentration through the Gibbs equation [23] . The Gibbs equation implies that the surface tension decreases with the logarithm of the protein concentration, up to reach a constant value for the critical assembly concentration, C* [23] [24] . At this critical concentration, the interface is fully covered by assembled proteins and the surface tension is minimized with the lowest packing energy, corresponding to the stable assembly state. However, at sufficiently high concentration ($C > C^*$), several proteins have been shown to self-assemble in the bulk solution in small asymmetric assemblies like oligomers [25] [26] [27] .

These spontaneous assemblies are strongly dependent on the anisotropic character of proteins both in their shape and in their interactions. Therefore, according to Gibbs equation, at high concentration, RodA hydrophobins may assemble not only at the air-growth medium interface but also in the bulk of the aqueous growth medium. Indeed, the presence of soluble oligomers rich in β-sheets was already reported in aqueous solutions of SC3 hydrophobins [28] [29] . The fact that very few defects were observed within rodletnanodomains imaged by AFM (Figure 1(a)) indicates that the elementary unit constitutive of rodlets is very small. Indeed, from a topological point of view, to pave a surface with lines without defects (as holes or dislocations) is only possible with small building blocks, such as oligomers. Therefore, RodA hydrophobins start probably to self-assemble in dimers in the surrounding medium through stacked β-sheets between anti-parallel molecules.

3.2. Modelling for the Monte Carlo Simulations

Some research groups have recently studied the peculiar case of phase behaviour of interacting rigid rods on two-dimensional square lattices by Monte Carlo simulations [30] [31] [32] . In this study, we propose to simulate the assembly of RodA hydrophobins on a surface in the monolayer regime with anisotropic interactions between particles using Monte Carlo method. In our model, each particle, representing a RodAdimer (Figure 1(b)), interacts with its four nearest- neighbours through anisotropic interactions modelled by a patchy model [30] [33] . Precisely, each particle possesses two bonding poles on two opposite sides inducing that two patches on different particles can only interact if they overlap. Thus, this anisotropy of the inter-particle interactions allows to mimic the formation of cross-β spine structure by favouring the particle alignment. Such patchy models have recently focused an increasing interest for simulations of protein crystallization and self-assembly of particles into chains [33] . Our Monte Carlo simulations would allow to either confirm or reject the assumption that the anisotropy of inter-hydrophobin interactions can control their assembly by constructing a possible scenario of rodlet formation covering the surface. As the RodA hydrophobin assembly takes spontaneously place in the surrounding medium during the conidia formation, the use of dynamic methods is not necessarily required to describe the surface organization of mature conidia [7] .

The surface of A. fumigatus is modelled as a two-dimensional square lattice of 150 × 150 adsorptive sites where periodic boundary conditions were applied in both horizontal and vertical directions. These conditions allow 1) to simulate an infinite surface where all sites are equivalent avoiding the border effects and 2) to represent the spherical geometry of A. fumigatus conidia. Particles represen- ting RodAdimers are adsorbed on the surface with two allowed orientations along the principal axis of the square lattice. Each site of the lattice is characterized by a three-state variable where the value 0 corresponds to an empty site, and the values ±1 represent the vertical and horizontal orientations, respectively. In order to represent the properties of a monolayer regime, each lattice site can only be single-occupied.

The following Monte Carlo procedure was used to simulate the adsorption process. Starting from an initial state where all sites of the square lattice are empty, a particle is adsorbed on one of the lattice sites at random and its orientation is also chosen randomly (horizontal or vertical orientation with probability 1/2 each). Then, another particle with its orientation chosen as random is adsorbed on an empty site of the lattice chosen as random. If there is no neighbouring particle, the orientation of the new particle is chosen at random. If it is not the case, the orientation of the particle is determined by calculating the Hamiltonian of the system by considering the interactions with its four nearest- neighbours. The Hamiltonian of the system is described in Equation 1

$$H = -J \sum_{<k,l>} \left| \vec{S}_{i,j} \vec{r}_{ij-kl} \right| \left| \vec{S}_{k,l} \vec{r}_{ij-kl} \right| \tag{1}$$

where J, is the interaction strength between two nearest-neighbour particles (kept constant in the simulations); , represents a sum over the nearest-neighbours sites identified by the row k and column l on the two-dimensi- onal square lattice; $\vec{S}_{i,j}$, is the occupation and orientation vector of site identified by the row i and column j (similar definitions are used for the occupation and orientation vector of nearest-neighbours sites identified by the row k and column l); $\vec{r}_{ij-kl}$, corresponds to the inter-particles vector connecting the sites (i, j) and (k, l).

The interaction strength between two nearest particles depends both on their relative orientations and on their orientations relative to the inter-particles vector (Equation (1)). Precisely, the bond energy is ?J if two particles are nearest neighbours on the lattice and both axes are parallel to the lattice vector connecting them (particles aligned), and is zero otherwise. Thus, for a given configuration, the energy is directly proportional to the number of shared bonds between particles. In the Monte Carlo procedure, the Hamiltonian is first calculated for the trial configuration with the random orientation, $H_{initial}$. Then, an attempt is made to change the orientation state and the corresponding Hamiltonian is calculated, H_{final}. The orientation change is determined with the probability given by the standard Metropolis acceptance criterion (Equation (2)):

$$P = \min\left[1, \exp\left(-\Delta H / k_B T\right)\right] \tag{2}$$

where $\Delta H = H_{final} - H_{initial}$, is the difference between the Hamiltonians of the final and initial states. If $\Delta H < 0$, the energy of the system is minimized, the orientation change is accepted and the process is repeated by choosing another site. If $\Delta H > 0$, the Boltzmann weight is compared to a random number comprised between 0 and 1. If this random number is lower than the Boltzmann weight, the orientation change is accepted; otherwise this change is rejected. At each time step, a new particle is added on the lattice and its orientation is either chosen at random if there is no particle in its neighbouring or chosen to minimize the total energy by considering the interactions with the nearest particles. It is important to note that the configurational entropy is implicitly taken into account in such Monte Carlo simulations with a jointly optimization of the total energy and configurational entropy of the system [31] .

The simulations were performed at constant temperature T, but the aggregation kinetics could also be simulated using Monte Carlo dynamics for different values of k_BT (where k_B is the Boltzmann constant) [7] . Monte Carlo simulations using a deposition-eva- poration algorithm of particles could also be an alternative to our model but the final result should not be significantly different [34] .

3.3. Monte Carlo Simulations of Rodletnano Domain Formation

Results of Monte Carlo simulations are shown on Figure 2 through four snapshots corresponding to different coverage ratio (χ) of adsorbed particles on the surface. For low coverage ratio (χ = 25%), the first adsorbed particles are either isolated on the square lattice or they initiate the particle assembly into short lines (Figure 2(a)). For χ = 50%, by adding more adsorbed particles on the lattice, the rod formation appears more clearly with longer linear assemblies (Figure 2(b)).

The first rod formed acts as a nucleation site to induce the formation of other rods parallel to the first one (repulsive interactions between rods).These longer rods correspond to the assembly of RodA hydrophobin dimers into rodlets through anisotropic interactions between stacked β-sheets, which conduct to a final linear cross-β spine structure. By increasing the coverage ratio up to 75%, the simulations show that the linear assemblies generate the formation of domains, which contain some parallel lines not totally complete (Figure 2(c)). When χ reaches 100%, the lattice is fully covered by domains that are composed of several parallel rods (Figure 2(d)).The boundaries of domains correspond to breaking symmetry coming from the intersection of lines of adjacent domains.

Based on the minimization of the total energy of the system with an underlying maximization of configurational entropy, our Monte Carlo simulations reveal that by taking only into account anisotropic interactions between the particles on a square lattice, several domains of parallel lines resulting from the linear assembly of particles were formed. The good agreement obtained between the AFM image of the protein assembly covering the surface and the final simulation, where the lines within domains modelled well the rodlets observed (Fig- ure 1 and Figure 2), shows that a sufficient condition for the self-assembly of hydrophobins into rodlets is to consider anisotropic interactions between hydro- phobins. The anisotropic interactions used in our patchy model possess a repulsive part (end-to-end dimers orientation) and an attractive part (side-to-side dimers orientation) modelling the interactions between hydrophobins through stacked β-sheets. However, it is not necessary to define precisely the nature of these anisotropic interactions in our simulations. Other simulations describing explicitly the anisotropic character of interactions with attractive/repulsive electrostatic interactions could lead to similar results [31] [32].

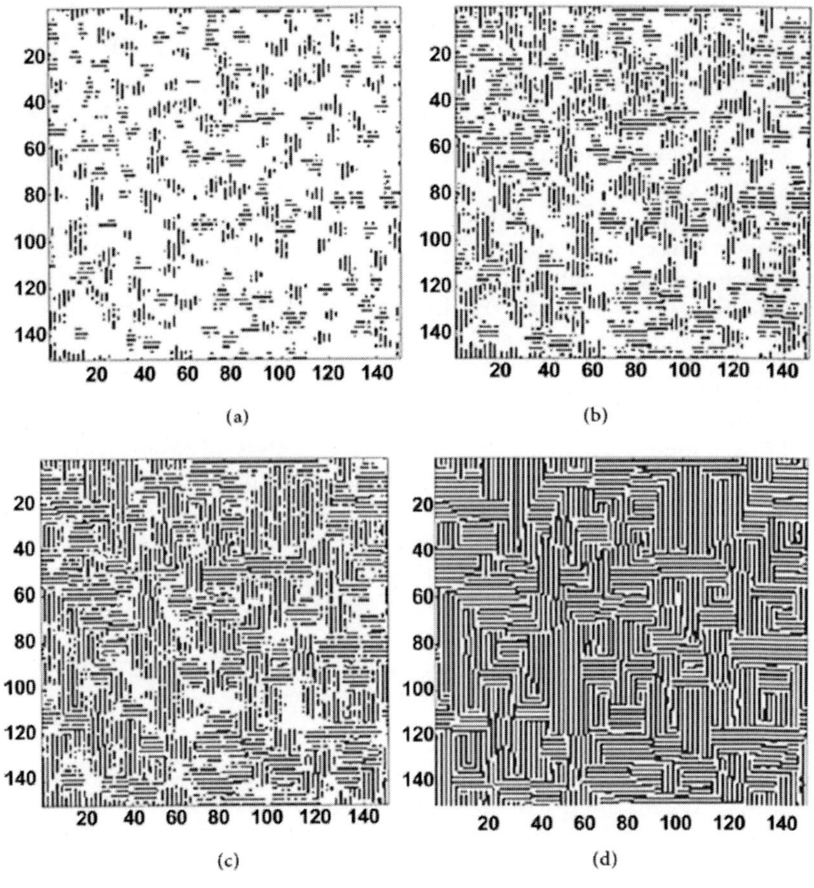

Figure 2. Snapshots of Monte Carlo simulations showing the domains formed from particles assembled into lines (which model RodA dimers assembly into rodlets before the formation of nanodomains) for different surface coverage ratios (a) 25%, (b) 50%, (c) 75%, (d) 100%.

From Figure 2, Monte Carlo simulations reveal a possible scenario to fully cover the surface of rodlet domains with 1) the formation of rodlets from the assembly of RodAhydrophobin dimers via the cross-β spine structure and 2) the subsequent formation of nanodomains. From Monte Carlo simulations, more detailed information can be obtained on the assembly mechanism by following the evolution of the rodlet number and rodlet length with the amount of adsorbed particles. In Figure 3(a), four distinct parts can be clearly distinguished on the curve of the average number of rodlets plotted as a function of the coverage ratio. In the first part, a linear increase in the average rodlet number up to 130 was observed for $\chi = 0.8\%$. This comes from the fact that each new particle added on the lattice generates the formation of a new rodlet. For χ from 0.8% to 40%, the average number of rodlets increased strongly from 130 to 1200. Indeed, new particles are still added on the lattice, which leads to the formation of new rodlets. However, in this part of the curve, new particles may also associate (i.e. assemble) to existing rodlets. For χ from 40% to 70%, a slight increase in the average rodlet number was noticed (up to 1500), which suggests that new particles contribute most likely to the increase in length of rodlets already formed rather than to formation of new rodlets.

In the last part of the curve, when the lattice becomes completely covered (χ from 70% to 100%), the average number of rodlets significantly decreased down to 850. It is most likely that new added particles contribute to connect rodlets already formed (Figure 2(c)), which decreases the average number of rodlets on the lattice. From Monte Carlo simulations, the evolution of the average length of rodlets can also be followed in function of the coverage ratio (Figure 3(b)).

Three parts are clearly visible. For χ up to 5%, the average rodlet length increased rapidly to get rodlets constituted of 4 units. In the second part, when χ increased from 5% to 85%, new added particles contribute to a continuous increase in the average rodlet length up to 13 units. Finally, the average length of rodlets strongly increased, leading to rodlets composed of 25 units, when the lattice was almost completely covered (χ from 85% to 100%). Indeed, as already suggested, the addition of new particles led to association of existing rodlets, thus resulting in important increase in the rodlet length.

The first monolayer of RodAhydrophobin dimers assembled into rodletsvia anisotropic interactions between stacked β-sheets on the conidia surface is most likely continuously completed by free dimers from growth medium. These free dimers spontaneously interact through their hydrophilic sides with the monolayer already assembled to decrease the monolayer surface energy. Finally, the formed bilayer constitutes the real rodlet as schematically shown on Figure 1(b).

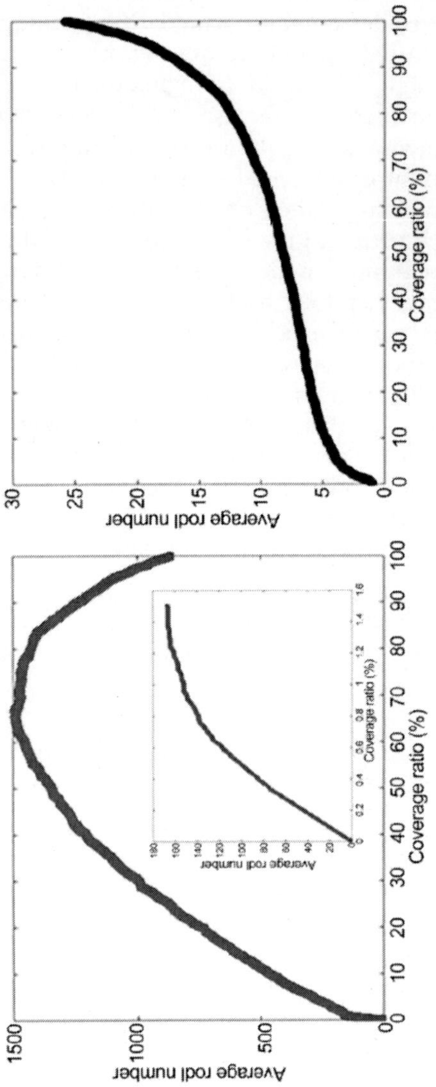

Figure 3. (a) Evolution of the average number of rodlets formed in function of the surface coverage ratio, deduced from Monte Carlo simulation. In the inset, the linear dependence observed for the first adsorbed particles indicating that they systematically initiate the formation of new rodlets; (b) Evolution of the average length of rodlets formed in function of the surface coverage ratio, obtained from Monte Carlo simulation.

4. CONCLUSION

In the present work, Monte Carlo simulations were used to support AFM investigations of the conidial surface in the human opportunistic pathogen A. fumigatus to understand the mechanism that leads RodA hydrophobins to self-ass- emble into rodlets. These small proteins, which possess an important anisotropic character via their interactions through stacked β-sheets, are ideal candidates to be modelled. Monte Carlo simulations allowed showing that anisotropic interactions between RodA hydrophobindimers (which are pre-assembled in the growth medium) are sufficient to drive them to assembly into parallel rodlets prior to the formation of nanodomains. These interactions possess an attractive part for side-to-side RodA hydrophobin dimers orientation modelling the interactions between hydrophobins through stacked β-sheets and a repulsive part for end-to- end dimers orientation. Such Monte Carlo simulations can be used as a complementary tool of experiments, offering a great support in the understanding of assembly mechanisms. Therefore, the complexity of mechanisms governing the assembly of peptides, colloids and proteins can be overcome.

ACKNOWLEDGEMENTS

The authors acknowledge the "Region des Pays de la Loire" for financial support in the frame of "Myco-AFM" research program.

REFERENCES

1. Chiti, F. and Dobson, C.M. (2006) Protein Misfolding, Functional Amyloid, and Human Disease. Annual Review of Biochemistry, 75, 333-366.
2. Uversky, V.N. and Eliezer, D. (2009) Biophysics of Parkinson's Disease: Structure and Aggregation of α-Synuclein. Current Protein & Peptide Science, 10, 483-499.
3. Andersson, A., Bohman, S., Borg, L.A.H., Paulsson, J.F., Schultz, S.W., Westermark, G.T. and Westermark, P. (2008) Amyloid Deposition in Transplanted Human Pancreatic Islets: A Conceivable Cause of Their Long-Term Failure. Experimental Diabetes Research, 2008, Article ID: 562985.
4. Auer, S., Dobson, C.M., Vendruscolo, M. and Maritan, A. (2008) Self-Templated Nucleation in Peptide and Protein Aggregation. Physical Review Letters, 101, Article ID: 258101.
5. Bianchi, E., Blaak, R. and Likos, C.N. (2011) Patchy Colloids: State of the Art and Perspectives. Physical Chemistry Chemical Physics, 13, 6397-6410.
6. Doppelbauer, G., Bianchi, E. and Kahl, G. (2010) Self-Assembly Scenarios of Patchy Colloidal Particles in Two Dimensions. Journal of Physics: Condensed Matter, 22, Article ID: 104105.
7. Irback, A., Jonsson, S.A., Linnemann, N., Linse, B. and Wallin, S. (2013) Aggregate Geometry in Amyloid Fibril Nucleation. Physical Review Letters, 110, Article ID: 058101.

8. Latgé, J.P. (1999) Aspergillus fumigatus and Aspergillosis. Clinical Microbiology Reviews, 12, 310-350.
9. Nivoix, Y., Velten, M., Letscher-Bru, V., Moghaddam, A., Natarajan-Amé, S., Fohrer, C., Lioure, B., Bilger, K., Lutun, P., Marcellin, L., Launoy, A., Bergerat, J.P. and Herbrecht, R. (2008) Factors Associated with Overall and Attributable Mortality in Invasive Aspergillosis. Clinical Infectious Diseases, 47, 1176-1184.
10. Tronchin, G., Pihet, M., Lopes-Bezerra, L.M. and Bouchara, J.-P. (2008) Adherence Mechanisms in Human Pathogenic Fungi. Medical Mycology, 46, 749-772.
11. Thau, N., Monod, M., Crestani, B., Rolland, C., Tronchin, G., Latgé, J.-P. and Paris, S. (1994) Rodletless Mutants of Aspergillus fumigates. Infection and Immunity, 62, 4380-4388.
12. Paris, S., Debeaupuis, J.-P., Crameri, R., Carey, M., Charlès, F., Prévost, M.C., Schmitt, C., Philippe, B. and Latgé, J.-P. (2003) Conidial Hydrophobins of Aspergillus fumigatus. Applied and Environmental Microbiology, 69, 1581-1588.
13. Aimanianda, V., Bayry, J., Bozza, S., Kniemeyer, O., Perruccio, K., RamuluElluru, S., Clavaud, C., Paris, S., Brakhage, A.A., Kaveri, S.V., Romani, L. and Latgé, J.-P. (2009) Surface Hydrophobin Prevents Immune Recognition of Airborne Fungal Spores. Nature, 460, 1117-1121.
14. Dague, E., Alsteens, D., Latgé, J.-P., Verbelen, C., Raze, D., Baulard, A.R. and Dufrêne, Y.F. (2007) Chemical Force Microscopy of Single Live Cells. Nano Letters, 7, 3026-3030.
15. Kwan, A.H., Winefield, R.D., Sunde, M., Matthews, J.M., Haverkamp, R.G., Templeton, M.D. and Mackay, J.P. (2006) Structural Basis for Rodlet Assembly in Fungal Hydrophobins. Proceedings of the National Academy of Sciences of the United States of America, 103, 3621-3626.
16. Kwan, A.H., Macindoe, I., Vukasin, P.V., Morris, V.M., Kass, I., Gupte, R., Mark, A.E., Templeton, M.D., Mackay, J.P. and Sunde, M. (2008) The Cys3-Cys4 Loop of the Hydrophobin EAS Is Not Required for the Formation and Surface Activity. Journal of Molecular Biology, 382, 708-720.
17. Macindoe, I., Kwan, A.H., Ren, Q., Morris, V.K., Yang, W., Mackay, J.P. and Sunde, M. (2012) Self-Assembly of Functional, Amphipathic Amyloid Monolayers by the Fungal Hydrophobin EAS. Proceedings of the National Academy of Sciences of the United States of America, 109, 804-811.
18. Nelson, R., Sawaya, M.R., Balbirnie, M., Madsen, A.O., Riekel, C., Grothe, R. and Eisenberg, D. (2005) Structure of the Cross-Beta Spine of Amyloid-Like Fibrils. Nature, 435, 773-778.
19. Goldschmidt, L., Teng, P.K., Riek, R. and Eisenberg, D. (2010) Identifying the Amylome, Proteins Capable of Forming Amyloid-Like Fibrils. Proceedings of the National Academy of Sciences of the United States of America, 107, 3487-3492.
20. Dague, E., Alsteens, D., Latgé, J.-P. and Dufrêne, Y.F. (2008) High-Resolution Cell Surface Dynamics of Germinating Aspergillus fumigatus Conidia. Biophysical Journal, 94, 656-660.
21. Zykwinska, A., Pihet, M., Radji, S., Bouchara, J.-P. and Cuenot, S. (2014) Self-Assembly of Proteins into a Three-Dimensional Multilayer System: Investigation of the Surface of the Human Fungal Pathogen Aspergillus fumigatus. Biochimica et Biophysica Acta, 1844, 1137-1144.
22. Wosten, H.A.B., Schuren, F.H.J. and Wessels, J.G.H. (1994) Interfacial Self-Assembly of a Hydrophobin into an Amphipathic Protein Membrane Mediates Fungal Attachment to Hydrophobic Surfaces. EMBO Journal, 13, 5848-5854.
23. Gibbs, J.W. (1931) The Collected Works of J. W. Gibbs. Longmans, Green, New York, Vol. 1, 219.
24. Adamson, A.W. and Gast, A.P. (1997) Physical Chemistry of Surfaces. 4th Edition, John Wiley, New York.
25. Fung, S.Y., Keyes, C., Duhamel, J. and Chen, P. (2003) Concentration Effect on the Aggregation of a Self-Assembling Oligopeptide. Biophysical Journal, 85, 537-548.
26. Liu, X.Y. (2000) Molecular Modelling for Non-Ideal Mixing of Amphiphilic Molecules. Langmuir, 18, 14-25.
27. Sipe, J.D. and Cohen, J.D. (2000) Review: History of the Amyloid Fibril. Journal of Structural Biology, 130, 88-98.

28. Wang, X., Graveland-Bikker, J.F., de Kruif, C.G. and Robillard, G.T. (2004) Oligomerization of Hydrophobin SC3 in Solution: From Soluble State to Self-Assembly. Protein Science, 13, 810-821.
29. Zykwinska, A., Guillemette, T., Bouchara, J.-P. and Cuenot, S. (2014) Spontaneous Self-Assembly of SC3 Hydrophobins into Nanorods in Aqueous Solution. Biochimica et Biophysica Acta, 1844, 1231-1237.
30. Tavares, J.M., Holder, B. and Telo da Gama, M.M. (2009) Structure and Phase Diagram of Self-Assembled Rigid Rods: Equilibrium Polydispersity and Nematic Ordering in Two Dimensions. Physical Review E, 79, Article ID: 021505. Longone, P., Linares, D.H. and Ramirez-Pastor, A.J. (2010) Critical Behaviour of Attractive Rigid Rods on Two-Dimensional Lattices. Journal of Chemical Physics, 132, Article ID: 184701.
31. Ghosh, A. and Dhar, D. (2007) On the Orientational Ordering of Long Rods on a Lattice. Europhysics Letters, 78, 20003.
32. Doye, J.P.K., Louis, A.A., Lin, I.-C., Allen, L.R., Noya, E.G., Wilber, A.W., Kok, H.C. and Lyus, R. (2007) Controlling Crystallization and Its Absence: Proteins, Colloids and Patchy Models. Physical Chemistry Chemical Physics, 9, 2197-2205.
33. Schwen, E.M., Mazilu, I. and Mazilu, D.A. (2015) A Stochastic Model of Particle Deposition and Evaporation for Ionic Self-Assembly of Thin Films. Journal of Physics, 574, Article ID: 012043.

CHAPTER 5

Conjugate Gradient Method to Solve Fluid Structure Interaction Problem

Mamadou Diop[1], Ibrahima Mbaye[2]

[1]*Thies University, Thies, Senegal*

[2]*Department of Mathematics and Computer Science, Thies University, Thies, Senegal*

ABSTRACT

In this paper, we propose a method to solve coupled problem. Our computational method is mainly based on conjugate gradient algorithm. We use finite difference method for the structure and finite element method for the fluid. Conjugate gradient method gives suitable numerical results according to some papers.

Keywords: Fluid-Structure Interaction, Beam, Stokes, Finite Element, Finite Difference Method, Conjugate Gradient Method

1. INTRODUCTION

Problem involving fluid structure interaction occurs in a wide vatiety of engi- neering problems and therefore has attracted the interest of many investiga- tions from different engineering disciplines. As a result, much efforts has gone into the development of general computational method for fluid structure systems [1] [2] [3] [4] [5] [6] .

Amongst the computational methods for fluid structure interaction problem, we cite the fixed point method, the Newton method, the Quasi-Newton method, the fictitious domain method. In this work we present a method based on the conjugate gradient algorithm. In effect, the fluid interaction problems occur in biomedical fluids areas for example blood flow interaction with elastic veins. Thus, this paper aims at showing that, we can combine the finite difference method, the finite element method and the conjugate gradient method to solve fluid structure interaction problem. On the one hand, we use finite difference method to approximate the structure model in order to have a linear systems, On the other hand, we solve the stokes equation by the finite element method. Moreover, conjugate gradient method will be intruduced to compute the displacement of the structure. Thus, the velocity v and the pressure p of the fluid are done in the deformed domain. In addition, the fluid represented by the blood is modelled by two dimensional Stokes equation for steady flow and the structure represented by the body vessel is modelled by the one dimensional beam equation.

2. POSITION OF PROBLEM

2.1. Domain Fluid

The fluid domain noted $\Omega_F^u \subset \mathbb{R}^2$ is represented in the Figure 1.

Where, the border $\partial\Omega_F^u = \Sigma_1 \cup \Sigma_2 \cup \Sigma_3 \cup \Gamma_2$.

Γ_2 is the interface between the fluid and the elastic structure

Σ_1 is The inflow

Σ_2 is a rigid border

Σ_3 is the outflow

L is the domain length

H is the domain height

u is the displacement of the structure

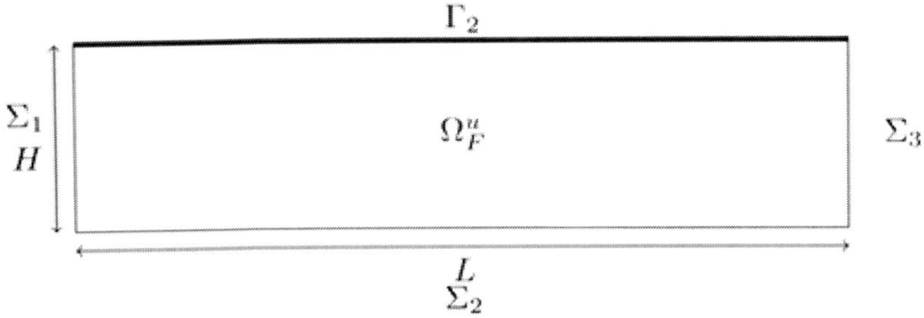

Figure 1. Reference domain.

2.2. Fluid Properties

The fluid is considered to be Newtonian, incompressible and its state is describ- ed by the velocity $v = (v_1, v_2)$ and the pressure p. The balance equations are

$$-\mu\Delta v + \nabla p = f^F, \text{ in } \Omega_F^u \tag{1}$$

$$\nabla \cdot v = 0, \text{ in } \Omega_F^u \tag{2}$$

$$v = G, \text{ on } \Sigma_1 \cup \Sigma_2 \tag{3}$$

$$v = 0, \text{ on } \Gamma_2 \tag{4}$$

$$-pI_d n + \frac{\partial v}{\partial n} = 0, \text{ on } \Sigma_3 \tag{5}$$

μ: the fluid viscosity

I_d the identity matrix

$f^F = (f_1^F, f_2^F)$ the volume force of the fluid

n is a unit normal vector

$G=(G_1,G_2)$ the velocity profil in Σ_1

2.3. Structure Properties

The structure is assumed by elastic beam. We note $u:[0,L]\rightarrow\mathbb{R}$ the displacement of the structure, it is modelled by the beam equation

$$Du^{(4)}(x)=p(x,H+u(x)),\forall x\in[0,L], \tag{6}$$

with the boundary conditions,

$$u(0)=u(L)=0 \tag{7}$$

$$u'(0)=u'(L)=0 \tag{8}$$

where,

$D=\dfrac{E\times h^3}{12(1-\nu^2)}$

E is the Young modulus

h elastic structure thickness

ν the Poisson's coefficient

Remark: In Equation (6) we assume that only the pressure force is acting on the interface and also u is the transversal displacement [3] .

3. COUPLED PROBLEM

The coupled problem is to find (u,v,p) such that:

$$\begin{cases} Du^{(4)}(x) = p(x, H+u(x)) \forall x \in [0, L] \\ u(0) = u(L) = 0 \\ u'(0) = u'(L) = 0 \\ -\mu\Delta v + \nabla p = f^F, \text{ in } \Omega_F^u \\ \nabla \cdot v = 0, \text{ in } \Omega_F^u \\ v = G, \text{ on } \Sigma_1 \cup \Sigma_2 \\ v = 0, \text{ on } \Gamma_2 \\ -pI_d n + \dfrac{\partial v}{\partial n} = 0, \text{ on } \Sigma_3 \end{cases}$$

In order to solve this coupled problem, we transform its continuous problem into a discreet problem by using finite difference method and finite element method.

3.1. Approximation by Taylor Development

Assumption: We consider u as a small displacement.

Thus, the Taylor formula gives

$$p(x, H+u(x)) \approx p(x, H) + u(x)\frac{\partial p(x, H)}{\partial y}, \tag{9}$$

the Equation (6) becomes:

$$Du^{(4)}(x) - u(x)\frac{\partial p(x, H)}{\partial y} = p(x, H), \tag{10}$$

we pose $\alpha(x) = -\frac{\partial p(x, H)}{\partial y}$, finally we have,

$$Du^{(4)}(x) + \alpha(x)u(x) = p(x, H). \tag{11}$$

To discretize the Equation (11), we introduce a space step $\Delta x = \frac{L}{N+1}$. We denote by u_i the value of the discrete solution at $x_i = i \times \Delta x$

for $i \in \{0,1,\cdots,N+1\}$. We must also discretize The boundary conditions . A centred formula gives

$$u_1 = u_{-1} \text{ et } u_{N+2} = u_N \tag{12}$$

and the boundary conditions $u(0) = u(L) = 0$ become

$$u_0 = u_{N+1} = 0 \tag{13}$$

we rewrite the Equation (11) in the discreet form

$$D\frac{u_{i-2} - 4u_{i-1} + 6u_i - 4u_{i+1} + u_{i+2}}{\Delta x^4} + \alpha_i u_i = P(x_i, H) \quad i = 1,2,3\cdots N \tag{14}$$

Then, the continuous problem becomes the following algebraic equation $AU = P$, where

$$A = \begin{pmatrix} \frac{7D}{\Delta x^4}+\alpha_1 & \frac{-4D}{\Delta x^4} & \frac{D}{\Delta x^4} & 0 & 0 & \cdots & 0 \\ \frac{-4D}{\Delta x^4} & \frac{6D}{\Delta x^4}+\alpha_2 & \frac{-4D}{\Delta x^4} & \frac{D}{\Delta x^4} & 0 & \cdots & 0 \\ \frac{D}{\Delta x^4} & \frac{-4D}{\Delta x^4} & \frac{6D}{\Delta x^4}+\alpha_3 & \frac{-4}{\Delta x^4} & \frac{D}{\Delta x^4} & \cdots & 0 \\ \vdots & \ddots & \ddots & \ddots & \ddots & \ddots & \vdots \\ 0 & \cdots & \frac{D}{\Delta x^4} & \frac{-4D}{\Delta x^4} & \frac{6D}{\Delta x^4}+\alpha_{N-2} & \frac{-4D}{\Delta x^4} & \frac{D}{\Delta x^4} \\ 0 & \cdots & 0 & \frac{D}{\Delta x^4} & \frac{-4D}{\Delta x^4} & \frac{6D}{\Delta x^4}+\alpha_{N-1} & \frac{-4D}{\Delta x^4} \\ 0 & \cdots & 0 & 0 & \frac{D}{\Delta x^4} & \frac{-4D}{\Delta x^4} & \frac{7D}{\Delta x^4}+\alpha_N \end{pmatrix}$$

$$P = \begin{pmatrix} p_1(x,H) \\ p_2(x,H) \\ \vdots \\ p_N(x,H) \end{pmatrix} \text{ et } U = \begin{pmatrix} u_1 \\ u_2 \\ \vdots \\ u_N \end{pmatrix}$$

Proposition 1. Note that A is symmetric positive definite under this assumption $\alpha(x) \geq 0$.

Proof. We will prove that $\alpha(x)\geq 0$ for all $x\in[0,L]$.

For all $(x,y)\in[0,L]\times[0,H]$ we have $\frac{\partial p(x,y)}{\partial y}=f_2^F(x,y)+\Delta v_2(x,y)$, for $y=H$ and $x\in[0,L]$ we have $v_2(x,H)=0$, then $\frac{\partial p(x,H)}{\partial y}=f_2^F(x,H)$. By choosing $f_2(x,H)=-g\times\rho^s\times h$ [3] where g is the gravity force and ρ^s the structure density, we obtain $\frac{\partial p(x,H)}{\partial y}\leq 0$. Finally, we deduce $\alpha(x)\geq 0$ for all $x\in[0,L]$.

3.2. Coupled Approximate Problem

Since A is symmetric positive definite, so we can use the conjugate gradient method to solve the following coupled problem. Find $(v,p)\in\left(H_0^1\left(\Omega_F^u\right)\right)^2\times L^2\left(\Omega_F^u\right)$ and U so that

$$\begin{cases} AU=P \\ \mu\int_{\Omega_F^u}\nabla v:\nabla w\mathrm{d}x-\int_{\Omega_F^u}p\,\mathrm{div}(v)\mathrm{d}x=\int_{\Omega_F^u}f^F w\mathrm{d}x, & \forall w\in\left(H_0^1\left(\Omega_F^u\right)\right)^2 \\ \int_{\Omega_F^u}\mathrm{div}(v)q\mathrm{d}x=0, & \forall q\in L^2\left(\Omega_F^u\right) \end{cases}$$

4. NUMERICAL METHOD

To solve numerically the coupled problem we use the following conjugate gradient algorithm.

Proposition 2. Let A be a symmetric positive definite matrix, and $u_0\in\mathbb{R}$. Let (u_k,r_k,z_k) be three sequences defined by the induction relations

$z_0=r_0=P-Au_0$, and for $k\geq 0\begin{cases} u_{k+1}=u_k+\lambda_k z_k \\ r_{k+1}=r_k-\lambda_k Az_k \\ z_{k+1}=r_{k+1}+\alpha_k z_k \end{cases}$

with

$$\alpha_k = \frac{\|r_{k+1}\|^2}{\|r_k\|^2} \text{ and } \lambda_k = \frac{\|r_k\|^2}{(z_k, Az_k)}$$

Then, $(u_k)_{0\le k\le k_0+1}$ is the sequence of approximate solutions of the the conjugate gradient method [7] .

Description of the computational method

Step 1: It computes in the initial field the velocity and the pressure.

Step 2: It uses the conjugate gradient algorithm to find the structure deformation u .

Step 3: It computes again the pressure and the velocity in the deformed domain

5. NUMERICAL RESULTS

Let the real noted test defined by $\text{test} = \|P\|^2 \times \varepsilon$. We define the stopping criterion of iterations for the conjugate gradient algorithm by $\text{rhoold} > \text{test}$ and $k < n$ where, k the number of iterations and $\text{rhoold} = \|z_0\|^2$.

We assume that the velocity on the boundary fluid domain is [3] :

$$G = \begin{cases} G_1(x_1,x_2) = v_1 = 30\left(1 - \frac{x_2^2}{H}\right) \text{ and } G_2(x_1,x_2) = v_2 = 0, \text{ on } \Sigma_1 \\ G_1(x_1,x_2) = v_1 = 30 \text{ on } G_2(x_1,x_2) = v_2 = 0, \text{ on } \Sigma_2 \\ G_1(x_1,x_2) = v_1 = 0 \text{ and } G_2(x_1,x_2) = v_2 = 0, \text{ on } \Sigma_3 \end{cases}$$

We take parameters for fluid and structure in [3] [4] . (Tables 1-3)

The Table 3 shows that, if we take the tolerance $\text{test} = 1.95\times10^{-3}$ we have the convergence of the algorithm after $k = 10$ iterations and $\text{rhoold} = 5.20\times10^{-21} < \text{test}$ and the norm of the displacement is $\|u\|_\infty = 3.3\times10^{-3}$.

Freefem ++ [8] is used for the numerical tests. Figures 2-7 following display the structure displacement, the pressure and the velocity.

Table 1. Parameters of the strcuture.

Paramerters	Values
E	$0.75 \times 10^{6}\ \mathrm{g/cm^{2} \cdot s^{2}}$
h	0.1 cm
ν	0.3
ρ^{s}	$1.1\ \frac{\mathrm{g}}{\mathrm{cm^{3}}}$

Table 2. Parameters of the fluid.

Paramerters	Values
μ	$0.035\ \frac{\mathrm{g}}{}$
L	3 cm
H	0.5 cm
g	$9.81\ \frac{\mathrm{cm}}{\mathrm{s^{2}}}$
$f^{F} = \left(f_1^{F}, f_2^{F}\right)$	$\left(0, -g \times \rho^{s} \times h\right)$

Table 3. Results related to the algorithm.

Paramerters	Values
ε	10^{-4}
test	1.95×10^{-3}
rhoold	5.20×10^{-21}
k	10
$\|u\|_{\infty}$	3.3×10^{-3}

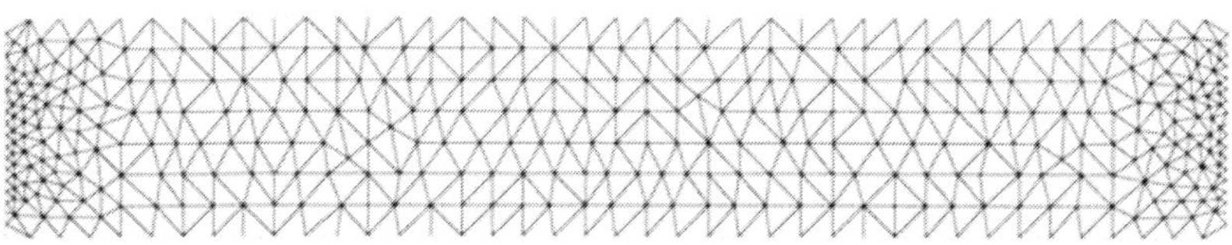

Figure 2. Initial grid.

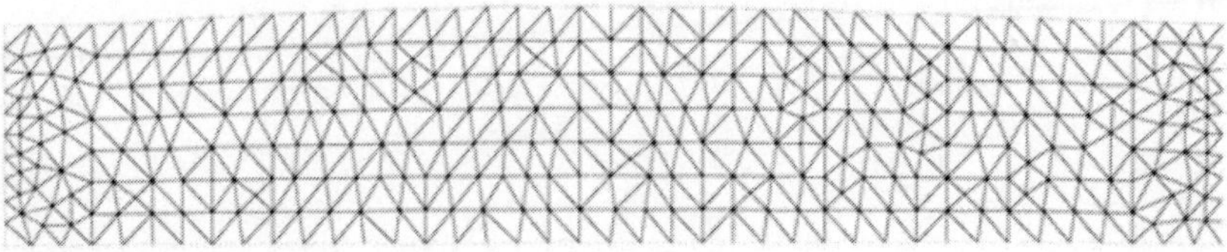

Figure 3. Final grid.

Figure 4. Pressure profile in the initial domain.

Figure 5. Pressure profile in the deformed domain.

Figure 6. Velocity profile in the initial domain.

Figure 7. Velocity profile in the deformed domain.

6. CONCLUSION

In this paper, we present a method to solve a steady coupled problem. Our method is based namely on the conjugate gradient algorithm, it takes simulta- neously into account three unknown parameters so that each of them depends on the others. To get the results it is necessary to solve the fluid in the initial domain with the finite element method in order to determine the displacement of the structure by the conjugate gradient method and finally to deduce the velocity and the pressure. The velocity, the pressure and the displacement profile compared to [2] [3] [5] appear good. In this work, the only thing to skill is to reduce the number of iterations and to apply this strategy on the unsteady problem.

REFERENCES

1. 1. Mbaye, I. (2012) Fourier Series Development Resulting from an Unsteady Coupled Problem. International Journal of Contemporary Mathematical Sciences, 7, 1695-1706.
2. Mbaye, I. (2013) Fourier Series Development for Solving a Steady Fluid Structure Interaction Problem. International Journal of Contemporary Mathematical Sciences, 8, 75-84.
3. Murea, C.M. (2005) The BFGS Algorithm for Nonlinear Least Squares Problem Arising from Blood Flow in Arteries. Computers & Mathematics with Applications, 49, 171-186. [Citation Time(s):6]
4. Mbaye, I. and Murea, C. (2007) Numerical Procedure with Analytic Derivative for Unsteady Fluid-Structure Interaction. Communications in Numerical Methods in Engineering, 24, 1257-1275.
5. Mbaye, I. (2012) Controllability Approach for a Fluid Structure Interaction Problem. Applied Mathematics, 3, 213-216.
6. Turek, S. and Hron, J. (2006) Proposal for Numerical Benchmarking of Fluid-Structure Interaction between an Elastic Object and Laminar Incompressible Flow. Computational Science and Engineering, 53, 371-385.
7. Allaire, G. (2007) Numerical Analysis and Optimization an Introduction to Mathematical Modelling and Numerical Simulation. Oxford University Press Inc., New York.
8. Hecht, F. and Pironneau, O. (2003) A Finite Element Software for PDE: FreeFem++.

CHAPTER 6

Linear Prolate Functions for Signal Extrapolation with Time Shift

Daniela Valente, Michael Cada, Jacek Ilow

Department of Electrical and Computer Engineering, Dalhousie University, Halifax, Canada

ABSTRACT

We propose a low complexity iterative algorithm for band limited signal extrapolation. The extrapolation method is based on the decomposition of finite segments of the signal via truncated series of real-valued linear prolate functions. Our theoretical derivation shows that given a truncated series (up to a selectable value) of prolate functions, it is possible to extrapolate the band limited function elsewhere if each extrapolated portion of the function is subject only to moderate truncation errors that we quantify in this paper. The effects of different sources of errors have been analyzed via extensive simulations. We have investigated a property of the signal decomposition formula based on linear prolate functions whereby the integration interval does not need to be symmetric with respect to the origin while time-shifted prolate functions are used in the series.

Keywords: Real-Valued Band Limited Eigenvectors, Signal Decomposition, Signal Extrapolation

1. INTRODUCTION

In the early 60's David Slepian and his colleagues discovered the bandlimited function that is maximally concentrated, in the mean-square sense, within a given time interval; this function is the prolate spheroidal wave function (PSWF) of zero-order.

The linear prolate functions (LPFs) are the one-dimensional version of the prolate spheroidal functions and they form sets of bandlimited functions which are orthogonal and complete over a finite interval. Moreover, unlike other functions, they are also complete and orthogonal over the infinite interval. An additional property is that the finite Fourier transform (FT) of a linear prolate function is proportional to the same prolate function. Although there are other functions which are their own infinite Fourier transform, only the prolate functions enjoy the property for the finite transform: this property uniquely defines the prolate functions [1] . Associated with each function, there is an eigenvalue $\lambda_n(c)$ and a free parameter c which is a useful descriptor of sys- tem performance [2] . Some of the mentioned mathematical properties make the prolate functions easily applicable to optics [3] . In particular, we are interested in the problem of determining a bandlimited function from the knowledge of a finite segment of the function, since it is relevant in many practical situations from application to filters in communication systems [4] to optical systems when, for example, due to intrinsic instrumental limits, only limited observation data are available.

Specifically, in the research area of bandlimited signal extrapolation, there have been contributions with iterative and non-iterative algorithms for extrapolation of signals in the LCT (linear canonical transform) domain that is a generalization of the Fourier transform. The challenges of convergence of algorithms based on the Gerchberg-Papoulis (GP) algorithm [5] and an applica- tion to high frequencies have been extensively investigated [6] . However, appro- aches based on the use of the prolate spheroidal wave functions [7] need to provide efficient ways to compute the prolate functions.

In this paper, we benefit from a proprietary algorithm developed theoretically and implemented numerically by Cada [8] , for accurate generation of linear prolate functions with desired high precision to use LPFs for signal extrapolation. In what follows, we introduces the basics of signal expansion using the linear prolate functions in Section 2; Section 3 presents our approach to signal extrapolation based on LPFs. In Section 4 and Section 5 simulation results, error analysis and numerical examples are presented and discussed. Finally conclu- sions are drawn in Section 6.

2. SIGNAL EXPANSION

As sets of bandlimited functions, orthogonal on the finite interval and orthonormal on the infinite interval, the linear prolate functions $\psi_n(c,t)$ can be successfully used for the expansion of a generally complex, bandlimited function $f(t)$:

$$f(t)=\sum_{n=0}^{\infty}a_n\psi_n(c,t) \tag{1}$$

the representation is valid for all t, the bandwidth parameter is $c=t_0\Omega_0$ where Ω_0 represents the finite bandwidth or a cutoff frequency, and t_0 is the time interval. The function $f(t)$ is supposed to be Ω_0-bandlimited. Adopting the criterion of a minimized mean-square error, the expansion coefficients a_n in (1) are given by:

$$a_n=\int_{-\infty}^{\infty}f(t)\psi_n(c,t)\mathrm{d}t \tag{2}$$

There is an alternative way to derive the coefficients a_n using only the values of $f(t)$ within the finite observation interval $[-t_0,t_0]$ and introducing the eigenvalues $\lambda_n(c)$. After multiplying (1) by $\psi_m(c,t)$, integrating as reported in [1] and from the orthogonality properties of the LPFs which are valid on both the finite and infinite interval, one can obtain:

$$a_n=[\lambda_n(c)]^{-1}\int_{-t_0}^{t_0}f(t)\psi_n(c,t)\mathrm{d}t \tag{3}$$

The latest expression for $\{a_n\}$ together with (1) states that when the bandlimited function $f(t)$ is known over a finite interval of extend $2t_0$ then $f(t)$ is theoretically known everywhere if one can accurately calculate the coefficients a_n for $n\to\infty$, the functions $\psi_n(c,t)$ and the eigenvalues $\lambda_n(c)$. $\lambda_n(c)$ can be regarded as the index of energy concentration of each function ψ_n in $[-t_0,t_0]$. Therefore:

$$f(t) \square \sum_{n=0}^{N} [\lambda_n(c)]^{-1} \psi_n(c,t) \int_{-t_0}^{t_0} f(t')\psi_n(c,t') dt' \tag{4}$$

Accurate computing of ψ_n, λ_n and a_n for $n > 2c/\pi$ (where $N_{crit} = 2c/\pi$ is known as the critical value) turns the orthogonal expansion expression presented in (1) into a signal extrapolation problem. Indeed, for any LPFs set with a fixed c, the energy concentration of the functions within $[-t_0, t_0]$ decreases as the order n increases and for $n = N_{crit}$, the signal's maximum concentration reaches the boundary of the observation interval. Hence the summation of $\{\lambda_n(c)\}$ is mostly determined by the first N_{crit} terms whose individual value is very close to 1, and the series $\sum_{n=0}^{\infty} \lambda_n(c)$ converges to a finite value $(2c/\pi)$, as extensively analyzed in [9] . Accurate estimation of the overlap integral $\int_{-t_0}^{t_0} f(t)\psi_n(c,t) dt$ for high orders of n becomes then a challenging problem of high-precision numerical integration with an absolute necessity of having $\psi_n(c,t)$ with a high precision as well [8] .

3. SIGNAL EXTRAPOLATION

Our main objective in signal extrapolation using linear prolate functions aims to take advantage of a generalized expression stated in [3] , never exploited so far, for the coefficients in (3) which enables the finite interval $2t_0$ to not be nece- ssarily symmetric with respect to the origin. Hence, for a general interval $T - t_0 \le t \le T + t_0$, relation (4) becomes:

$$f(t) \square \sum_{n=0}^{N} [\lambda_n(c)]^{-1} \psi_n(c,t-T) \int_{T-t_0}^{T+t_0} f(t')\psi_n(c,t'-T) dt' \tag{5}$$

Substituting $y = t' - T$, the following expression for $f(t)$ is obtained:

$$f(t) \square \sum_{n=0}^{N} [\lambda_n(c)]^{-1} \psi_n(c,t-T) \int_{-t_0}^{t_0} f(y+T)\psi_n(c,y) dy \tag{6}$$

with

$$a_n(T) = \frac{1}{\lambda_n(c)} \int_{-t_0}^{t_0} f(y+T)\psi_n(c,y)\,\mathrm{d}y \tag{7}$$

Thanks to the significant generalization for the calculation of coefficients $\{a_n\}$, one can use (6) to perform the signal extrapolation on $f(t)$ beyond the observation interval $[-t_0, t_0]$ using an iterative approach.

Specifically, we start from the assumption that the function $f(t)$ is perfectly known in the interval $[-t_0, t_0]$. We use (6) with $T = 0$ to extrapolate the signal by an interval $\Delta t^{(0)}$ using already proposed algorithms to obtain accurate calculations of $a_n(0)$ till $N = N^{(0)} > 2c/\pi$. Instead of pursuing the more challenging computing for $N \square 2c/\pi$, Formula (6) is re-applied for $T = \Delta t^{(0)}$ to extrapolate the signal by an additional interval $\Delta t^{(T)}$ via accurate calcula- tions of $a_n(T)$ to $N^{(T)} > 2c/\pi$. The procedure is then repeated for the i-th iteration and up to the number of iterations that has been set. Specifically, at each iteration i, we form the function which becomes the input for iteration $i+1$:

$$\begin{aligned} f_i(t) &= \tilde{f}_i(t) + \left[f_{i-1}(t) - \tilde{f}_i(t)\right] p_{\Delta t}(t) \\ &= \begin{cases} f_{i-1}(t), -t_0 + i\Delta t \le t \le t_0 + (i-1)\Delta t \\ \tilde{f}_i(t), t_0 + (i-1)\Delta t < t \le t_0 + i\Delta t \end{cases} \end{aligned} \tag{8}$$

with $\tilde{f}_i(t)$ being defined in (6). For the sake of simplicity Δt is chosen to be the same at every iteration. Also,

$$p_{\Delta t}(t) = \begin{cases} 1, & -t_0 + i\Delta t \le t \le t_0 + (i-1)\Delta t \\ 0, & \text{otherwise} \end{cases}$$

4. NUMERICAL RESULTS

A LPFs set with bandwidth parameter $c = 20\pi$ and $t_0 = 1$ is used as the orthogonal basis for the proposed extrapolation method. The functions are discretized in time at a sampling rate of 0.001 for numerical implementation and each discrete sample has a high numerical precision greater than 100 digits. The software Mathematica characterized by high precision computing has been used for the simulations. Extrapolation is carried out on the Ω_0-bandlimited test function shown in Figure 1.

$$f(t) = \cos\left(2\pi t - \left(\frac{\pi}{11}\right)\right) + \cos\left(\frac{2\pi t}{7}\right) - \cos\left(\frac{3\pi t}{2}\right)$$

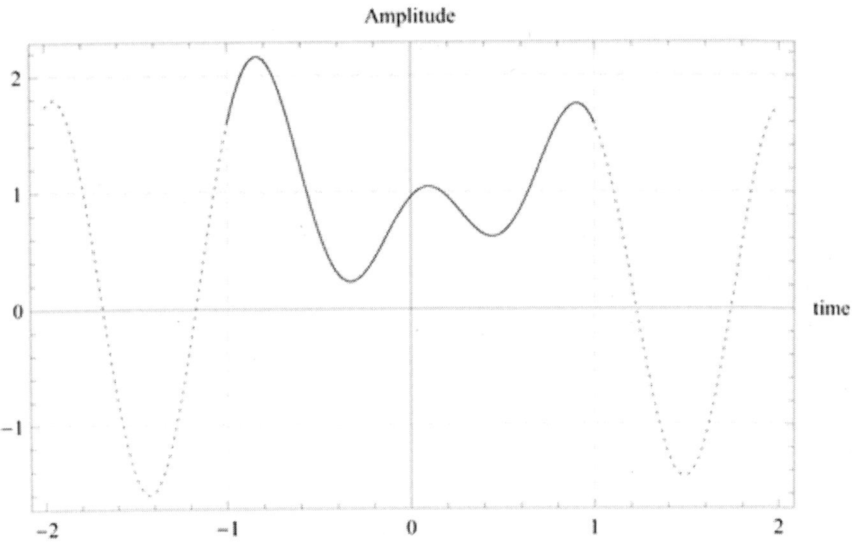

Figure 1. $f(t)$ in the interval $[-1,1]$.

4.1. Perfect Knowledge of $f(t)$ in the Integration Interval

In order to test the proposed approach for signal extrapolation as described in Section 3, we consider the ideal case first. This assumption means that at each iteration of the extrapolation, the function $f(t+T)$ in the integral in (6) is known in Mathematica user-defined precision. Figure 2 shows reconstruction/ extrapolation after the first iteration when the time shift is $T = 0$, the truncation value is $N = 97$ and the Mathematica built-in interpolation and integration functions are used to calculate the overlap integral. In Figure 3, we show results for signal reconstruction/extrapolation after the second iteration with an actual application of our time shift approach; specifically $T=-t_0/2=-0.5$ and $N=97$. In both cases, the extrapolation succeeds to extrapolate the signal over an additional interval which is up to the 60% of half of the time range where the function is known. The presence of the truncation error is discussed in Section 5.

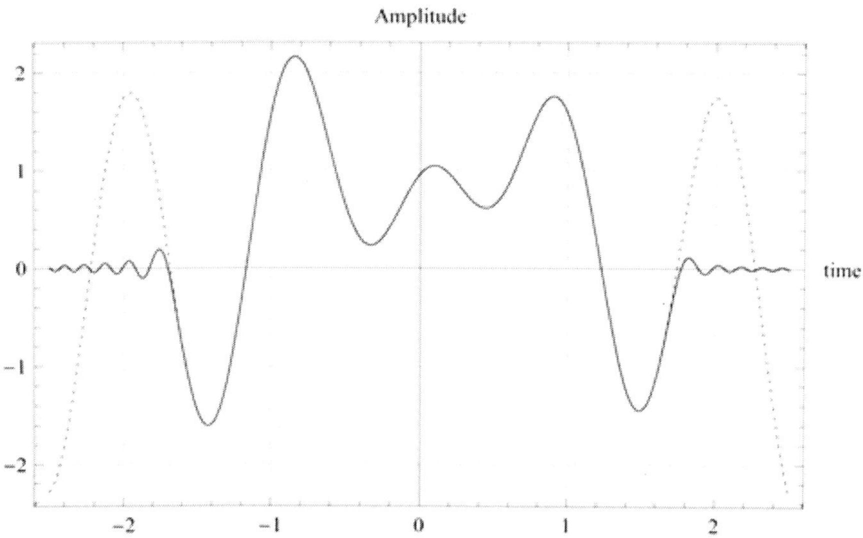

Figure 2. Extrapolated $f(t)$, no time shift.

4.2. Estimate of $f(t)$ in the Integration Interval

We consider a more realistic case when the function $f(t+T)$ in the integral in (6) is known in Mathematica user-defined precision only for $T=0$. Figure 4(b) shows signal reconstruction/extrapolation after 16 iterations with a total time shift $|T|=(i-1)\cdot\Delta t=0.3$ and $N=97$.

Figure 4(c) shows results after 36 iterations with a total time shift $|T|=0.7$ and $N=97$. Each progressive shift is equal to $\Delta t=0.02$. At every iteration $i+1$, we use the function $\tilde{f}_i(t)$ as the new input to (6) to make the integral calculation successful.

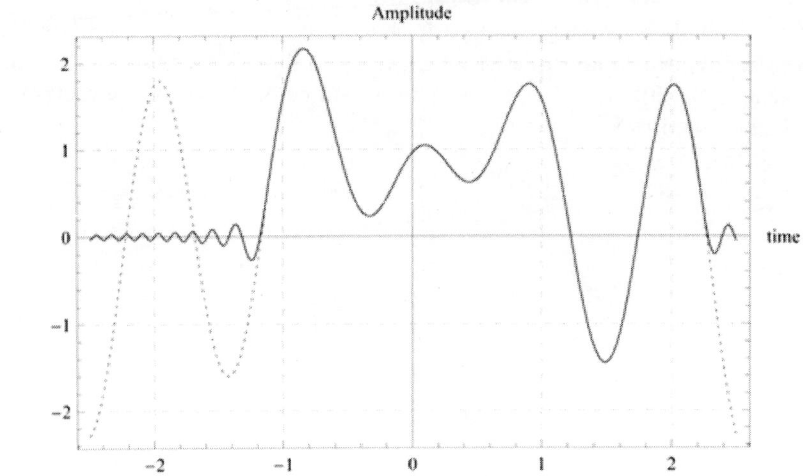

Figure 3. Extrapolated $f(t)$, shift$^{[-0.5,1.5]}$.

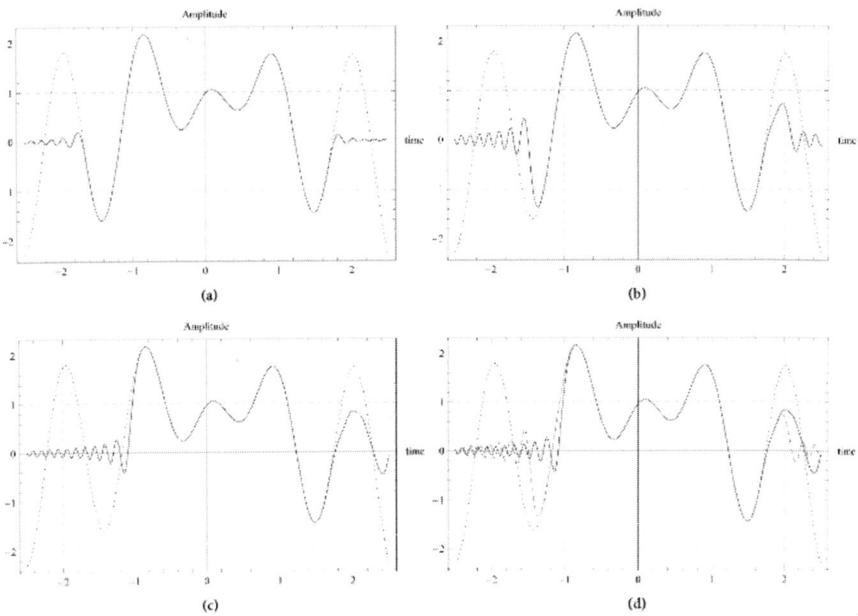

Figure 4. Extrapolation outputs. (a) $f(t)$ known in $[-1,1]$, $i=1$; (b) $f(t)$ known in $[-0.7,1.3]$, $i=16$; (c) $f(t)$ known in $[-0.3,1.7]$, $i=36$; (d) Reconstruction cases in (b) and (c) (solid line).

The piecewise polynomial interpolation method present- ed in [10] has been applied for the accurate computing of the overlap integral and the LPFs set with bandwidth parameter $c = 20\pi$ has been used. Indeed, in terms of the normalized mean-squared error (NMSE), the method in [10] performs superiorly when compared to the iterative approach proposed in [6] and the generalized PSWFs (prolate spheroidal wave functions) expansion method proposed in [7] . Specifically, for comparison purposes, Figure 4(a) is obtained by setting $i = 1$ and hence $T = 0$, and reproduces results presented in [10] .

In Figure 5, details are shown for the extrapolation of the portion of the function in the time interval $[1.65, 1.9]$. Despite the effect of accumulated errors, it verifies that given the same truncation value N, the shift-approach ($i = 36$) outperforms the reference approach ($i = 1$) when the extrapolation capability of the reference approach vanishes.

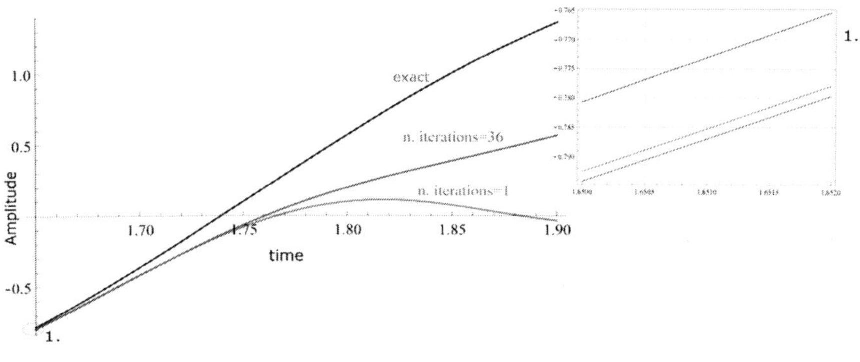

Figure 5. Detailed curves.

The difference quotients in Table 1 calculated between time instant 1.73 and time instant 1.78 are indices of the curves slope and show that the shift-approach follows better the slope of the exact function. In Figure 6, extrapolation details are shown for the portion of the function in the time interval $[1.85, 2.2]$. Despite the effect of accumulated errors, the shift-approach for $i = 36$ outper- forms the shift-approach for $i = 16$. The difference quotients in Table 2 calcu- lated between time instants 1.90 and 2.00 show that the shift-approach for $i = 36$ follows better the slope of the exact function.

Table 1. Difference quotient.

$[f(x_0+h)-f(x_0)]/h$	exact	36 iterations ($\|T\|=0.7$)	16 iterations ($\|T\|=0.3$)	1 iteration (no shift)
$[f(1.74)-f(1.73)]/0.01$	9.40661	6.83248	6.83238	6.36935
$[f(1.75)-f(1.74)]/0.01$	9.46608	6.49618	6.49601	5.82323
$[f(1.76)-f(1.75)]/0.01$	9.4923	6.11953	6.11922	5.18117
$[f(1.77)-f(1.76)]/0.01$	9.48503	5.71649	5.71596	4.45234
$[f(1.78)-f(1.77)]/0.01$	9.44413	5.30341	5.30248	3.651

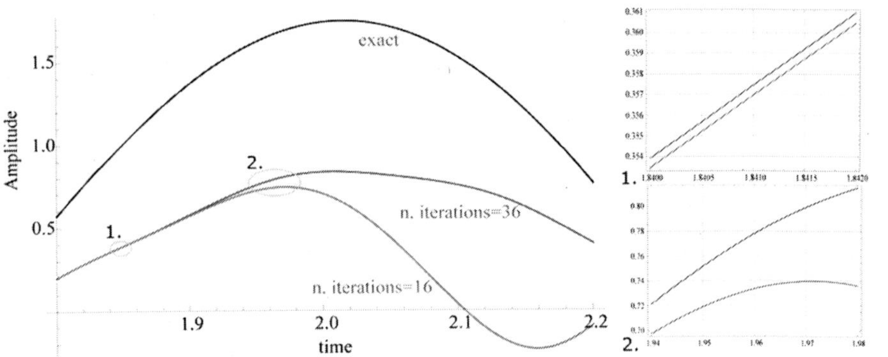

Figure 6. Detailed curves.

5. ERROR ANALYSIS

The proposed method is subject to an inherent series truncation error. Its mean squared error expression is the following, after an extrapolation interval T_e:

$$E_T = \int_{-t_0}^{+t_0+T_e} |f(t) - f_N(t)|^2 \, dt = \int_{-t_0}^{+t_0} |f(t) - f_N(t)|^2 \, dt + \int_{+t_0}^{+t_0+T_e} |f(t) - f_N(t)|^2 \, dt \tag{9}$$

$$= \sum_{n=N+1}^{\infty} a_n^2 \lambda_n(c) + \int_{+t_0}^{+t_0+T_e} \left| f(t) - f_N(t) \right|^2 \mathrm{d}t \tag{10}$$

Table 2. Difference quotient.

$\left[f(x_0+h)-f(x_0)\right]/h$	exact	36 iterations ($\lvert T\rvert=0.7$)	16 iterations ($\lvert T\rvert=0.3$)
$\left[f(1.91)-f(1.90)\right]/0.01$	6.00265	3.8751	3.61211
$\left[f(1.92)-f(1.91)\right]/0.01$	5.54131	3.8203	3.45436
$\left[f(1.93)-f(1.92)\right]/0.01$	5.05783	3.67135	3.16975
$\left[f(1.94)-f(1.93)\right]/0.01$	4.55379	3.41957	2.74234
$\left[f(2.00)-f(1.99)\right]/0.01$	1.19338	0.532228	−2.40967

The first term in the summation in (9) represents the error in the fit of $f_N(t)$ (defined in (6)) to $f(t)$ within the interval $[-t_0, t_0]$. Specifically, as reported in [1] , the calculation of $\int_{-t_0}^{+t_0} \left| f(t) - f_N(t) \right|^2 \mathrm{d}t = \int_{-t_0}^{+t_0} \left(\sum_{n=N+1}^{\infty} a_n \psi_n(c,t) \right)^2 \mathrm{d}t$ into the sum form in (10) follows from the fact that $\psi_n(c,t)$ are orthogonal on the interval $[-t_0, t_0]$. The term $\int_{+t_0}^{+t_0+T_e} \left| f(t) - f_N(t) \right|^2 \mathrm{d}t$ is the truncation error in

the extrapolation interval and then depends on the quantity T_e. It is reasonable to consider the first term not critical for truncation values N above $2c/\pi$, when the energy factor $\lambda_n(c)$ rapidly approaches zero (an example is shown in Figure 7 for $c = 20\pi$). The last statement has to satisfy the condition that for a sufficiently large n, the integral $\left(\int_{-t_0}^{t_0} f(t) \psi_n(c,t) \mathrm{d}t \right)^2$ tends to zero faster than the corresponding $\lambda_n(c)$ at the denominator of the products $a_n^2 \lambda_n(c)$. This consideration motivated our work and indirectly highlights again that calculating accurate coefficients a_n for large n is critical since both overlap integrals and eigenvalues become very small quantities. This has been a known problem since the 60's of the last century. The critical aspect is an accurate calculation of $\psi_n(c,t)$, which is now possible [8] . Under the assumption that the function $f(t)$ is initially known in $[-t_0, t_0]$, the sum of truncation errors, computer roundoff and analog to digital conversion errors makes the practical implementation of the proposed iterative approach subject to a total error that has been measured as normalized mean-square error (NMSE) between the original function and the extrapolated signal f_e as:

$$\text{NMSE} = \frac{\|f_e - f\|^2}{\|f\|^2} = \frac{\int_{-t_0+T_e}^{t_0+T_e} |f_i(t) - f(t)|^2 \, dt}{\int_{-t_0+T_e}^{t_0+T_e} |f(t)|^2 \, dt} \tag{11}$$

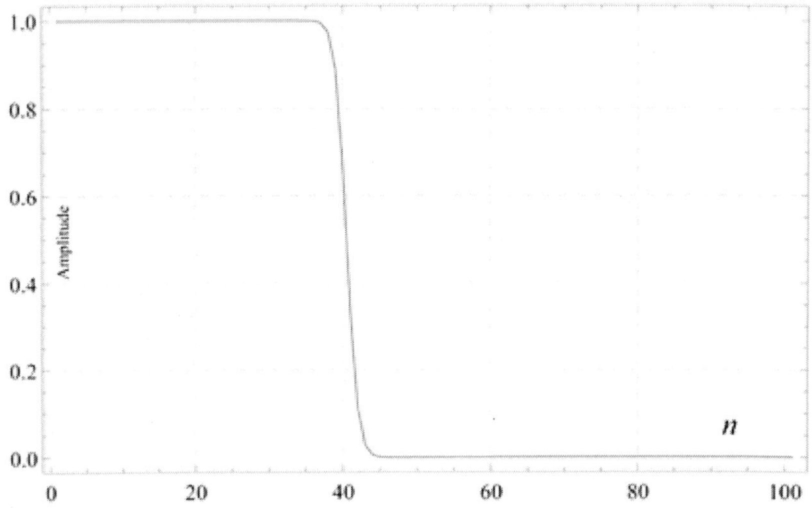

Figure 7. Eigenvalues $\lambda_n(c)$ vs. order n for $c = 20\pi$ and $n_{crit} = 2c/\pi = 40$.

It is clear from Table 3 that the zero-shift still gives the best results up to the point it is capable to extrapolate correctly (e.g., $T_e = 0.65$); increasing N will extend T_e and it will still be the best extrapolation up to that point. However, if N is a limit, as in practice always is, our shifting method performs better. This can be interpreted as a consequence of the multiplication by the quantity $\psi_n(c, t-T)$ in (6) which means moving the energy of the LPFs accordingly to the shift of the function. Hence results confirm that our approach works in principle and, as pointed out in [10], the chosen algorithm for computing the overlap integral shows its sensitivity to reduced accuracy and noise when applied to an already extrapolated portion of the function. It is also important observing that the reported NMSE values have been calculated for $f_i(t)=\tilde{f}_i(t)$ following notation in (8). The extrapolation can be optimized by reconstructing the extra- polated function $f_i(t)$ as concatenation of the known function with segments of optimum estimates. However, we still observe numerical inaccuracies oc- curring at the points of concatenations, which is presently under investigation.

Table 3. Normalized mean-square error (NMSE).

T_e	NMSE-36 iterations ($\|T\|=0.7$)	NMSE-16 iterations ($\|T\|=0.3$)	NMSE-1 iteration (no shift)
0.65	1.432420×10^{-7}	1.432421×10^{-7}	1.02351×10^{-7}
0.78	1.99422×10^{-4}	1.99432×10^{-4}	2.28493×10^{-4}

6. CONCLUSION

In this paper, we have proposed and implemented a low complexity iterative algorithm for bandlimited signal extrapolation based on orthogonal projections over real-valued eigenvectors: the linear prolate functions. The method is valid for an arbitrary large range of frequencies with immediate applications in signal processing. The main contribution of our work is a theoretical derivation such that given a truncated series (up to a selectable value) of prolate functions, it is possible to extrapolate the bandlimited function (initially known in a limited time interval) elsewhere if each extrapolated portion of the function is subject only to moderate series truncation errors. These errors are controllable by the depth of extrapolation at each iteration. By doing so and with the aim of finding an alternative solution to the initial problem of implementing an accurate summation of infinite terms, we have investigated a property of the signal de- composition formula based on LPFs according to which the integration interval does not need to be symmetric with respect to the origin while time-shifted prolate functions are used in the summation. Also, we have investigated the effects of different sources of errors by implementing and analyzing the iterative algorithm as a generalization of the special case presented in [10] . Our method has shown to outperform concurrent approaches in terms of the normalized mean-square error of the extrapolated signal.

REFERENCES

1. Slepian, D. and Pollak, H.O. (1961) Prolate Spheroidal Wave Functions, Fourier Analysis, and Uncertainty—I. The Bell System Technical Journal, 40, 43-63.
2. Moore, I.C. and Cada, M. (2004) Prolate Spheroidal Wave Functions, an Introduction to the Slepian Series and Its Properties. Applied and Computational Harmonic Analysis, 16, 208-230.
3. Frieden, B.R. (1971) Evaluation, Design and Extrapolation Methods for Optical Signals Based on Use of the Prolate Functions. Progress in Optics, 9, 311-407.
4. Soman, S. and Cada, M. (2017) Design and Simulation of a Linear Prolate Filter for a Baseband Receiver. Journal of Information Technology and Software Engineering, 7, 197. https://www.omicsgroup.org/journals/design-and-simulation-of-a-linear-prolate-filter-for-a-baseband-receiver-2165-7866-1000197.pdf
5. Papoulis, A. (1975) A New Algorithm in Spectral Analysis and Band-Limited Extrapolation. IEEE Transactions on Circuits and Systems, 22, 735-742.
6. Shi, J., Sha, X., Zhang, Q. and Zhang, N. (2012) Extrapolation of Bandlimited Signals in Linear Canonical Transform Domain. IEEE Transactions on Signal Processing, 60, 1502-1508.
7. Zhao, H., Ran, Q.W., Ma, J. and Tan, L.Y. (2010) Generalized Prolate Spheroidal Wave Functions Associated with Linear Canonical Transform. IEEE Transactions on Signal Processing, 58, 3032-3041.
8. Cada, M. (2012) Private Communication. Department of Electrical and Computer Engineering, Dalhousie University, Halifax, Canada.
9. Xiao, H. (2001) Prolate Spheroidal Wave Functions, Quadrature, Interpolation, and Asymptotic Formulae. Ph.D. Dissertation, Department of Computer Science, Yale University, New Haven.
10. Devasia, A. and Cada, M. (2013) Bandlimited Signal Extrapolation Using Prolate Spheroidal Wave Functions. IAENG International Journal of Computer Science, 40, 291-300.

CHAPTER 7

Multi-Objective Mathematical Model for the Optimal Time to Harvest Sugarcane

Surattana Sungnul[1,2], Wisanlaya Pornprakun[1,2], Santipong Prasattong[1], Chanasak Baitiang[1]

[1]*Department of Mathematics, King Mongkut's University of Technology North Bangkok, Bangkok, Thailand*

[2]*Centre of Excellence in Mathematics, CHE, Bangkok, Thailand*

ABSTRACT

In this paper, the sugarcane and sugar industry in Thailand is studied. The government determines the sugarcane prices which is based on the two main factors: 1) weight and 2) commercial cane sugar (standard value equal 10 C.C.S.). Usually, the C.C.S. will increase with time and the weight will de- crease. The main purpose of this research is to find the optimal harvest time to maximize revenue and minimize gathering cost. The mathematical model is first formulated under the regulations of the Office of the Cane and Sugar Board (OCSB). The ε-constraints method is then applied to solve the multi- objective mathematical model. The optimal harvest times in the four regions of Thailand (Northern, Central, Eastern, North-Eastern) for crop years 2012/ 13, 2013/14 and 2014/15 are obtained for comparison.

Keywords: Multi-Objective Mathematical Model, Sugarcane, ε-Constraint Method

1. INTRODUCTION

Sugarcane is an important crop in Thailand because sugarcane is a raw material of the sugar industry, which is one of the top five agricultural products that makes revenue for the country each year. In the last ten years, the link between agriculture and industry and the structure of sugar markets have both changed. These changes have affected the amount and direction of the global sugar trade [1] . Thailand is an exporter with export volume ranked second in the world and and it also has a large role in the Asian market. Because Thailand has the advan- tage of location and an oversupply of sugar it can respond quickly to changes in demand in the Asian market. The forecast volume of exports in the next ten years is 11 million tons, which is an increase of 69 percent over the average volume of exports in the past ten years. Thus, sugarcane is an important eco- nomic crop in Thailand.

A survey by the Promotion Center of the Cane and Sugar Industry has reported that, in the crop year 2012/13 the average cost of production of sugarcane including the cost of transportation was 1196.31 baht per ton (transport cost was 143.61 baht per ton). This average cost of production was an increase of 246.31 baht per ton over the cost of 950 baht per ton in crop year 2011/12. This increase was because of increases in land rents and prices of labour, fertilizer and ma- terials. In crop year 2012/13, sugarcane production for the whole country was at the level of 94.64 million tons from sugarcane plantations of 9.33 million rai for an average yield of 10.14 tons per rai. This production represented a decrease from crop year 2011/12 when the annual production was 97.98 million tons. The decrease occurred because of drought during the crop year 2012/13 when the rains came later than usual.

Many researchers have discussed the renewable energy which can be produced from the biomass of agricultural products. For example, the biomass can be used to produce ethanol which can then increase the profit and reduce the cost of the agricultural product. In 2008, Helenice de Oliveira Florentino [2] studied multi-objective optimization of the economics of sugarcane harvest bio- mass in Brazil. The aim of this work was to develop a model to optimize plant variety selection, to minimize cost of the residual biomass transfer process, to evaluate the economics of using this material, and to address sucrose production subject to planting area constraints, and considering distance from planting area to processing center. For this, 0 - 1 multiple objective linear programming tech- niques were used. The results showed the viability of the model when selecting varieties and the increased profit that could be obtained from residual biomass use. In 2010, Maximiliano Salles Scarpari and Edgar Gomes Ferreira de Beauclair [3] used linear programming to develop an optimized plan for sugarcane farming in Brazil. The program language used was General Algebraic Modelling System (GAMS) as this system was seen to be an excellent tool to allow profit maximization and harvesting time schedule optimization in the sugar mill stu- died.

The results supported this optimized planning model as being a very useful tool for sugarcane management. In 2012, Francisco Regis Abreu Gomes [4] studied a bi-objective mathematical model for choosing sugarcane varieties with harvest residual biomass in energy co-generation. This study developed a bi- objective mathematical model for choosing sugarcane varieties that result in maximum revenue from electricity sales and minimum gathering cost of sugar- cane by harvesting residual biomass.

The approach used to solve the proposed model was based on the ε-constraints method. Experiments were performed using real data from sugarcane varieties and costs and showed effectiveness of the model and method proposed. These experiments showed the possibility of increasing net revenue from electricity sale, i.e., discounting the cost increase with residual biomass gathering, by up to 98.44%.

At present, the Thai government determines the price of sugarcane each year depending on its quality (sweetness) and weight. In general, the sweetness in- creases over time while the weight decreases. In this paper, a model is developed to find the optimal harvest time of sugarcane in order to maximize the agri- culturists revenue and to minimize the cost.

The mathematical model is for- mulated in a multi-objective optimization framework under the Office of Cane and Sugar Board (OCSB) conditions. The ε-constraints method is used to solve the multi-objective optimization problem. The results for production of fresh sugarcane (not burnt before cutting) and fired sugarcane (was burnt before cutting) in crop years 2012/13, 2013/14 and 2014/15 are compared and analyzed.

2. MULTI-OBJECTIVE OPTIMIZATION

A multi-objective optimization problem has a number of objective functions which are to be minimized or maximized. As in a single-objective optimization problem, the problem usually has a number of constraints which any feasible solution (including the optimal solution) must satisfy. In its general form, the mul- ti-objective optimization problem (MOOP) can be stated as follows (see, e.g., [5]):

$$
\begin{aligned}
\text{Minimize} \quad & f_m(x), m = 1,2,\cdots,M \\
\text{subject to} \quad & g_j(x) \geq 0, j = 1,2,\cdots,J \\
& h_k(x) = 0, k = 1,2,\cdots,K \\
& x_i^{(L)} \leq x_i \leq x_i^{(U)}, i = 1,2,\cdots,n
\end{aligned}
\tag{1}
$$

A solution x is a vector of n decision variables : $x = (x_1, x_2, \cdots, x_n)^T$. The last set of constraints are called variable bounds as they restrict each decision variable x_i to take a value within a lower bound $x_i^{(L)}$ and an upper bound $x_i^{(U)}$. The feasible region D for the MOOP is the set of vectors x that satisfy all constraints. If each objective function $f_m(x)$ is denoted by z_m and the vector of all objective functions is denoted by z, then the objective function space can be defined as:

$$
Z = \left\{ z = (z_1, z_2, \cdots, z_M) \middle| z_m = f_m(x), \forall x \in D, m = 1,2,\cdots,M \right\}
\tag{2}
$$

In this work, the ε-constraints method [5] was used to solve the MOOP. This method consists of reformulating a multi-objective problem by choosing the most important objective while maintaining other objectives constrained by upper bounds defined by a decision maker. For example, if $f_\mu(x)$ is selected as the most important objective, then the problem can be reformulated as follows:

$$
\begin{aligned}
\text{Minimize} \quad & f_\mu(x) \\
\text{subject to} \quad & f_m(x) \leq \varepsilon_m, m = 1,2,\cdots,M; m \neq \mu \\
& g_j(x) \geq 0, j = 1,2,\cdots,J \\
& h_k(x) = 0, k = 1,2,\cdots,K \\
& x_i^{(L)} \leq x_i \leq x_i^{(U)}, i = 1,2,\cdots,n
\end{aligned}
\tag{3}
$$

where ε_m is an upper bound of objective m, $m = 1,2,\cdots,M; m \neq \mu$ and D is the set of feasible solutions to the ε-constrained problem.

3. MATHEMATICAL MODEL

In this section, the mathematical model is formulated in a multi-objective op- timization framework under the Office of the Cane and Sugar Board (OCSB) conditions. The main purpose is to find the optimal harvest time of sugarcane in order to maximize the agriculturists' revenue and minimize the cost. This pro- blem is a bi-objective optimization problem to maximize revenue and minimize cost. For the ε-constraint formulation, the maximum of revenue from the sugarcane sale will be chosen as the objective function and the gathering cost of production of the sugarcane will be bounded above by an ε-constraint.

Revenue of selling: The government determines the sugarcane prices which are based on two main factors; 1) weight and 2) commercial cane sugar (C.C.S.).

1) Revenue from the weight of sugarcane: The OCSB classifies sugarcane going into factories into 2 types; a) fresh sugarcane and b) fired sugarcane. As determined by the government, agriculturists who sell fired sugarcane will have 20 baht/ton deducted from the price of sugarcane based on weight. The factory will share this amount of money between agriculturists who sell fresh sugarcane and increase their price based on weight by a maximum of 70 baht/ton. The price of fired sugarcane based on weight $P_w(B)$ is therefore defined by

$$P_w(B) = P_w - 20, \tag{4}$$

where P_w is the basic price of sugarcane based on weight set by the govern- ment (baht/ton). The price of fresh sugarcane based on weight $P_w(A)$ is de- fined by

$$P_w(A) = P_w + \frac{20a_{j,k}(B)}{a_{j,k}(A)}, \tag{5}$$

where $a_{j,k}(A)$ is the amount of fresh sugarcane (tons) from planted area j in the harvest at time k and $a_{j,k}(B)$ is the amount of fired sugarcane (tons) from planted area j in the harvest at time k. Reasonable values for the total planted areas of sugarcane were estimated from OCSB data. The actual values for $a_{j,k}$ were computed by the optimization program.

Therefore the revenue from weight of sugarcane is defined by

$$P_1(i) = P_w(i)\,a_{j,k}(i), \quad i = A, B \tag{6}$$

2) Revenue from C.C.S. of sugarcane. The price per ton based on C.C.S is defined by

$$P_2(i) = \left[P_c\left(1 + 0.06y_{j,k}\right)\right]a_{j,k}(i), \quad i = A, B \tag{7}$$

where P_c is the price per ton determined by the government for sugarcane with 10 C.C.S.

$y_{j,k} = \text{C.C.S.} - 10$, where C.C.S. is the average C.C.S. from sugarcane in planted area j in the harvest at time k. The factor 0.06 is the rate of change of price per 1 C.C.S. change from the base level of 10.

Therefore, revenue $\left[RV_{j,k}(i)\right]$ from the sale of sugarcane from planted area j in the harvest at time k is determined by adding Equation (6) and Equation (7) as shown in Equation (8)

$$RV_{j,k}(i) = P_1(i) + P_2(i) = \left\{P_w(i) + \left[P_c\left(1 + 0.06y_{j,k}\right)\right]\right\}a_{j,k}(i), \quad i = A, B \tag{8}$$

Gathering cost of production: The gathering cost of production can be separated into two parts: 1) average cost of production on the farm and 2) cost of transport. The total gathering cost $GC_{j,k}(i)$ baht/ton of sugarcane production from planted area j in the harvest at time k is given by

$$GC_{j,k}(i) = \left(C_j + C_{T_j}\right)a_{j,k}(i), \quad i = A, B \tag{9}$$

where C_j is the average total cost of production of sugarcane on farms in area j and C_{T_j} is cost of transport to factories for sugarcane produced in area j. The average total cost of production C_j consists of fixed costs such as farm rent and depreciation of equipment, and variable costs such as labor, materials and interest rates.

3.1. The Mathematical Model of Fresh Sugarcane

The multi-objective mathematical model of fresh sugarcane is described by Equa- tions (10)-(13). The objective function Equation (10) maximizes revenue from the fresh sugarcane selling. The constraint Equation (11) represents the second objective of the problem which is to minimize gathering cost of production, with upper bound given by ε_e. The decision variables $X_{j,k}$ are defined by $X_{j,k}=1$ means that planted area j is harvested at time k and $X_{j,k}=0$ means that j is not harvested at k. In the constraint set, Equation (12) and Equation (13) ensure that in each area the sugarcane is harvested only at one time k.

$$\text{Maximize} \sum_{j=1}^{m}\sum_{k=1}^{n} RV_{j,k}(A)\cdot X_{j,k} \tag{10}$$

$$\text{subject to} \sum_{j=1}^{m}\sum_{k=1}^{n} GC_{j,k}(A)\cdot X_{j,k} \le \varepsilon_e \tag{11}$$

$$\sum_{k=1}^{n} X_{j,k} = 1; \quad \forall j = 1,2,\cdots,m \tag{12}$$

$$X_{j,k} \in \{0,1\}; \quad \forall j = 1,2,\cdots,m; \quad \forall k = 1,2,\cdots,n \tag{13}$$

3.2. The Mathematical Model of Fired Sugarcane

The multi-objective mathematical model of fired sugarcane is described by Equations (14)-(17). The objective function Equation (14) maximizes revenue from the fired sugarcane selling. The constraint Equation (15) represents the second objective of the problem which is to minimize gathering cost of pro- duction, with upper bound given by ε_e. As for the fresh sugarcane model, the decision variables $X_{j,k}$ are defined by $X_{j,k}=1$ means that planted area j is harvested at time k and $X_{j,k}=0$ means that j

is not harvested at k. In the constraint set, Equation (16) and Equation (17) ensure that in each area the sugarcane is harvested only at one time k.

$$\text{Maximize } \sum_{j=1}^{m}\sum_{k=1}^{n} RV_{j,k}(B)\cdot X_{j,k} \tag{14}$$

$$\text{subject to } \sum_{j=1}^{m}\sum_{k=1}^{n} GC_{j,k}(B)\cdot X_{j,k} \le \varepsilon_e \tag{15}$$

$$\sum_{k=1}^{n} X_{j,k} = 1;\quad \forall j = 1,2,\cdots,m \tag{16}$$

$$X_{j,k} \in \{0,1\};\quad \forall j = 1,2,\cdots,m;\quad \forall k = 1,2,\cdots,n \tag{17}$$

Method of determining value of ε_e. The model will be solved for p values of ε_e defined as follows:

$$\varepsilon_e = \varepsilon_{e-1} + \Delta\varepsilon;\quad e = 1,2,\cdots,p, \tag{18}$$

where $\Delta\varepsilon = \frac{UB - LB}{p-1}$ and LB and UB are lower and upper bounds defined as follows.

$$\begin{aligned}
&LB = \varepsilon_1 \left(\text{summation of minimum of gathering cost in each area } j\right),\\
&UB = \varepsilon_p \left(\text{summation of maximum of gathering cost in each area } j\right) \text{and}\\
&\qquad p \text{ is number of experiments.}
\end{aligned} \tag{19}$$

3.3. Technologies and Equipment Used

The multi-objective mathematical models for fresh and fired sugarcane were solved using the linear programming and mixed-integer programming package GLPK (GNU Linear Programming Kit) version 4.52.

3.4. Data Used in Experiments

In this work, the optimal times to harvest sugarcane were determined for crop years 2012/13, 2013/14 and 2014/15. Examples of the real data used in the ex- periments for crop year 2012/13 are presented in Table 1 & Table 2, respectively [6] . Table 1 shows quantity of fresh and fired sugarcane delivered to factories in the four regions of Thailand, Table 2 the price of the sugarcane and the average total cost of production, Table 3shows the average C.C.S. of the sugarcane. The data in Table 3 were used to determine the values of the $y_{j,k}$ in the C.C.S. price $P_2(i)$ for the sugarcane in Equation (7).

Table 1. The quantity of sugarcane ($\times 10^5$ tons) delivered into factories in four regions of Thailand for crop year 2012/13.

Time	Northern		Central		Eastern		North-Eastern	
(k)	Fresh	Fired	Fresh	Fired	Fresh	Fired	Fresh	Fired
15-30/11/12	3.51	4.33	1.15	2.17	0.31	1.17	3.02	3.72
1-15/12/12	6.80	13.74	3.77	8.79	0.53	2.99	17.98	18.03
16-31/12/12	10.12	21.16	11.40	24.84	0.95	4.18	27.87	24.48
1-15/01/13	10.45	21.42	14.48	29.75	1.29	4.31	22.60	24.29
16-31/01/13	10.60	23.27	16.05	33.41	1.47	3.70	28.20	27.31
1-14/02/13	8.80	20.68	11.64	25.15	1.41	3.55	22.35	28.09
15-28/02/13	8.29	21.04	12.50	30.32	1.16	3.75	19.25	28.84
1-15/03/13	6.78	23.25	11.77	30.43	1.05	3.62	15.95	32.13
16-31/03/13	4.87	22.07	8.51	22.99	0.80	4.28	9.06	29.04
1-15/04/13	0.55	4.50	1.83	4.04	0.35	2.98	1.99	11.95
16-30/04/13	0.01	0.16	0.00	0.00	0.15	1.62	0.71	4.36
1-16/05/13	0.00	0.00	0.00	0.00	0.07	1.16	0.10	0.88

Table 2. The price of sugarcane and average total cost of production (baht/ton) in crop year 2012/13.

$P_w(A)$ (baht/ton)	$P_w(B)$ (baht/ton)	P_c (baht/ton)	C_{Northern} (baht/ton)	C_{Central} (baht/ton)	C_{Eastern} (baht/ton)	$C_{\text{North-Eastern}}$ (baht/ton)
198.465	140	999.2	1113.49	1114.85	1077.65	963.45

A comparison of the quantities of fresh and fired sugarcane delivered to the sugar mills in the three crop years 2012/2013, 2013/2014 and 2014/2015 are shown in Figure 1. It can be seen that the percentages of fresh and fired sugar- cane delivered are approximately 30% and 70%, respectively.

4. RESULTS AND DISCUSSION

The results of crop years 2012/13, 2013/14 and 2014/15 are presented in this section. The results are given for both fresh sugarcane and fired sugarcane for the four regions of Thailand.

4.1. Results for Crop Year 2012/13

The mathematical model for fresh sugarcane is described by objective function Equation (10) subject to constraints Equations (11)-(13) and the mathematical model for fired sugarcane is described by objective function Equation (14) subject to constraints Equations (15)-(17). The data from Tables 1-3 were used in the optimization. As stated earlier, the data in Table 3 were used to determine the values of the $y_{j,k}$ and the price $P_2(i)$ in Equation (7). The results are as follows.

1) Fresh Sugarcane :

The value of ε_e was calculated from the formula in Equation (18). The values of the lower bound $LB = 1.46 \times 10^8$ baht and upper bound $UB = 5.84 \times 10^9$ baht were estimated for each region. In the computations, 100 values of ε_e were used equally spaced between LB and UB in order to determine how the maximum revenue and the optimal cutting time changed as the upper bound on the gathering cost changed.

The results of revenue from the fresh sugarcane sales, gathering cost and the optimal harvest time for the four regions in Thailand; Northern, Central, Eastern and North-Eastern in each ε_e are presented in Table 4. It was found that the revenue increased at a rate greater than the increase of cost as shown in Figure 2(a).

For example, for ε_2, the revenue would be 0.199×10^9 baht, the gathering cost would be 0.196×10^9 baht and the optimal harvest times in each area would be as follows: 1) Northern should be harvested in 16-30 April 2013, 2)

Table 3. Average C.C.S. of sugarcane in crop year 2012/13.

Time (k)	Northern	Central	Eastern	North-Eastern
15-30/11/12	8.5	8.07	8.87	9.69
1-15/12/12	8.79	8.59	9.24	10.77
16-31/12/12	9.27	9.14	9.7	11.11
1-15/01/13	9.7	9.54	10.01	11.4
16-31/01/13	10.13	9.95	10.29	11.72
1-14/02/13	10.44	10.2	10.45	11.96
15-28/02/13	10.71	10.45	10.62	12.17
1-15/03/13	10.92	10.66	10.76	12.34
16-31/03/13	11.1	10.82	10.85	12.49
1-15/04/13	11.19	10.89	10.93	12.56
16-30/04/13	11.2	10.89	10.94	12.57
1-16/05/13	11.2	10.89	10.93	12.56

Table 4. The optimal harvest time for fresh sugarcane in 2012/13 with $p = 100$ ex- periments.

ε	Revenue	Gathering Cost	Optimal Time for Harvesting			
	($\times 10^9$ baht)	($\times 10^9$ baht)	Northern	Central	Eastern	North-Eastern
ε_1	0.148	0.146	16-30/04/13	15-30/11/12	1-16/05/13	1-16/05/13
ε_2	0.199	0.196	16-30/04/13	15-30/11/12	1-15/12/12	1-16/05/13
ε_3	0.288	0.253	16-30/04/13	1-15/04/13	1-15/04/13	1-16/05/13
ε_4	0.371	0.312	16-30/04/13	1-15/04/13	1-15/04/13	16-30/04/13
ε_5	0.439	0.359	16-30/04/13	15-30/11/12	1-15/04/13	1-15/04/13
⋮	⋮	⋮	⋮	⋮	⋮	⋮
ε_{98}	6.907	5.724	16-31/01/13	16-31/01/13	1-15/04/13	16-31/01/13
ε_{99}	6.963	5.772	16-31/01/13	16-31/01/13	16-31/03/13	16-31/01/13
ε_{100}	7.041	5.844	16-31/01/13	16-31/01/13	16-31/01/13	16-31/01/13

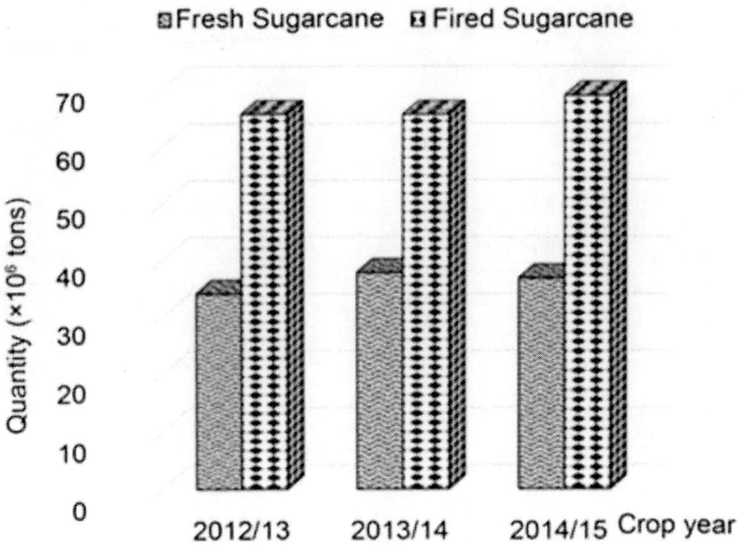

Figure 1. Comparison between quantities of fresh and fired sugarcane delivered to sugar mills.

Central should be harvested in 15-30 November 2012, 3) Eastern should be harvested in 1-15 December 2012 and 4) North-Eastern should be harvested in 1-16 May 2013.

2) Fired Sugarcane :

The values of the LB and UB for gathering cost of the fired sugarcane were estimated as described for the fresh sugarcane. With these values, 100 values of ε_e were used in the range $\varepsilon_e; 4.69\times10^8 \le \varepsilon_e \le 9.87\times10^9$. The calculated revenues from the fired sugarcane sales, gathering costs and the optimal harvest time for the four regions for each ε_e are shown in Table 5. It was again found that the revenue increased at a rate greater than the increase of cost as shown in Figure 2(b).

For example, for ε_5, It was found that the optimal harvest times in each area were as follows: 1) Northern should be harvested in 16-30 April 2013, 2) Central should be harvested in 15-30 November 2012, 3) Eastern should be harvested in 1-16 May 2013 and 4) North-Eastern should be harvested in 16-30 April 2013. The maximum values of revenue from the fired sugarcane selling would be 9.45×10^8 baht and the minimum values of gathering cost would be 8.05×10^8 baht.

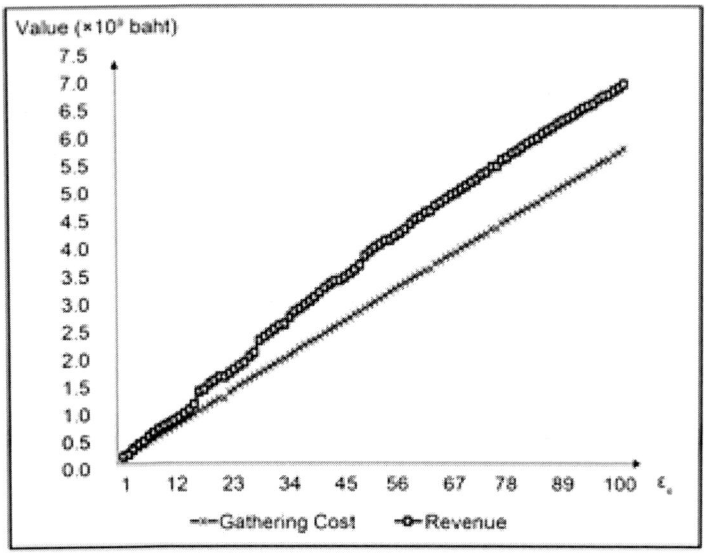

(a)

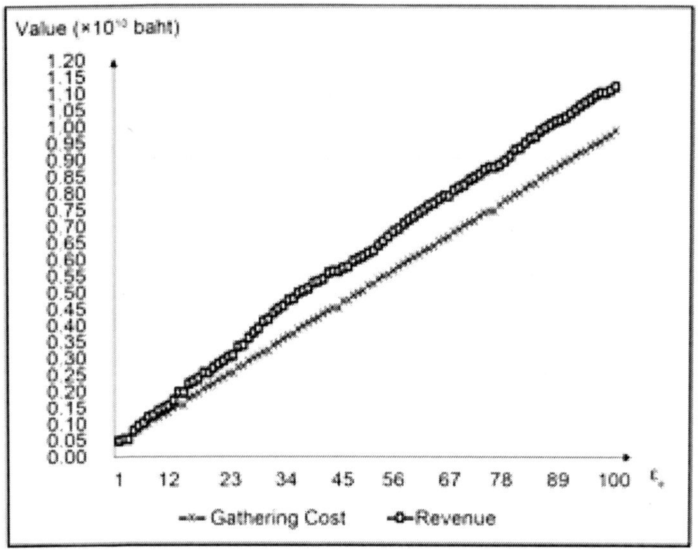

(b)

Figure 2. Comparison between revenue from sugarcane sales and gathering cost of production for 2012/2013. (a) Fresh Sugarcane. (b) Fired Sugarcane.

Table 5. The optimal harvest time for fired sugarcane in 2012/13 with $p=100$ ex- periments.

ε	Revenue	Gathering Cost	Optimal Time for Harvesting			
	($\times 10^9$ baht)	($\times 10^9$ baht)	Northern	Central	Eastern	North-Eastern
ε_1	0.493	0.469	16-30/04/13	15-30/11/12	1-16/05/13	1-16/05/13
ε_2	0.549	0.519	16-30/04/13	15-30/11/12	16-30/04/13	1-16/05/13
ε_3	0.549	0.519	16-30/04/13	15-30/11/12	16-30/04/13	1-16/05/13
ε_4	0.808	0.727	16-30/04/13	1-15/04/13	16-30/04/13	1-16/05/13
ε_5	0.945	0.805	16-30/04/13	15-30/11/12	1-16/05/13	16-30/04/13
⋮	⋮	⋮	⋮	⋮	⋮	⋮
ε_{98}	10.995	9.667	16-31/03/13	16-31/01/13	1-15/03/13	1-15/03/13
ε_{99}	11.076	9.739	16-31/03/13	16-31/01/13	16-31/03/13	1-15/03/13
ε_{100}	11.192	9.869	1-15/03/13	16-31/01/13	16-31/03/13	1-15/03/13

Table 6. The optimal harvest time for fresh sugarcane in 2013/14 with $p=100$ ex- periments.

ε	Revenue	Gathering Cost	Optimal Time for Harvesting			
	($\times 10^9$ baht)	($\times 10^9$ baht)	Northern	Central	Eastern	North-Eastern
ε_1	0.096	0.083	25-30/11/13	1-15/04/14	16-30/04/14	1-9/05/14
ε_2	0.149	0.136	25-30/11/13	1-15/04/14	1-15/12/13	1-9/05/14
ε_3	0.249	0.210	1-15/04/14	1-15/04/14	16-30/04/14	1-9/05/14
ε_4	0.315	0.272	25-30/11/13	1-15/04/14	16-31/03/14	16-30/04/14
ε_5	0.409	0.342	1-15/04/14	1-15/04/14	1-15/12/13	16-30/04/14
⋮	⋮	⋮	⋮	⋮	⋮	⋮
ε_{98}	7.937	6.684	16-31/01/14	16-31/01/14	16-31/03/14	16-31/01/14
ε_{99}	7.937	6.684	16-31/01/14	16-31/01/14	16-31/03/14	16-31/01/14
ε_{100}	8.083	6.828	16-31/01/14	16-31/01/14	16-31/01/14	16-31/01/14

4.2. Results for Crop Year 2013/14

The mathematical model for fresh sugarcane is described by objective function (10) subject to constraints (11)-(13) and the mathematical model of fired sugarcane is described by objective function (14) subject to constraints (15)-(17). Real data for crop year 2013/14 were used for the C.C.S. and the prices $P_1(i)$ for the weight, $P_2(i)$ for the C.C.S. and the gathering cost.

The results of revenues from the fresh and fired sugarcane sales, gathering costs and the optimal harvest time for the four regions in Thailand; Northern, Central, Eastern and North-Eastern for each ε_e are presented in Table 6 and Table 7, respectively. It was found that in both cases the revenue increased at a rate greater than the increase of cost as shown in Figure 3(a) and Figure 3(b).

Table 7. The optimal harvest time for fired sugarcane in 2013/14 with $p = 100$ ex- periments.

ε	Revenue	Gathering Cost	Optimal Time for Harvesting			
	($\times 10^9$ baht)	($\times 10^9$ baht)	Northern	Central	Eastern	North-Eastern
ε_1	0.165	0.153	16-30/04/14	1-15/04/14	1-9/05/14	25-30/11/13
ε_2	0.271	0.251	16-30/04/14	1-15/04/14	1-15/12/13	1-9/05/14
ε_3	0.399	0.338	1-15/04/14	1-15/04/14	1-9/05/14	1-9/05/14
ε_4	0.481	0.428	1-15/04/14	1-15/04/14	1-15/12/13	1-9/05/14
ε_5	0.577	0.507	1-15/04/14	1-15/04/14	1-15/04/14	1-9/05/14
⋮	⋮	⋮	⋮	⋮	⋮	⋮
ε_{98}	11.955	10.219	1-15/03/14	15-28/02/14	1-15/03/14	16-31/03/14
ε_{99}	11.955	10.219	1-15/03/14	15-28/02/14	1-15/03/14	16-31/03/14
ε_{100}	12.194	10.711	16-31/12/13	16-31/01/14	1-15/03/14	16-31/03/14

4.3. Results for Crop Year 2014/15

The mathematical model for fresh sugarcane is described by objective function (10) subject to constraints (11)-(13) and the mathematical model for fired sugarcane is described by objective function (14) subject to constraints (15)-(17). Real data for crop year 2014/15 were used for the

C.C.S. and the prices $P_1(i)$ for the weight, $P_2(i)$ for the C.C.S. and the gathering cost.

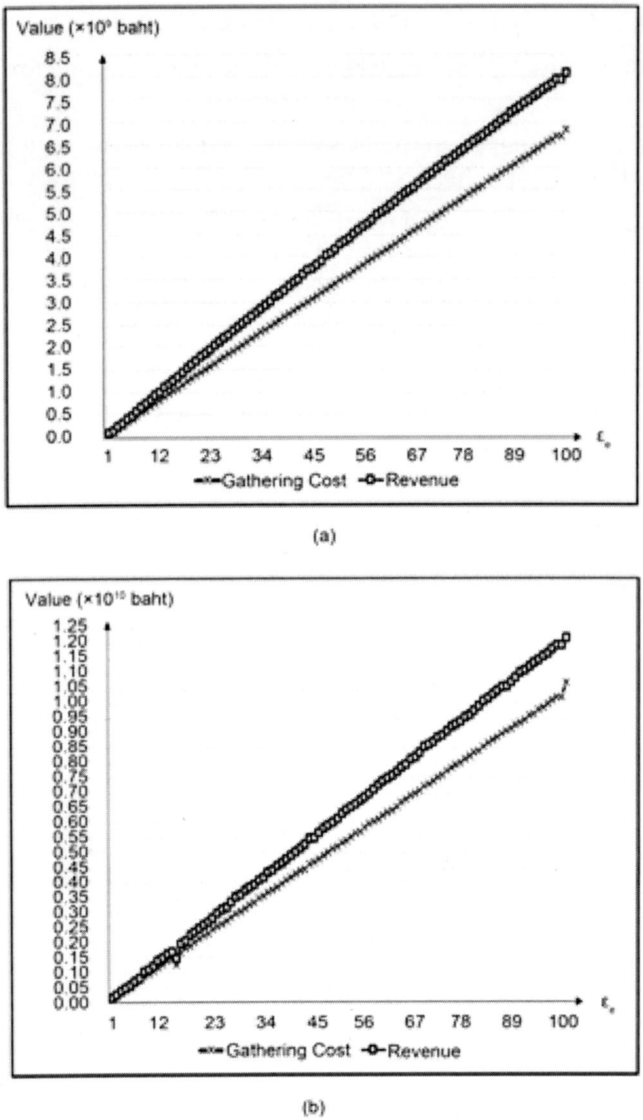

Figure 3. Comparison between revenue from sugarcane sales and gathering cost of production for 2013/2014. (a) Fresh Sugarcane; (b) Fired Sugarcane.

Table 8. The optimal harvest time for fresh sugarcane in 2014/15 with $p=100$ ex- periments.

ε	Revenue	Gathering Cost	Optimal Time for Harvesting			
	($\times 10^9$ baht)	($\times 10^9$ baht)	Northern	Central	Eastern	North-Eastern
ε_1	0.145	0.151	-	-	-	-
ε_2	0.145	0.151	-	-	-	-
ε_3	0.212	0.224	-	-	-	-
$\vdots$	$\vdots$	$\vdots$	$\vdots$	$\vdots$	$\vdots$	$\vdots$
ε_{24}	1.588	1.556	16-30/04/15	1-15/04/15	1-15/04/15	16-31/03/15
ε_{25}	1.655	1.630	16-30/04/15	1-15/04/15	16-31/03/15	16-31/03/15
ε_{26}	1.698	1.689	16-30/04/15	1-15/04/15	1-15/12/14	16-31/03/15
ε_{27}	1.765	1.754	16-30/04/15	1-15/04/15	15-28/02/15	16-31/03/15
ε_{28}	1.815	1.812	16-30/04/15	1-15/04/15	1-14/02/15	16-31/03/15
$\vdots$	$\vdots$	$\vdots$	$\vdots$	$\vdots$	$\vdots$	$\vdots$
ε_{99}	6.075	6.201	-	-	-	-
ε_{100}	6.143	6.273	-	-	-	-

Table 9. Optimal harvesting times for fresh sugarcane.

Crop year	Northern	Central	Eastern	North-Eastern
2012/2013	16-30 April 2013	1-15 April 2013	1-15 April 2013	16-31 January 2013
2013/2014	1-15 March 2014	15-28 February 2014	1-15 April 2014	16-31 March 2014
2014/2015	16-30 April 2015	1-15 April 2015	1-15 April 2015	16-31 January 2015

1) Fresh Sugarcane.

The results of revenues from the fresh sugarcane sales, gathering costs, values of ε_e and the optimal harvest time for the four regions in Thailand: Northern, Central, Eastern and North-Eastern for each ε_e are presented in Table 8. As shown in Figure 4(a), it was found that for this crop year only 20% of the ε_e values gave revenue greater than the gathering cost.

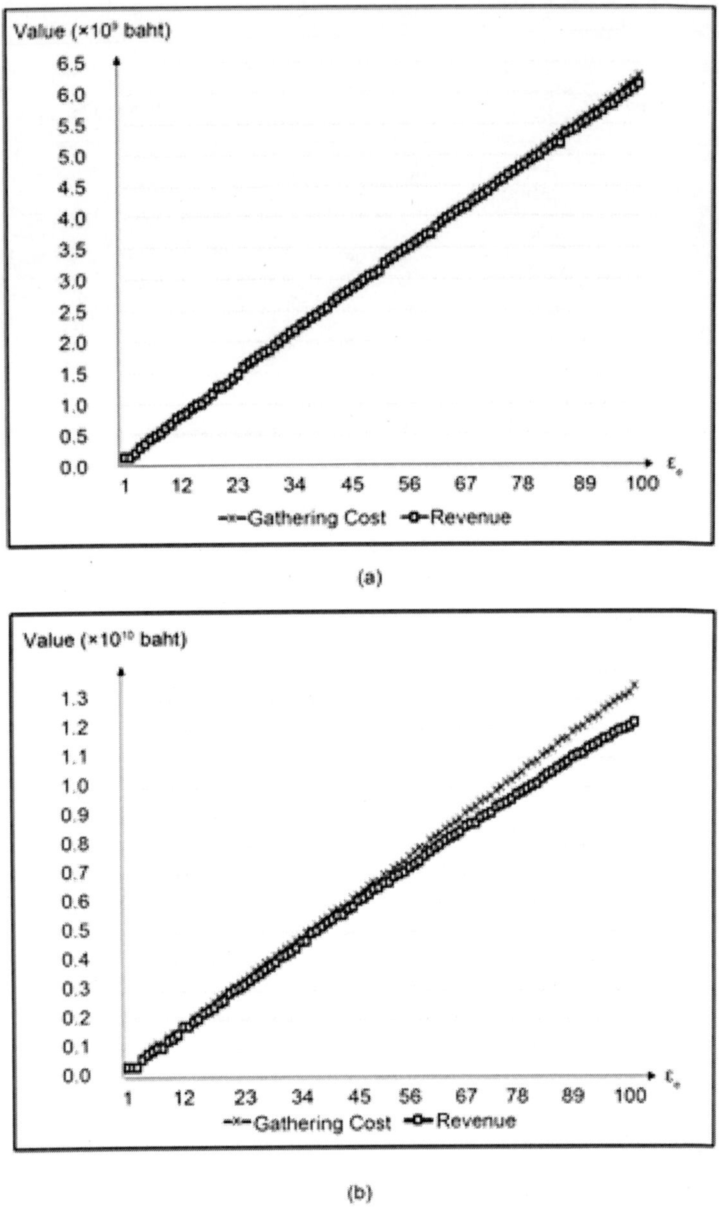

Figure 4. Comparison between revenue from sugarcane sales and gathering cost of production for 2014/2015. (a) Fresh Sugarcane; (b) Fired Sugarcane.

2) Fired Sugarcane.

For this crop year it was found, as shown in Figure 4(b), that the revenue was less than the gathering cost for all ε_e. Therefore there was no optimal harvest time in this case.

4.4. Summary of Results

A summary of the optimal harvest times for fresh and fired sugarcane for the crop years 2012/13, 2013/14 and 2014/15 are shown in Table 9 and Table 10.

Table 10. Optimal harvesting times for fired sugarcane.

Crop year	Northern	Central	Eastern	North-Eastern
2012/2013	16-30 April 2013	1-15 April 2013	1-16 May 2013	1-15 March 2013
2013/2014	16-31 March 2014	1-15 March 2014	1-9 May 2014	16-30 April 2014
2014/2015	16-30 April 2015	1-15 April 2015	1-15 April 2015	16-31 March 2015

5. CONCLUSIONS

In this paper, a multi-objective mathematical model has been presented for the calculation of the optimal times to harvest fresh and fired sugarcane in Thailand for the crop years 2012/13, 2013/14 and 2014/15. The two objectives in the model were to maximize revenue and minimize cost.

The ε-constraint method was used to change the multi-objective optimization model into a single-ob- jective model to maximize revenue subject to an ε-constraint of an upper limit on the cost. The GNU Linear Programming Kit version 4.52 was used to compute the optimal harvest time, maximum revenue and minimum gathering cost for a range of 100 values of ε-constraint. The results of the computations are sum- marized in Table 9 and Table 10.

The results show that the methods used in this paper can be used to predict optimal harvesting times. In this model, the main factor in determining the optimal harvest time is the change in C.C.S. with time as shown in Table 3. However, there are a number of other factors that are not included in the model that could change the optimal harvest times. These factors include: 1) changes in fixed and variable production costs during a year, 2) changes in delivery costs during a year, 3) decisions by the sugar mills on their requirements for sugar cane.

ACKNOWLEDGEMENTS

This research was partially funded by King Mongkut's University of Technology North Bangkok (Contract No. KMUTNB-GOV-59-23) and Centre of Excellence in Mathematics, The Commission on Higher Education, Thailand.

REFERENCES

1. Naranong, V., et al. (2013) A Study of Reform the Structure of Thailand's Sugar and Cane Industry. Thailand Development Research Institute.
2. Florentino, H.O., Moreno, E.V. and Sartori, M.P. (2008) Multiobjective Optimization of Economic Balances of Sugarcane Harvest Biomass. Science Agricultural, 65, 561-564.
3. Scarpari, M.S. and Ferreira, E.G. (2010) Optimized Agricultural Planning of Sugarcane Using Linear Programming. Revista Investigacion Operacional, 31, 126-132.
4. Gomes, F.R.A. (2012) Bi-Objective Mathematical Model for Choosing Sugarcane Varieties with Harvest Residual Biomass in Energy Cogeneration. International Journal of Agricultural and Biological Engineering, 5, 50-58.
5. Deb, K. (2001) Multi-Objective Optimization Using Evolutionary Algorithms. John Wiley and Sons, Chichester.
6. Office of the Cane and Sugar Board. http://www.ocsb.go.th

CHAPTER 8

A Family of Combined Iterative Methods for Solving Nonlinear Equations

Shuping Chen[1] and Youhua Qian[2]

[1]*College of Mathematics, Xiamen University of Technology, xiamen 361024, P. R. China*

[2]*College of Mathematics, Physics and Information Engineering, Zhejiang Normal University, Jinhua, 321004, P. R. China*

ABSTRACT

In this article we construct some higher-order modifications of Newton's method for solving nonlinear equations, which is based on the undetermined coefficients. This construction can be applied to any iteration formula. It can be found that per iteration the resulting methods add only one additional function evaluation, their order of convergence can be increased by two or three units. Higher order convergence of our methods is proved and corresponding asymptotic error constants are expressed. Numerical examples, obtained using Matlab with high precision arithmetic, are shown to demonstrate the convergence and efficiency of the combined iterative methods. It is found that the combined iterative methods produce very good results on tested examples, compared to the results produced by the existing higher order schemes in the related literature.

Keywords: Newton's method, combined iterative methods, nonlinear equations, order of convergence, computational efficiency

1. INTRODUCTION

A variety of problems in different fields of science and technology require to find the solution of nonlinear equations. Iterative methods such as 's method are the most used technique. In this paper, we consider a new family of combined iterative methods to find a simple root ξ of a nonlinear equation $f(x)=0$, where f is a real function $f: I \subseteq R \rightarrow R$, defined in an open interval I.

The well-known numerical method for the calculation of ξ is the classical Newton's method as given by

$$x_{n+1} = x_n - \frac{f(x_n)}{f'(x_n)}, n = 0, 1, \cdots, \tag{1}$$

where x_0 is an initial approximation that is sufficiently close to ξ. The convergence order of the classical 's method is quadratic for simple roots and linear for multiple roots.

Recently, a number of authors including [1-6][1] have derived new variants of Newton's methods that offer higher order convergence. These methods are frequently composed of more than two formulas and derived in different ways.

In [6] some fifth order modifications of Newton's method which is extending a general form of third order method are considered. In a similar way, some sixth-order class of modified Ostrowski's methods [7] that improves the order of convergence of Ostrowski's method are presented as

Follows

$$\begin{cases} y_n = x_n - \dfrac{f(x_n)}{f'(x_n)}, \\ z_n = y_n - \dfrac{f(x_n)}{f(x_n) - 2f(y_n)} \dfrac{f(y_n)}{f'(x_n)}, \\ x_{n+1} = z_n - H(v_n) \dfrac{f(z_n)}{f'(x_n)}, \end{cases} \tag{2}$$

where $v_n = \frac{f(y_n)}{f(x_n)}$, and $H(t)$ is a real-valued function for $H(0)=1$, $H'(0)=2$ and $H''(0)<\infty$.

Described below in equation (3) is the three-point sixth-order method [8], which requires only two derivatives and two functions:

$$\begin{cases} y_n = x_n - \dfrac{f(x_n)}{f'(x_n)}, \\ z_n = x_n - \dfrac{2f(x_n)}{f'(x_n)+f'(y_n)}, \\ x_{n+1} = z_n - \dfrac{f'(x_n)+f'(y_n)}{3f'(y_n)-f'(x_n)} \cdot \dfrac{f(z_n)}{f'(x_n)}. \end{cases} \tag{3}$$

This is a three-step method. The first and the second equations of equation (3) compose a third-order method developed by the authors in [9].

It is worth noting that equations (2) and (3) use the technique that consists in applying of new function on the existing iterative schemes. Motivated by the activities in this direction, in this paper, our special attention is paid to the development of a general class of higher-order combined iterative methods. The important feature of our methods is only to add the evaluation of the function at another point. However, their convergence order can be improved and increased above the original level.

This paper is organized as follows. In section 2, the new methods are formulated, and the local convergence theorem is established. Some concrete iterative methods are discussed in section 3. In section 4, the new methods are verified through a number of numerical examples, comparisons of results are also reported to show the effectiveness of the present approach. Finally, the paper ends with conclusions in section 5.

2. THE METHODS AND ANALYSIS OF CONVERGENCE

Let $\phi(x; f(x), f'(x), f'(y))$ be a function from $R \to R$ with the information f, f' at x and y, this means that the functions $f(x)$, $f'(x)$ and $f'(y)$ at each iteration step are required to evaluate in the computation of ϕ. Now we consider the modification of 's method as given below

$$\begin{cases} z_n = \phi\left(x_n; f(x_n), f'(x_n), f'(y_n)\right), \\ x_{n+1} = z_n - \dfrac{f(z_n)}{f'(z_n)}, \end{cases} \tag{4}$$

where $z_n = \phi\left(x; f(x), f'(x), f'(y)\right)$ represents any iterative method whose order of convergence is at least m.

In recent years, a type of multipoint iterative methods have been proposed [9-11][9]. These methods can be viewed as obtained by approximating $f'(z_n)$ with the expressions $a_1(x_n), \cdots, a_k(x_n)$, where the function $f'(z_n)$ is defined as

$$f'(z_n) = \phi\left(x_n; a_1(x_n), \cdots, a_k(x_n)\right). \tag{5}$$

Analogously, in order to derive the new methods, we consider the expression

$$\begin{aligned} f'(z_n) &= Af'(x_n) + Bf'(y_n) \\ &\quad + Cf(z_n) + Df'\left(\frac{x_n + y_n}{2}\right), \end{aligned} \tag{6}$$

for application of the method of undetermined coefficients.

Expand the terms $f'(z_n)$, $f'(y_n)$, $f'\left(\frac{x_n + y_n}{2}\right)$ and $f(z_n)$ about the point x_n up to the third derivatives and collect terms. Upon comparing the coefficients of the derivatives of f at x_n, we get

$$A + B + \alpha C + D = 1, \tag{7}$$

$$C = 0, \tag{8}$$

$$B\beta + \frac{C\alpha^2}{2} + \frac{1}{2}D\beta = \alpha, \tag{9}$$

$$\frac{B\beta^2}{2}+\frac{C\alpha^3}{6}+\frac{1}{8}D\beta^2=\frac{\alpha^2}{2}, \tag{10}$$

where $\alpha = z_n - x_n$ and $\beta = y_n - x_n$.

Solving equations (7)-(10), we have

$$A=\frac{\beta^2-3\alpha\beta+2\alpha^2}{\beta^2}, \tag{11}$$

$$B=\frac{-\alpha\beta+2\alpha^2}{\beta^2}, \tag{12}$$

$$C=0, \tag{13}$$

$$D=\frac{4\left(\alpha\beta-\alpha^2\right)}{\beta^2}. \tag{14}$$

(11)-(14) into equation (6), we obtain

$$f'(z_n)=\frac{\left(\beta^2-3\alpha\beta+2\alpha^2\right)}{\beta^2}f'(x_n) -\frac{\left(\alpha\beta-2\alpha^2\right)}{\beta^2}f'(y_n) +\frac{4\left(\alpha\beta-\alpha^2\right)}{\beta^2}f'\left(\frac{x_n+y_n}{2}\right). \tag{15}$$

By using the arithmetic mean of $f'(x_n)$ and $f'(y_n)$ instead of the midpoint value $f'\left(\frac{x_n+y_n}{2}\right)$ in equation (15), we have

$$f'(z_n)=\frac{\left(\beta^2-\alpha\beta\right)}{\beta^2}f'(x_n)+\frac{\alpha\beta}{\beta^2}f'(y_n). \tag{16}$$

Substituting equation (16) into the second step of equation (4), we propose a higher-order family of combined iterative methods in the following form:

$$\begin{cases} z_n = \phi(x; f(x), f'(x), f'(y)), \\ x_{n+1} = z_n - \dfrac{\beta^2 f(z_n)}{(\beta^2 - \alpha\beta) f'(x_n) + \alpha\beta f'(y_n)}. \end{cases} \tag{17}$$

For this family of methods, the following result can be established.

Theorem 2.1. Let us suppose that $f(x)$ is a sufficiently differentiable function and $f(x)$ has a simple zero ξ. If the initial guess x_0 is close enough to ξ and iteration function $z_n = \phi(x_n; f(x_n), f'(x_n), f'(y_n))$ satisfies condition

$$z_n - \xi = Ae_n^m + O\left(e_n^{m+1}\right). \quad (18$$

If $m < 3$, then the sequence $\{x_k\}$ generated by equation (17) is of order at least $2m$.

If $m \geq 3$, then the sequence $\{x_k\}$ generated by equation (17) is of order at least $m+3$.

Proof. Let $e_n = x_n - \xi$ and $d_n = z_n - \xi$. Using the Taylor expansion and taking into account $f(\xi) = 0$, we arrive at

$$f(x_n) = f'(\xi)\Big[e_n + c_2 e_n^2 + c_3 e_n^3 + c_4 e_n^4 + c_5 e_n^5 + c_6 e_n^6 + c_7 e_n^7 + O\left(e_n^8\right)\Big], \tag{19}$$

$$f'(x_n) = f'(\xi)\Big[1 + 2c_2 e_n + 3c_3 e_n^2 + 4c_4 e_n^3 + 5c_5 e_n^4 + 6c_6 e_n^5 + 7c_7 e_n^6 + O\left(e_n^7\right)\Big], \tag{20}$$

where $c_k = \dfrac{f^k(\alpha)}{k!f'(\alpha)}$.

By simple calculations, we have

$$\begin{aligned}
y_n &= x_n - \frac{f(x_n)}{f'(x_n)} \\
&= \xi + c_2 e_n^2 + 2\left(c_3 - c_2^2\right)e_n^3 \\
&+ \left(4c_2^3 - 7c_2c_3 + 3c_4\right)e_n^4 \\
&+ \left(-8c_2^4 + 20c_2^2c_3 - 6c_3^2 - 10c_2c_4 + 4c_5\right)e_n^5 \\
&+ \left(16c_2^5 - 52c_2^3c_3 + 28c_4c_2^2 - 13c_5c_2 + 33c_3^2c_2\right. \\
&\left.+5c_6 - 17c_3c_4\right)e_n^6 + \left(-32c_2^6 + 128c_2^4c_3 - 72c_2^3c_4\right. \\
&- 126c_3^2c_2^2 + 36c_5c_2^2 + 92c_2c_3c_4 \\
&+ 6c_7 - 16c_2c_6 - 12c_4^2 \\
&\left.+27c_3^3 + 5c_3c_5 - 9c_3^3 - 27c_3c_5\right)e_n^7 + O\left(e_n^8\right),
\end{aligned} \tag{21}$$

$$\begin{aligned}
f'(y_n) = f'(\xi)\Big[1 &+ 2c_2^2e_n^2 - 4c_2\left(c_2^2 - c_3\right)e_n^3 \\
&- \left(11c_2^2c_3 - 8c_2^4 - 6c_2c_4\right)e_n^4 \\
&- 4c_2\left(4c_2^4 - 7c_2^2c_3 + 5c_2c_4 - 2c_5\right)e_n^5 \\
&+ 2c_2\left(16c_2^5 - 40c_2^3c_3 + 30c_2^2c_4 - 13c_2c_5\right. \\
&\left.+12c_2c_3^2 + 5c_6 - 8c_3c_4\right)e_n^6 + O\left(e_n^7\right)\Big],
\end{aligned} \tag{22}$$

$$f(z_n) = f'(\xi)\left[d_n + c_2d_n^2 + O\left(d_n^3\right)\right], \tag{23}$$

$$\alpha = z_n - x_n = -e_n + (z_n - \xi), \tag{24}$$

$$\begin{aligned}
\beta &= y_n - x_n \\
&= -e_n + c_2 e_n^2 + 2\left(c_3 - c_2^2\right)e_n^3 \\
&+ \left(4c_2^3 - 7c_2c_3 + 3c_4\right)e_n^4 \\
&+ \left(-8c_2^4 + 20c_2^2c_3 - 6c_3^2 - 10c_2c_4 + 4c_5\right)e_n^5 \\
&+ \left(16c_2^5 - 52c_2^3c_3 + 28c_4c_2^2 - 13c_5c_2\right. \\
&\left.+33c_3^2c_2 + 5c_6 - 17c_3c_4\right)e_n^6 \\
&+ \left(-32c_2^6 + 128c_2^4c_3 - 72c_2^3c_4\right. \\
&- 126c_3^2c_2^2 + 36c_5c_2^2 + 92c_2c_3c_4 \\
&- 16c_2c_6 - 12c_4^2 + 6c_7 \\
&\left.+ 27c_3^3 + 5c_3c_5 - 9c_3^3 - 27c_3c_5\right)e_n^7 + O\left(e_n^8\right).
\end{aligned} \tag{25}$$

Hence, we obtain

$$\begin{aligned}
\beta^2 &= e_n^2 - 2c_2e_n^3 + \left(5c_2^2 - 4c_3\right)e_n^4 \\
&+ 6\left(3c_2c_3 - 2c_2^3 - c_4\right)e_n^5 \\
&+ \left(28c_2^4 - 62c_2^2c_3 + 26c_2c_4 + 16c_3^2 - 8c_5\right)e_n^6 \\
&+ \left(-64c_2^5 + 188c_2^3c_3 - 88c_4c_2^2 + 34c_2c_5\right. \\
&\left.- 106c_2c_3^2 - 10c_6 + 46c_3c_4\right)e_n^7 \\
&+ \left(144c_2^6 - 528c_2^4c_3 + 264c_2^3c_4 + 471c_2^2c_3^2\right. \\
&- 114c_2^2c_5 - 300c_2c_3c_4 + 42c_2c_6 + 33c_4^2 \\
&\left.- 12c_7 - 60c_3^3 + 60c_3c_5\right)e_n^8 + O\left(e_n^9\right),
\end{aligned} \tag{26}$$

$$
\begin{aligned}
\alpha\beta = e_n\Big[& (1+c_2d_n)e_n - \left(c_2 - 2\left(c_3 - c_2^2\right)d_n\right)e_n^2 \\
& - \left(2\left(c_3 - c_2^2\right) - \left(4c_2^3 - 7c_2c_3 + 3c_4\right)d_n\right)e_n^3 \\
& - \left(4c_2^3 - 7c_2c_3 + 3c_4\right)e_n^4 \\
& - \left(8c_2^4 - 20c_2^2c_3 + 6c_3^2 + 10c_2c_4 - 4c_5\right)d_ne_n^4 \\
& + \left(8c_2^4 - 20c_2^2c_3 + 6c_3^2 + 10c_2c_4 - 4c_5\right)e_n^5 \\
& + \left(16c_2^5 - 52c_2^3c_3 + 28c_2^2c_4 - 13c_2c_5\right. \\
& \left. + 33c_2c_3^2 + 5c_6 - 17c_3c_4\right)d_ne_n^5 \\
& - \left(16c_2^5 - 52c_2^3c_3 + 28c_2^2c_4\right. \\
& \left. + 33c_2c_3^2 + 5c_6 - 17c_3c_4\right)e_n^6 \\
& + \left(32c_2^6 - 128c_2^4c_3 + 126c_2^2c_3^2 - 36c_2^2c_5 + 16c_2c_6\right. \\
& + 12c_4^2 - 6c_7 - 13c_2c_5 + 72c_2^3c_4 - 92c_2c_3c_4 \\
& \left. - 18c_3^3 + 22c_3c_5\right)e_n^7 - d_n\Big] + O\left(e_n^9\right).
\end{aligned} \tag{27}
$$

We then have

$$
\begin{aligned}
\beta^2 - \alpha\beta = {} & e_nd_n - c_2d_ne_n^2 - \left(c_2 + 2\left(c_3 - c_2^2\right)d_n\right)e_n^3 \\
& + \left(\left(3c_2^2 - 2c_3\right) - \left(4c_2^3 - 7c_2c_3 + 3c_4\right)d_n\right)e_n^4 \\
& + \left(\left(11c_2c_3 - 8c_2^3 - 3c_4\right)\right. \\
& \left. + \left(8c_2^4 - 4c_5 - 20c_2^2c_3 + 6c_3^2 + 10c_2c_4\right)d_n\right)e_n^5 \\
& + \left(20c_2^4 - 4c_5 - 42c_2^2c_3 + 16c_2c_4 + 10c_3^2\right)e_n^6 \\
& - \left(48c_2^5 - 136c_2^3c_3 + 60c_2^2c_4 - 21c_2c_5 + 73c_2c_3^2\right. \\
& \left. + 5c_6 - 29c_3c_4\right)e_n^7 + \left(112c_2^6 - 400c_2^4c_3 + 192c_2^3c_4\right. \\
& + 345c_2^2c_3^2 - 78c_2^2c_5 - 208c_2c_3c_4 + 26c_2c_6 + 21c_4^2 \\
& \left. - 6c_7 - 42c_3^3 + 38c_3c_5\right)e_n^8 + O\left(e_n^9\right).
\end{aligned} \tag{28}
$$

Thus, from equations (20), (22), (27) and (28), we attain

$$\begin{aligned}\left(\beta^2-\alpha\beta\right)f'(x_n)=f'(\xi)\Big[&e_nd_n+c_2d_ne_n^2\\&-(c_2-c_3d_n)e_n^3+\left(\left(c_2^2-2c_3\right)+c_4d_n\right)e_n^4\\&+\left(\left(4c_2c_3-2c_2^3-3c_4\right)+c_5d_n\right)e_n^5\\&+\left(4c_2^4-11c_2^2c_3+6c_2c_4+4c_3^2-4c_5\right)e_n^6\\&-\left(8c_2^5+5c_6-28c_2^3c_3+16c_2^2c_4-8c_2c_5\right.\\&\left.+20c_2c_3^2-12c_3c_4\right)e_n^7\\&+\left(16c_2^6-68c_2^4c_3+40c_2^3c_4+73c_2^2c_3^2-21c_2^2c_5\right.\\&-6c_7-12c_3^3-58c_2c_3c_4+10c_2c_6+9c_4^2\\&\left.+16c_3c_5\right)e_n^8+O\left(e_n^9\right)\Big],\end{aligned} \tag{29}$$

$$\begin{aligned}\alpha\beta f'(y_n)=f'(\xi)\Big[&-d_ne_n+(1+c_2d_n)e_n^2\\&-\left(c_2-2\left(c_3+2c_2^2\right)d_n\right)e_n^3\\&+\left(\left(4c_2^2-2c_3\right)+\left(10c_2^3-11c_2c_3+3c_4\right)d_n\right)e_n^4\\&+\left(\left(11c_2c_3-10c_2^3-3c_4\right)\right.\\&\left.-\left(12c_2^2c_3-6c_3^2-10c_2c_4+4c_5\right)d_n\right)e_n^5\\&+\left(24c_2^4-39c_2^2c_3+6c_3^2+16c_2c_4-4c_5\right)e_n^6\\&-\left(56c_2^5-121c_2^3c_3+60c_2^2c_4\right.\\&\left.-21c_2c_5+41c_2c_3^2+5c_6-17c_3c_4\right)e_n^7\\&+\left(128c_2^6-358c_2^4c_3+196c_2^3c_4+212c_2^2c_3^2\right.\\&-78c_2^2c_5-6c_7-132c_2c_3c_4+26c_2c_6\\&\left.+12c_4^2-18c_3^3+22c_3c_5\right)e_n^8+O\left(e_n^9\right)\Big]\end{aligned} \tag{30}$$

Using equations (29) and (30), we have

$$\begin{aligned}
&\left(\beta^2-\alpha\beta\right)f'\left(x_n\right)+\alpha\beta f'\left(y_n\right)\\
&=f'(\xi)\Big[e_n^2-2c_2e_n^3+\left(5c_2^2-4c_3\right)e_n^4+2c_2d_ne_n^2\\
&-\left(6c_4+12c_2^3-15c_2c_3\right)e_n^5\\
&+\left(28c_2^4-50c_2^2c_3+10c_3^2+22c_2c_4-8c_5\right)e_n^6\\
&+\left(3c_3-4c_2^2\right)d_ne_n^3+\left(10c_2^3-11c_2c_3+4c_4\right)d_ne_n^4\\
&-\left(12c_2^2c_3-6c_3^2-10c_2c_4+3c_5\right)d_ne_n^5\\
&-\Big(10c_6-29c_3c_4+64c_2^5-149c_2^3c_3\\
&+76c_2^2c_4-29c_2c_5+61c_2c_3^2\Big)e_n^7\\
&+\Big(144c_2^6-426c_2^4c_3+236c_2^3c_4+285c_2^2c_3^2-99c_2^2c_5\\
&-190c_2c_3c_4+36c_2c_6+21c_4^2\\
&-12c_7-30c_3^3+38c_3c_5\Big)e_n^8+O\left(e_n^9\right)\Big]
\end{aligned} \tag{31}$$

In the same way, we attain

$$\begin{aligned}
&\beta^2f\left(z_n\right)\\
&=f'(\xi)e_n^2d_n\Big[1-2c_2e_n+\left(5c_2^2-4c_3\right)e_n^2\\
&+6\left(3c_2c_3-2c_2^3-c_4\right)e_n^3\\
&+\left(28c_2^4-62c_2^2c_3+26c_2c_4+16c_3^2-8c_5\right)e_n^4\\
&+\Big(188c_2^3c_3-64c_2^5-88c_4c_2^2+34c_2c_5\\
&\quad-106c_2c_3^2-10c_6+46c_3c_4\Big)e_n^5\\
&+c_2d_n-2c_2^2e_nd_n+\left(5c_2^3-4c_2c_3\right)e_n^2d_n\\
&+6\left(3c_2^2c_3-2c_2^4-c_2c_4\right)e_n^3d_n\\
&+\Big(144c_2^6-528c_2^4c_3+264c_2^3c_4+471c_2^2c_3^2\\
&\quad-114c_2^2c_5-300c_2c_3c_4+42c_2c_6+33c_4^2\\
&-12c_7-60c_3^3+60c_3c_5\Big)e_n^6+O\left(e_n^7\right)\Big]
\end{aligned} \tag{32}$$

Dividing equation (32) by equation (31), we get

$$\frac{\beta^2 f(z_n)}{\left(\beta^2 - \alpha\beta\right) f'(x_n) + \alpha\beta f'(y_n)} = d_n\left[1 + 3c_2c_3e_n^3 - c_2d_n + O\left(e_n^4\right)\right]. \tag{33}$$

Since z_n is of order at least m, there exists a constant A such that

$$d_n = z_n - \xi = Ae_n^m + O\left(e_n^{m+1}\right). \tag{34}$$

Thus,

$$\begin{aligned} e_{n+1} &= z_n - \xi - \frac{\beta^2 f(z_n)}{\left(\beta^2 - \alpha\beta\right) f'(x_n) + \alpha\beta f'(y_n)} \\ &= d_n - d_n\left[1 + 3c_2c_3e_n^3 - c_2d_n + O\left(e_n^4\right)\right] \\ &= -3c_2c_3d_ne_n^3 + c_2d_n^2 + O\left(e_n^{n+4}\right). \end{aligned} \tag{35}$$

Hence, if $m < 3$ then using equations (34) and (35) reduce to

$$e_{n+1} = c_2A^2e_n^{2m} + O\left(e_n^{2m+1}\right), \tag{36}$$

which implies that the method defined by equation (17) is of order at least $2m$. Furthermore, if $m \geq 3$, then using equations (34) and (35), we obtain

$$e_{n+1} = -3c_2c_3Ae_n^{m+3} + O\left(e_n^{m+4}\right). \tag{37}$$

Therefore, the method defined by equation (17) is of order at least $m+3$.

This completes the proof.

3. THE CONCRETE ITERATIVE METHODS

This section describes some interesting studies based on different form of $z_n = \phi\left(x_n; f\left(x_n\right), f'\left(x_n\right), f'\left(y_n\right)\right)$. Throughout the rest of this article, y_n is defined by equation (21).

3.1. Some Fourth-Order Methods

Case 3.1. For the function ϕ defined by

$$\phi\left(x; f\left(x\right), f'\left(x\right), f'\left(y\right)\right) = x - \frac{f\left(x\right)}{f'\left(x\right)}, \tag{38}$$

then we obtain the two-step Newton's method

$$\begin{cases} z_n = x_n - \dfrac{f\left(x_n\right)}{f'\left(x_n\right)}, \\ x_{n+1} = z_n - \dfrac{f\left(z_n\right)}{f'\left(z_n\right)}. \end{cases} \tag{39}$$

It is easy to see that equation (39) is the well-known two-point fourth-order double-Newton method [12, 13].

Case 3.2. If we take a second order method from [14]

$$z_n = x_n - \frac{f\left(x_n\right)}{f'\left(x_n\right) + \alpha' f\left(x_n\right)}, \tag{40}$$

where α' is a parameter, then we obtain the new four-order method

$$\begin{cases} z_n = x_n - \dfrac{f\left(x_n\right)}{f'\left(x_n\right) + \alpha' f\left(x_n\right)}, \\ x_{n+1} = z_n - \dfrac{\beta^2 f\left(z_n\right)}{\left(\beta^2 - \alpha\beta\right) f'\left(x_n\right) + \alpha\beta f'\left(y_n\right)}. \end{cases} \tag{41}$$

These two methods have order four and use four functional evaluations per step, and their efficiency index [15] is the same with the known methods of order two. However, numerical examples show that these modified methods may be efficient enough and have better performance as compared to the known methods of order two.

3.2. Some Sixth-Order Methods

Case 3.3. If we take a third-order variant of Newton's method appeared in [9]

$$z_n = x_n - \frac{2f(x_n)}{f'(y_n)+f'(x_n)}. \tag{42}$$

Then, we get the new sixth-order method

$$\begin{cases} z_n = x_n - \dfrac{2f(x_n)}{f'(y_n)+f'(x_n)}, \\ x_{n+1} = z_n - \dfrac{\beta^2 f(z_n)}{\left(\beta^2 - \alpha\beta\right)f'(x_n)+\alpha\beta f'(y_n)}. \end{cases} \tag{43}$$

Case 3.4. If we use the cubically convergent iterative scheme in [16]

$$x_{n+1} = x_n - \frac{f(x_n)\left(f'(x_n)+f'(y_n)\right)}{2f'(x_n)f'(y_n)}. \tag{44}$$

Then, the following expressions can be resulted

$$\begin{cases} z_n = x_n - \dfrac{f(x_n)\left(f'(x_n)+f'(y_n)\right)}{2f'(x_n)f'(y_n)}, \\ x_{n+1} = z_n - \dfrac{\beta^2 f(z_n)}{\left(\beta^2 - \alpha\beta\right)f'(x_n)+\alpha\beta f'(y_n)}. \end{cases} \tag{45}$$

The new methods (43) and (45) have the efficiency index equal to $6^{\frac{1}{4}} \approx 1.5651$, which is better than $\sqrt{2} \approx 1.4142$ of Newton's method and $3^{\frac{1}{3}} \approx 1.4422$ of the methods (42) and (44).

3.3. The Seventh-Order Method

Case 3.5. If we take the fourth-order Jarratt method [17] defined by

$$\begin{cases} y_n = x_n - \dfrac{2}{3}\dfrac{f(x_n)}{f'(x_n)}, \\ z_n = x_n - \dfrac{3f'(y_n)+f'(x_n)}{6f'(y_n)-2f'(x_n)}\dfrac{f(x_n)}{f'(x_n)}. \end{cases} \tag{46}$$

Then, we obtain the new seventh-order method

$$\begin{cases} y_n = x_n - \dfrac{2}{3}\dfrac{f(x_n)}{f'(x_n)}, \\ z_n = x_n - \dfrac{3f'(y_n)+f'(x_n)}{6f'(y_n)-2f'(x_n)}\dfrac{f(x_n)}{f'(x_n)}, \\ x_{n+1} = z_n - \dfrac{\beta^2 f(z_n)}{(\beta^2-\alpha\beta)f'(x_n)+\alpha\beta f'(y_n)}. \end{cases} \tag{47}$$

Case 3.6. If we take the fourth-order method presented by Khattri and Abbasbandy in [18]

$$\begin{cases} y_n = x_n - \dfrac{2}{3}\dfrac{f(x_n)}{f'(x_n)}, \\ z_n = y_n - \left[1+\left(\dfrac{21}{8}\right)\left(\dfrac{f'(y_n)}{f'(x_n)}\right)+\left(-\dfrac{9}{2}\right)\left(\dfrac{f'(y_n)}{f'(x_n)}\right)^2 \right. \\ \qquad \left. +\left(\dfrac{15}{8}\right)\left(\dfrac{f'(y_n)}{f'(x_n)}\right)^3\right]\dfrac{f(x_n)}{f'(x_n)}. \end{cases} \tag{48}$$

Then, we obtain the new seventh-order method

$$\begin{cases} y_n = x_n - \dfrac{2}{3}\dfrac{f(x_n)}{f'(x_n)}, \\ z_n = y_n - \left[1 + \left(\dfrac{21}{8}\right)\left(\dfrac{f'(y_n)}{f'(x_n)}\right) + \left(-\dfrac{9}{2}\right)\left(\dfrac{f'(y_n)}{f'(x_n)}\right)^2 \right. \\ \left. \qquad + \left(\dfrac{15}{8}\right)\left(\dfrac{f'(y_n)}{f'(x_n)}\right)^3\right]\dfrac{f(x_n)}{f'(x_n)}, \\ x_{n+1} = z_n - \dfrac{\beta^2 f(z_n)}{(\beta^2 - \alpha\beta)f'(x_n) + \alpha\beta f'(y_n)}. \end{cases} \quad (49)$$

The methods defined by equations (47) and (49) require two function and two first derivative evaluations per iteration. Each method improves the order of convergence of the fourth-order method from four to seven with additional function evaluation at the point iterated by the fourth-order method. We have that the methods obtained by the formulae (47) and (49) have the efficiency index equal to $7^{\frac{1}{4}} \approx 1.6266$, which is better than the fourth-order methods $4^{\frac{1}{3}} \approx 1.5874$. It should be pointed out that many new higher-order methods can be obtained by considering the different choices of z_n in (17).

4. NUMERICAL EXAMPLES

In this section, the results of some numerical tests are given to demonstrate the convergence efficiencies of various iterative schemes. We employ the present methods (39), (41)$^{(\alpha'=1)}$, (43), (45), (47) and (49) denoted by INW, IKT, IWM, IOM, IJA and IKA, respectively to solve some nonlinear equations and compare with Newton's method (NW), the method (KT) developed by Kanwar et al. [14], the method (WM) developed by Weerakoon et al. [9], the method (OM) by Özban [16], the Jarratt method (JA) and the method (KA) developed by Khattri et al. [18]. The equations $f(x)=0$ was solved using the following test functions with corresponding starting values x_0:

$$f_1(x) = \sin^2 x - x^2 + 1, \quad \xi_1^* = 1.4044916482153412260,$$
$$f_2(x) = x^2 - e^x - 3x + 2, \quad \xi_2^* = 0.25753028543986076046,$$
$$f_3(x) = \cos x - xe^x + x^2, \quad \xi_3^* = 0.63915409633200758106,$$
$$f_4(x) = \cos x - x, \quad \xi_4^* = 0.7390851332151606416 6.$$

Numerical computations have been carried out using variable precision arithmetic, with 1000 digits, in Matlab . The stopping criterion is taken as $|x_{n+1} - x_n| + |f(x_{n+1})| < 10^{-100}$. In cases where the exact solution was not available, we used the approximation ξ^*, which was also calculated with 1000 digits. For simplicity, only 20 digits are displayed.

Table 1. Numerical comparisons of the existing methods and the present combined iterative methods

	Method	N	$\lvert f(x_{n+1}) \rvert$	CPU	COC
f_1, $x_0 = 1$					
	NW	9	3.4e-101	0.046	1.9998515889198 38
	INW	5	3.4e-101	0.031	3.9914935280052 24
	KT	11	3.7e-168	0.047	2.0000000001524 57
	IKT	6	1.6e-234	0.203	3.9416041797456 91
	WM	7	7.5e-266	0.088	3.0453285919194 75
	IWM	5	1.2e-566	0.111	6.2626999321361 88
	OM	5	1.1e-186	0.531	3.0105884468160 82
	IOM	3	6.5e-123	0.063	6.5702533927094 60
	JA	6	1.4e-334	0.152	4.2487089847933 18
	IJA	4	6.0e-426	0.060	6.8831539989591 28
	KA	8	3.0e-244	0.203	3.9449471853762 85
	IKA	6	9.0e-728	0.125	6.8968993023260 81
f_1, $x_0 = 2.3$					
	NW	9	1.7e-104	0.031	1.9997625515356 68
	INW	5	1.7e-104	0.031	3.9927058769489 16
	KT	10	6.9e-110	0.031	2.0000000067284 49
	IKT	6	3.8e-296	0.078	3.9739311085623 68
	WM	6	5.8e-106	0.031	2.9775481645616 99
	IWM	5	1.0e-520	0.046	5.4287063276118 60
	OM	6	1.3e-213	0.031	3.0047059877604 14
	IOM	4	1.7e-129	0.054	5.7560540923451 67
	JA	6	1.4e-311	0.076	3.9918531104247 56
	IJA	5	6.0e-426	0.078	6.9105915314008 21
	KA	6	1.0e-234	0.094	3.9292326152766 95
	IKA	5	1.9e-1000	0.124	6.9915998991280 05

f_2, $x_0 = 0$					
	NW	8	8.9e-201	0.031	2.001221658759151
	INW	5	1.9e-402	0.031	4.036830839017272
	KT	8	4.9e-124	0.046	2.000000001565404
	IKT	5	1.4e-289	0.047	4.072048404246373
	WM	5	7.8e-106	0.033	2.989730807986798
	IWM	4	5.4e-271	0.046	5.998509512335663
	OM	5	4.3e-112	0.172	2.999984519710846
	IOM	4	3.5e-276	0.125	5.992519937559335
	JA	5	1.0e-286	0.057	3.987739193299696
	IJA	4	2.8e-827	0.061	6.997013442884977
	KA	5	1.6e-292	0.109	4.220027174017621
	IKA	4	5.3e-833	0.094	7.095300609185996
f_2, $x_0 = 1$					
	NW	8	1.7e-189	0.032	1.997942151132725
	INW	5	7.1e-380	0.032	3.725609846208910
	KT	9	4.1e-138	0.031	2.000000000185425
	IKT	5	2.7e-207	0.063	3.878358818775769
	WM	6	1.4e-201	0.062	3.038062042190221
	IWM	4	1.2e-201	0.031	5.825290885992024
	OM	6	3.4e-206	0.047	3.001050663973445
	IOM	4	3.1e-202	0.047	5.843786348256205
	JA	5	2.4e-258	0.062	3.722160566330791
	IJA	4	3.2e-633	0.062	6.833672609585321
	KA	5	5.6e-264	0.110	3.981598688393819
	IKA	4	5.5e-643	0.110	7.023459826814717
f_3, $x_0 = 1$					
	NW	9	1.3e-151	0.047	2.001106997169595
	INW	5	1.3e-151	0.031	3.999685578348938
	KT	9	1.3e-108	0.062	2.000000012840528
	IKT	5	7.2e-122	0.093	3.998627364224366
	WM	6	2.9e-131	0.063	3.000546669049605
	IWM	4	3.9e-133	0.062	5.866966779516045
	OM	6	3.4e-186	0.234	3.000513674703602
	IOM	4	2.8e-129	0.062	5.983351099780604
	JA	6	3.4e-425	0.109	3.990537088703305
	IJA	4	9.3e-444	0.106	6.985005730502670
	KA	6	3.7e-302	0.172	3.991952875605968
	IKA	5	0	0.125	6.825746339976237
f_3, $x_0 = 0.5$					
	NW	8	8.9e-122	0.046	2.003980933345205
	INW	5	2.6e-243	0.047	4.016633132132276
	KT	9	8e-184	0.047	1.999999964933080

The computational order of convergence (COC) was given by (see [19])

$$COC = \frac{\ln\left(|x_{k+1} - x_k| / |x_k - x_{k-1}|\right)}{\ln\left(|x_k - x_{k-1}| / |x_{k-1} - x_{k-2}|\right)}. \tag{50}$$

Table 1 summarizes the results obtained by using the mentioned methods in order to estimate a root of nonlinear equations. For every function we specify the initial estimate x_0, the number of iterations N required to meet the stopping criteria, the value of $|f(x_{n+1})|$ in the last iteration and the value of COC.

We also compare our methods with the fourth-order King's method [20] (KM), the Ostrowski's method [21] (OST), the sixth order methods given by Chun and Ham [7] (CM1 and CM2), Parhi and Gupta [8](PM), and the eighth order method given by Grau-Sánchez [22] (MG) (see Table 2).

Table 2. Numerical comparisons of the methods of KM, OST, PM, CM1, CM2, MG and the present methods

	Method	N	$\lvert f(x_{n+1})\rvert$	CPU
f_1, $x_0 = 1$				
	INW	5	3.4e-101	0.031
	IKT	6	1.6e-234	0.203
	KM	10	6.3e-301	0.251
	OST	5	1.0e-109	0.062
	IWM	5	1.2e-566	0.111
	IOM	3	6.5e-123	0.063
	CM1	5	1.4e-226	0.098
	CM2	6	3.6e-245	0.090
	PM	5	1.2e-566	0.068
	IIA	4	6.0e-426	0.060
	IKA	6	9.0e-728	0.125
	MG	5	7.6e-345	0.121
f_1, $x_0 = 2.3$				
	INW	5	1.7e-104	0.031
	IKT	6	3.8e-296	0.078
	KM	6	1.0e-259	0.078
	OST	6	8.3e-389	0.063
	IWM	5	1.0e-520	0.046
	IOM	4	1.7e-129	0.054
	CM1	5	1.6e-391	0.113
	CM2	5	1.4e-372	0.085
	PM	5	1.0e-520	0.075
	IIA	5	6.0e-426	0.078
	IKA	5	1.9e-1000	0.124
	MG	4	1.6e-130	0.320
f_2, $x_0 = 0$				
	INW	5	1.9e-402	0.031
	IKT	5	1.4e-289	0.047
	KM	5	1.9e-376	0.059
	OST	5	1.1e-352	0.063
	IWM	4	5.4e-271	0.046
	IOM	4	3.5e-276	0.125
	CM1	4	8.2e-313	0.081
	CM2	4	4.2e-291	0.067
	PM	4	5.4e-271	0.052
	IIA	4	2.8e-827	0.061
	IKA	4	5.3e-833	0.094
	MG	4	1.5e-387	0.133

f_2, $x_0 = 1$				
	INW	5	7.1e-380	0.032
	IKT	5	2.7e-207	0.063
	KM	5	3.8e-281	0.080
	OST	5	6.6e-258	0.047
	IWM	4	1.2e-201	0.031
	IOM	4	3.1e-202	0.047
	CM1	4	3.4e-194	0.074
	CM2	4	1.4e-194	0.062
	PM	4	1.2e-201	0.059
	IJA	4	3.2e-633	0.062
	IKA	4	5.5e-643	0.110
	MG	4	3.5e-326	0.129
f_3, $x_0 = 1$				
	INW	5	1.3e-151	0.031
	IKT	5	7.2e-122	0.093
	KM	5	1.5e-103	0.196
	OST	5	9.5e-187	0.094
	IWM	4	3.9e-133	0.062
	IOM	4	2.8e-129	0.062
	CM1	4	9.0e-124	0.102
	CM2	4	4.3e-118	0.098
	PM	4	4.0e-133	0.131
	IJA	4	9.3e-444	0.106
	IKA	5	0	0.125
	MG	4	1.1e-205	0.190
f_3, $x_0 = 0.5$				
	INW	5	2.6e-243	0.047

In Table 1 it is seen that our combined iterative methods generally arrive at the iterated solution with less number of iterations than the corresponding second, cubic and the fourth methods, so that the proposed methods improve the computational efficiency of the existing iterative methods. Our examples show that the combined iterative methods sometimes require more CPU time per iteration, compared to the existing methods. Although our methods require more time per iteration, they yield better numerical results. The numerical results in Table 1 show that for almost all of the test functions, our methods are well in accordance with the theory developed in section 2.

The test results in Table 2 show that for most of the functions we tested, the methods introduced in the present presentation for numerical tests have equal or better performance compared to the other methods of the same order. In each of these 8 test cases, the INW method outperformed the KM method in every case and it outperformed the OST method in 5 out of the 8 cases. Our IKT method outperformed KM method in 6 out of the 8 cases. We also implemented the three 6th order schemes of [7, 8] using these 8 cases, and found that the IWM and IOM methods outperformed the PM, CM1 and CM2 methods in 7 out of the 8 cases.

In [22] a eighth-order method, denoted with MG was considered. The IJA method outperformed the MG method in 7 out of the 8 cases, and the IKA method outperformed the MG method in 4 out of the 8 cases. Besides, we can see that the local convergence property of the new methods depending on the structure of the tested functions and the choice of initial approximations.

5. CONCLUSION

The new modified Newton's methods presented in this paper offer an increase rate of convergence over the existing methods. Unlike other higher-order methods, the distinct feature of such methods is only to add the evaluation of the function at another point, while their order of convergence can be improved effectively. Our new combined iterative methods are relatively simple and robust, more high-order convergence methods can be constructed by using the family of methods (17).

Computational results for test functions, presented in Table 1 and Table 2, show that our methods are efficient and show at least equal or better performance as compared with other higher order (4th order, 6th order and 8th order) schemes or Newton's method itself. Our methods show similar good performance for other functions.

ACKNOWLEDGEMENTS

The authors gratefully acknowledge the support of the National Natural Science Foundations of China (NNSFC) through grants No.11302184, 11202189, the Scientific Research Fund of Zhejiang Provincial Education Department through grant No. Y201121157 and the Scientific Research Foundation of Xiamen University of Technology through grant No. YKJ13007R.

REFERENCES

1. Zhou, X.J., "A class of Newton's methods with third-order convergence," Applied Mathematics Letters, 20 (9). 1026-1030. Sep.2007.
2. Cordero, A., Jordán, C. and Torregrosa, J.R.,"One-point Newton-type iterative methods: A unified point of view," Journal of Computational and Applied Mathematics, 275. 366-374. Feb. 2015.
3. Neta, B. and Scott, M., "On a family of Halley-like methods to find simple roots of nonlinear equations,"Applied Mathematics and Computation, 219 (15). 7940-7944. Mar. 2013.
4. Soleymani, F., Khattri, S.K. and Vanani, S.K., "Two new classes of optimal Jarratt-type fourth-order methods," Applied Mathematics Letters, 25 (5). 847-853. May. 2012.

5. Wang, X. and Liu, L.P., "Two new families of sixth-order methods for solving non-linear equations," Applied Mathematics and Computation, 213 (1). 73-78. Jul. 2009.
6. Kou, J.S., Li, Y.T. and Wang, X.H., "Some modifications of Newton's method with fifth-order convergence," Journal of Computational and Applied Mathematics, 209 (2). 146-152. Dec. 2007.
7. Chun, C. and Ham, Y., "Some sixth-order variants of Ostrowski root-finding methods," Applied Mathematics and Computation, 193 (2). 389-394. Nov. 2007.
8. Parhi, S.K. and Gupta, D.K., "A sixth order method for nonlinear equations," Applied Mathematics and Computation, 203 (1). 50-55. Sep. 2008.
9. Weerakoon, S. and Fernando, T.G.I., "A variant of Newton's method with accelerated third-order convergence," Applied Mathematics Letters, 13 (8). 87-93. Nov. 2000.
10. Petkovic, S.M., Neta B., Petkovic L.D. and Džunic, J., "Multiple methods for solving nonlinear equations: A survey," Applied Mathematics and Computation, 226. 635-660. Jan. 2014.
11. Steffensen, I.F., "Remarks on iteration," Scandinavian Actuarial Journal, 1993(1). 64-72. Dec. 2011.
12. Khattri, S.K. and Argyros, I.K., "How to develop fourth and seventh order iterative methods," Novi Sad Journal of Mathematics, 40 (2). 61-67. 2010.
13. Traub, J.F., Iterative methods for the solution of equations, Chelsea Publishing Company, New York, 1982.
14. Kanwar, V. and Tomar, S.K., "Modified families of Newton, Halley and Chebyshev methods," Applied Mathematics and Computation, 192 (1). 20-26. Sep. 2007.
15. Gautschi, W., Numerical Analysis: An introduction, Birkhäuser, 1997.
16. Özban, A. Y., "Some new variants of Newton's method," Applied Mathematics Letters, 17 (6). 677-682. Jun. 2004.
17. Jarratt, P., "Some efficient fourth order multipoint methods for solving equations," BIT Numerical Mathematics, 9(2). 119-124. Jun. 1969.
18. Khattri, S.K. and Abbasbandy, S., "Optimal fourth order family of iterative methods," Matematicki Vesnik, 63 (1). 67-72. Jan. 2011.
19. Cordero, A. and Torregrosa, J.R., "Variants of Newton's method using fifth-order quadrature formulas," Applied Mathematics and Computation, 190 (1). 686-698. Jul. 2007.
20. King, R.F, "A family of fourth-order methods for nonlinear equations," SIAM Journal on Numerical Analysis, 10 (5). 876-879. Oct. 1973.
21. Ostrowski, A.M., Solution of equations in Euclidean and Banach spaces, Academic, New York, 1960.
22. Grau-Sánchez, M., "Improving order and efficiency: Composition with a modified Newton's method," Journal of Computational and Applied Mathematics, 231 (2). 592-597. Sep. 2009.

CHAPTER 9

A New Entropy-based Intuitionistic Fuzzy Multi-attribute Decision Making Method

Lanping Li

Department of Basic Subjects, Hunan University of Finance and Economics, Changsha, China

ABSTRACT

The main task of this paper is to develop a new decision making method based on a novel entropy measure of intuitionistic fuzzy sets. First a novel intuitionistic fuzzy entropy is constructed, then based on this information measure, new weighting methods are proposed for the intuitionistic fuzzy decision making problems with the attribute weights are completely unknown or partly known. Further the intuitionistic fuzzy TOPSIS method is developed in this paper, and two examples are given to illustrate effectiveness and practicability of proposed method.

Keywords: Intuitionistic Fuzzy Number, Entropy, Multi-Attribute Decision Making, TOPSIS Method

1. INTRODUCTION

Because of the complexity and limitations of human understanding of the world, more and more management problems involve a lot of fuzzy concepts, such as, expected cost between 180 with 200 million, demand about 200 people. Therefore, since the fuzzy set was firstly proposed by Zadeh [1], fuzzy sets have been applied to various fields, especially the multiple attribute decision making problems based on fuzzy sets have been widely studied and applied to many field, such as supplier partner selection, military weapon system evaluation and selection of manufacturers [2-4]. Because of the complexity and the uncertainty of information of the actual management problem, in decision-making process decision makers often show some hesitation. In order to describe the hesitancy degree, Atanassov [5] proposed the concept of intuitionistic fuzzy sets. Intuitionistic fuzzy numbers can not only describe the general range of these fuzzy numbers, but also can describe the uncertainty of decision makers on the judgment of the decision makers. It also considers the degree of membership, non membership and hesitancy degree of three aspects of information, which can describe the fuzzy concept betwixt the more delicate describe the fuzzy nature of the objective world. It is the expansion and development of Zadeh fuzzy set theory [6]. In 1993, Gau and Buehrer [7] defined vague sets, and then Bustince and Burillo [8] pointed out that the notion of vague sets was the same with that of intuitionistic fuzzy set. In recent years, there are many intuitionistic fuzzy multi-attribute decision making (MADM) methods are developed to deal with decision making problems [9-11].

Entropy is an important measure to describe the degree of fuzzy set and the degree of uncertainty. Fuzzy entropy has been widely applied to pattern recognition, image processing, clustering analysis and MADM problem [12-14]. The research on intuitionistic fuzzy entropy has been concerned and studied by some scholars. For example, Burillo and Bustince [15] introduced the intuitionistic fuzzy entropy measure for measuring fuzziness degree or uncertain information of intuitionistic fuzzy sets. Zhang and Jiang [16] defined a measure of intuitionistic fuzzy entropy by generalizing of the Deluca and Termini logarithmic fuzzy entropy; Ye [17] proposed two intuitionistic fuzzy entropy measures using two triangular functions; Verma and Sharma [18] defined an exponential intuitionistic fuzzy entropy measure by generalizing of the Pal and Pal's exponential fuzzy entropy. Wang [19] proposed an intuitionistic fuzzy entropy measure using a cotangent function.

In this paper, we will put forward a new intuitionistic fuzzy entropy measure, and then we will develop a new entropy-based intuitionistic fuzzy MADM method in which the weights information are complete unknown and partially known.

The organization of this paper is: In Section 2, the basic definitions and notations of intuitionistic fuzzy set are given. Section 3 constructs a new intuitionistic fuzzy information measure and proves that it is an entropy measure. Section 4 puts forward an intuitionistic fuzzy MADM method in which the weights of attribute are obtained according to the proposed intuitionistic fuzzy entropy measure. Two examples are provided in Section 5. Finally, conclusions are given in Section 6.

2. PRELIMINARIES

Definition1 [20]. Let X be a given universal set, the intuitionistic fuzzy set A is defined as $A=\{<x_i,\mu_A(x_i),\upsilon_A(x_i)>|\, x_i\in X\}$, where the functions $\mu_A: X\mapsto[0,1]$ and $\upsilon_A: X\mapsto[0,1]$ are the membership degree and non-membership degree of x_i respectively, and they satisfy $0\le\mu_A(x_i)+\upsilon_A(x_i)\le 1$ for any $x_i\in X$. Further more, $\pi_A(x_i)=1-\mu_A(x_i)-\upsilon_A(x_i)$ is called the hesitancy degree of x_i. Conveniently, $<x_i,\mu_A(x_i),\upsilon_A(x_i)>$ is called the intuitionistic fuzzy number, abbreviated as $(\mu_A(x_i),\upsilon_A(x_i))$.

Definition2. Let $A=\{<x_i,\mu_A(x_i),\upsilon_A(x_i)>|\, x_i\in X\}$ and $B=\{<x_i,\mu_B(x_i),\upsilon_B(x_i)>|\, x_i\in X\}$ be two intuitionistic fuzzy sets, then the operation laws can be founded in [20]:

(1) $A\subseteq B$ iff $\mu_A(x_i)\le\mu_B(x_i),\upsilon_A(x_i)\ge\upsilon_B(x_i),\forall x_i\in X$;

(2) $A=B$ iff $A\subseteq B$ and $B\subseteq A$;

(3) The complementary set A^C of A is $A^C=\{<x_i,\upsilon_A(x_i),\mu_A(x_i)>|\, x_i\in X\}$;

(4) $A\preceq B$ called A less fuzzy than B, that is,

for $\forall x_i\in X$,

If $\mu_B(x_i)\le\upsilon_B(x_i)$, then $\mu_A(x_i)\le\mu_B(x_i),\upsilon_A(x_i)\ge\upsilon_B(x_i)$;

If $\mu_B(x_i)\ge\upsilon_B(x_i)$, then $\mu_A(x_i)\ge\mu_B(x_i),\upsilon_A(x_i)\le\upsilon_B(x_i)$.

Definition 3[21]. Let $A=\{<x_i,\mu_A(x_i),\upsilon_A(x_i)>|\, x_i\in X\}$ and $B=\{<x_i,\mu_B(x_i),\upsilon_B(x_i)>|\, x_i\in X\}$ be two intuitionistic fuzzy set, w_i is the weight of element x_i, then we can define the weighted Hamming distance measure between A and B as

$$d(A,B)=\frac{1}{2}\sum_{i=1}^{n}w_i(|\mu_A(x_i)-\mu_B(x_i)| + |\upsilon_A(x_i)-\upsilon_B(x_i)|+|\pi_A(x_i)-\pi_B(x_i)|) \quad (1)$$

The intuitionistic fuzzy entropy measure is given in Definition 4.

Definition 4 [15]. Let $IFS(X)$ is the set of all intuitionistic fuzzy sets defined in X. Then A map $E:IFS(X)\rightarrow[0,1]$ called the intuitionistic fuzzy entropy, if it satisfies the following four conditions:

(i) $E(A)=0$ if and only if A is a crisp set;

(ii) $E(A)=1$ if and only if $\mu_A(x_i)=\upsilon_A(x_i),\forall x_i\in X$;

(iii) $E(A)=E(A^C)$;

(iv) If $A\preceq B$, then $E(A)\leq E(B)$.

3. A NEW INTUITIONISTIC FUZZY ENTROPY

This section will first construct a new information measure and then prove that it is a entropy measure for intuitionistic set.

The new intuitionistic fuzzy information measure is constructed as follows:

$$E(A)=\frac{1}{n}\sum_{i=1}^{n}\frac{1-(\mu_A(x_i)-\upsilon_A(x_i))^2}{1+3(\mu_A(x_i)-\upsilon_A(x_i))^2} \quad (2)$$

Theorem. The measure given by Eq.(2) is an intuitionistic fuzzy entropy.

Proof. To prove the measure given by Eq.(2) is an intuitionistic fuzzy entropy, we only need to prove it satisfies the four cases (i)-(iv) in definition 4. Obviously, $0\leq E(A)\leq 1$.

For the case (i). Let A be a crisp set. That is, for $\forall x_i\in X$, $\mu_A(x_i)=0,\upsilon_A(x_i)=1$ or $\mu_A(x_i)=1,\upsilon_A(x_i)=0$. It is obvious that $E(A)=0$.

If $E(A)=0$, that is $E(A)=\frac{1}{n}\sum_{i=1}^{n}\frac{1-(\mu_A(x_i)-\upsilon_A(x_i))^2}{1+3(\mu_A(x_i)-\upsilon_A(x_i))^2}=0$,

then for $\forall x_i \in X$, we have

$$\sum_{i=1}^{n}\frac{1-(\mu_A(x_i)-\upsilon_A(x_i))^2}{1+3(\mu_A(x_i)-\upsilon_A(x_i))^2}=0,$$

It is obvious that $1-(\mu_A(x_i)-\upsilon_A(x_i))^2=0$, then we can conclude that $\mu_A(x_i)=0, \upsilon_A(x_i)=1$ or $\mu_A(x_i)=1, \upsilon_A(x_i)=0$. Therefore the set A is a crisp set.

For the case (ii). If $\mu_A(x_i)=\upsilon_A(x_i)$, $\forall x_i \in X$, then we can easily shown that $E(A)=1$ for the Eq. (2).

Now we suppose that $E(A)=1$, then for all $x_i \in X$, we have

$$\frac{1-(\mu_A(x_i)-\upsilon_A(x_i))^2}{1+3(\mu_A(x_i)-\upsilon_A(x_i))^2}=1,$$

Then we immediately obtain the conclusion $\mu_A(x_i)=\upsilon_A(x_i)$ for all $x_i \in X$.

For the case (iii). Because $A^C=\{<x_i,\upsilon_A(x_i),\mu_A(x_i)>| x_i \in X\}$,

we have

$$E(A^C)=\frac{1}{n}\sum_{i=1}^{n}\frac{1-(\mu_{A^c}(x_i)-\upsilon_{A^c}(x_i))^2}{1+3(\mu_{A^c}(x_i)-\upsilon_{A^c}(x_i))^2}$$
$$=\frac{1}{n}\sum_{i=1}^{n}\frac{1-(\upsilon_A(x_i)-\mu_A(x_i))^2}{1+3(\upsilon_A(x_i)-\mu_A(x_i))^2}$$
$$=\frac{1}{n}\sum_{i=1}^{n}\frac{1-(\mu_A(x_i)-\upsilon_A(x_i))^2}{1+3(\mu_A(x_i)-\upsilon_A(x_i))^2}=E(A).$$

For the case (iv). Construct the function

$$f(x,y)=\frac{1-(x-y)^2}{1+3(x-y)^2} \quad (3)$$

Where $x, y \in [0,1]$

We can easily derived the partial derivatives of $f(x,y)$ to x and to y, respectively

$$\frac{\partial f(x,y)}{\partial x} = \frac{-8(x-y)}{[1+3(x-y)^2]^2} \tag{4}$$

$$\frac{\partial f(x,y)}{\partial y} = \frac{8(x-y)}{[1+3(x-y)^2]^2} \tag{5}$$

When $x \leq y$, we have

$$\frac{\partial f(x,y)}{\partial x} \geq 0 \text{ and } \frac{\partial f(x,y)}{\partial x} \leq 0,$$

That is to say, $f(x,y)$ is an increasing function with x and a decreasing function with y. Then when the conditions $\mu_B(x_i) \leq \upsilon_B(x_i)$, $\mu_A(x_i) \leq \mu_B(x_i)$ and $\upsilon_A(x_i) \geq \upsilon_B(x_i)$ are satisfied, we have $f(\mu_A(x_i), \upsilon_A(x_i)) \leq f(\mu_B(x_i), \upsilon_B(x_i))$;

Similarly, it is easily proved that when $x \geq y$, we have $\frac{\partial f(x,y)}{\partial x} \leq 0$ and $\frac{\partial f(x,y)}{\partial x} \geq 0$, then $f(x,y)$ is a decreasing function with x and an increasing function with y.

Thus when the conditions $\mu_B(x_i) \geq \upsilon_B(x_i)$, $\mu_A(x_i) \geq \mu_B(x_i)$ and $\upsilon_A(x_i) \leq \upsilon_B(x_i)$ satisfied, we have $f(\mu_A(x_i), \upsilon_A(x_i)) \leq f(\mu_B(x_i), \upsilon_B(x_i))$.

Therefore, if $A \preceq B$, we have $\frac{1}{n}\sum_{i=1}^{n} f(\mu_A(x_i), \upsilon_A(x_i)) \leq \frac{1}{n}\sum_{i=1}^{n} f(\mu_B(x_i), \upsilon_B(x_i))$,

Then $E(A) \leq E(B)$.

The theorem is proved.

4. NEW ENTROPY-BASED INTUITIONISTIC FUZZY MADM METHOD

This section will propose a new decision making method based on the new constructed intuitionistic fuzzy entropy.

Now we consider a MADM problem including m non-inferior alternatives $A_i\,(i = 1, 2, \ldots, m)$ and n attributes $o_j\ (j = 1, 2, \ldots, n)$. Here we let $A = \{A_1, A_2, \cdots, A_m\}$ be the set of these alternatives and $O = \{o_1, o_2, \cdots, o_n\}$ be the set of these attributes. The task of this section is to choose the best desirable alternative from n non-inferior alternatives. Suppose that the evaluation values of alternatives A_i on attributes o_j are intuitionistic fuzzy numbers $\tilde{a}_{ij} = (\mu_{ij}, \upsilon_{ij})\ i = 1, 2, \ldots, m, j = 1, 2, \ldots, n$, Here μ_{ij} and υ_{ij} are the membership (satisfactory) degree and nonmembership (nonsatisfactory) degree of the alternative A_i on the attribute o_j, and for any i and j, μ_{ij} and υ_{ij} satisfy the following conditions:

$$0 \le \mu_{ij} \le 1,\quad 0 \le \upsilon_{ij} \le 1 \text{ and } 0 \le \mu_{ij} + \upsilon_{ij} \le 1.$$

Thus, a this intuitionistic fuzzy MADM problem can be expressed with the following decision matrix $\mathbf{D} = (\tilde{a}_{ij})_{m\times n}$ as follows:

$$\mathbf{D} = (\tilde{a}_{ij})_{m\times n} = \begin{array}{c} \\ A_1 \\ A_2 \\ \vdots \\ A_m \end{array}\!\!\begin{array}{c} \begin{array}{cccc} o_1 & o_2 & \cdots & o_m \end{array} \\ \begin{pmatrix} (\mu_{11}, \upsilon_{11}) & (\mu_{12}, \upsilon_{12}) & \cdots & (\mu_{1n}, \upsilon_{1n}) \\ (\mu_{21}, \upsilon_{21}) & (\mu_{22}, \upsilon_{22}) & \cdots & (\mu_{2n}, \upsilon_{2n}) \\ \vdots & \vdots & \vdots & \vdots \\ (\mu_{m1}, \upsilon_{m1}) & (\mu_{m2}, \upsilon_{m2}) & \cdots & (\mu_{mn}, \upsilon_{mn}) \end{pmatrix} \end{array}.$$

Let $\boldsymbol{w} = (w_1, w_2, \cdots, w_n)^{\mathrm{T}}$ be the weight vector of all attributes, where $0 \le w_j \le 1$ is the important degree of attributes o_j ($j = 1, 2, \cdots, n$), and $\sum_{j=1}^{n} w_j = 1$. The attribute weights information is usually unknown or partially known due to the limitation of time, knowledge etc. of decision makers in the decision making process. Therefore, the determination of attribute weights is an important issue which attracts great attention by many researchers [22-25]. This section will put forward two methods to determine the atstribute weights for the above-mentioned two cases.

4.1 Weighting Method with Unkown Attribute Weights Information

When the attribute weights are completely unknown, this section will propose a new entropy weighting method based on the new intuitionistic fuzzy entropy as follows:

Step1. Calcalate the intuitionistic fuzzy entropy of $\tilde{a}_{ij} = (\mu_{ij}, \upsilon_{ij})$:

$$E(\tilde{a}_{ij}) = \frac{1-(\mu_{ij}-\nu_{ij})^2}{1+3(\mu_{ij}-\nu_{ij})^2} \tag{6}$$

Step2. The weight of the attribute o_j is obtained as

$$w_j = \frac{1-e_j}{n-\sum_{j=1}^{n} e_j}, \quad j = 1, 2, \ldots, n \tag{7}$$

Here $e_j = \frac{1}{m}\sum_{i=1}^{m} E(\tilde{a}_{ij})$.

4.2 Weighting Method with Partially Known Attribute Weights Information

In actual decision process, attribute weights information sometimes is partially known under intuitionistic fuzzy environment. To determine the weights, Xu [26] constructed an optimization model on the basis pf Chen and Tan's score function [27] to slove the optimal weights. Wu and Zhang [28] solved the attribute weights by establishing a programming model according to minimum entropy principle. Inspired by reference [28], this paper will use the new intuitionistic fuzzy entropy measure to solve the attribute weights by establishing a programming model as follows:

$$\begin{aligned} \min\ E &= \sum_{i=1}^{m} E(A_i) = \sum_{i=1}^{m}\sum_{j=1}^{n} w_j \frac{1-(\mu_{ij}-\nu_{ij})^2}{1+3(\mu_{ij}-\nu_{ij})^2} \\ s.t.\ & \sum_{j=1}^{n} w_j = 1 \\ & \boldsymbol{w} \in \boldsymbol{H} \end{aligned} \tag{8}$$

Hence, the optimal attribute weights is the optimal solution $w^* = \arg\min E$ of the programming model (8).

4.3 The New Entropy-based Intuitionistic Fuzzy MADM Method

This subsection will put forward a new intuitionistic fuzzy MADM method based on the above-mentioned work combining with the concept of TOPSIS. The calculation steps are given as follows:

Step 1. Calculate the attribute weights according to Eq.(6) and Eq.(7) when the attribute weights are completely unknown or partially known;

Step 2. Determine the positive ideal solution and negative ideal solution.

The positive ideal solution is defined as follows:

$$A^* = ((\mu_1^*, \upsilon_1^*), (\mu_2^*, \upsilon_2^*), \cdots, (\mu_n^*, \upsilon_n^*)) \tag{9}$$

where $(\mu_j^*, \upsilon_j^*) = (1, 0)$ ($j = 1, 2, \cdots, n$).

The negative ideal solution is defined as follows:

$$A^- = ((\mu_1^-, \upsilon_1^-), (\mu_2^-, \upsilon_2^-), \cdots, (\mu_n^-, \upsilon_n^-)) \tag{10}$$

where $(\mu_j^-, \upsilon_j^-) = (0, 1)$ ($j = 1, 2, \cdots, n$).

Step 3. According to Definition 3, the weighted Hamming distance measures between alternative A_i with positive and negative ideal solutions are respectively calculated as follows:

$$\begin{aligned} d(A_i, A^*) &= \frac{1}{2}\sum_{j=1}^{n} w_j(|\mu_{ij} - \mu_j^*| + |\upsilon_{ij} - \upsilon_j^*| + |\pi_{ij} - \pi_j^*|) \\ &= \frac{1}{2}\sum_{j=1}^{n} w_j(|1 - \mu_{ij}| + |\upsilon_{ij}| + |1 - \mu_{ij} - \upsilon_{ij}|) \end{aligned} \tag{11}$$

$$\begin{aligned} d(A_i, A^-) &= \frac{1}{2}\sum_{j=1}^{n} w_j(|\mu_{ij} - \mu_j^-| + |\upsilon_{ij} - \upsilon_j^-| + |\pi_{ij} - \pi_j^-|) \\ &= \frac{1}{2}\sum_{j=1}^{n} w_j(|\mu_{ij}| + |1 - \upsilon_{ij}| + |1 - \mu_{ij} - \upsilon_{ij}|) \end{aligned} \tag{12}$$

Step 4. Calculate the relative closeness degree of each alternative.

The closeness degree C_i is defined as:

$$C_i = \frac{d(A_i, A^-)}{d(A_i, A^-) + d(A_i, A^+)} \tag{13}$$

Step 5. Rank the alternatives according to C_i in decreasing order. The best alternative is the closest to the positive ideal solution and the farthest to the negative ideal solution.

5. NUMERICAL EXAMPLES

In order to illustrate the effectiveness and feasibility of the proposed MADM method, two examples are given as follows:

Example 1. (The case of complete unknown attributes weights information) This example is adopted from Herrera and Herrera-Viedma [29], Ye [30]. Suppose that a company wants to invest a large amount of money to some project. There are four parallel alternatives to be selected: a car company (A_1), a food company (A_2), a computer company (A_3) and an arms company (A_4). To choose the best desirable alternative, the decision maker consider the following evaluation attributes: the risk analysis (o_1), the growth analysis (o_2) and the environmental impact analysis (o_3). Using statistical methods, we can obtain the satisfactory degree μ_{ij} and non-satisfactory degree υ_{ij} for the alternative A_i satisfying the attributes o_j respectively. The intuitionistic fuzzy decision matrix is illustrated in Table 1.

The calculation steps are given as follows:

Step1. According to the Eq. (7), the attribute weights vector is obtained as

$$w = (w_1, w_2, w_3)^T = (0.3384, 0.3306, 0.3309)^T$$

Table 1. Intuitionistic fuzzy decision matrix.

Alternative	Evaluation attribute		
	o_1	o_2	o_3
A_1	(0.45,0.35)	(0.50,0.30)	(0.20,0.55)
A_2	(0.65,0.25)	(0.65,0.25)	(0.55,0.15)
A_3	(0.45,0.35)	(0.55,0.35)	(0.55,0.20)
A_4	(0.75,0.15)	(0.65,0.20)	(0.35,0.15)

Step 2.The positive ideal solution A^* and negative ideal solution A^- are respectively

$$A^* = ((\mu_1^*, \upsilon_1^*), (\mu_2^*, \upsilon_2^*), (\mu_3^*, \upsilon_3^*)) = ((1,0), (1,0), (1,0))$$

$$A^- = ((\mu_1^-, \upsilon_1^-), (\mu_2^-, \upsilon_2^-), (\mu_3^-, \upsilon_3^-)) = ((0,1), (0,1), (0,1))$$

Step 3. The weighted Hamming distance measures of each alternative from positive and negative ideal solutions are calculated as

$$d(A_1, A^*) = 0.6162, d(A_2, A^*) = 0.3831,$$
$$d(A_3, A^*) = 0.4838, d(A_4, A^*) = 0.4154$$

$$d(A_1, A^-) = 0.6003, d(A_2, A^-) = 0.8169,$$
$$d(A_3, A^-) = 0.6996, d(A_4, A^-) = 0.8335$$

Step4. The relative closeness degrees are obtained as :

$$C_1 = 0.4935, C_2 = 0.6808,$$
$$C_3 = 0.5912, C_4 = 0.6674$$

Therefore, the ranking order of all alternatives is $A_2 \succ A_4 \succ A_3 \succ A_1$, and A_2 is the desirable alternative. The ranking order is in agreement with Ye [30].

Example 2. (The case of partially known attributes weights information) The example is adopted from Li [31], which considers an air-condition system selection problem. Suppose there are three air-condition systems

A_1, A_2, A_3 to be selected. The decision maker choses the following evaluation attributes: economical (o_1), function (o_2) and being operative (o_3). Using statistical methods, The satisfactory degree μ_{ij} and non-satisfactory degree υ_{ij} for the alternative A_i satisfying the attribute o_j respectively. The intuitionistic fuzzy decision matrix is illustrated in Table 2.

Table 2. Intuitionistic fuzzy decision matrix.

Alternative	Evaluation attribute		
	o_1	o_2	o_3
A_1	(0.75,0.1)	(0.6,0.25)	(0.8,0.2)
A_2	(0.8,0.15)	(0.68,0.2)	(0.45,0.5)
A_3	(0.4,0.45)	(0.75,0.05)	(0.6,0.3)

Assume the weights information of the three attribute is partially known, and the weights satisfies the set

$$H = \{0.25 \le w_1 \le 0.75, 0.35 \le w_2 \le 0.60, 0.30 \le w_3 \le 0.35\}.$$

The calculation steps of the proposed MADM method is given as follows:

Step1. According to the Eq. (8), we can establish the following programming model:

$$\min E = 0.6697w_1 + 0.6654w_2 + 0.8172w_3$$
$$s.t. \begin{cases} 0.25 \le w_1 \le 0.75 \\ 0.35 \le w_2 \le 0.60 \\ 0.30 \le w_3 \le 0.35 \\ w_1 + w_2 + w_3 = 1 \end{cases}$$

Then We the optimum attribute weight vector is solved by using Matlab software, and the optimal weight vector is obtained as $w = (0.25, 0.45, 0.30)^T$.

Step 2. The positive ideal solution A^+ and negative ideal solution A^- are respectively

$$A^{*} = ((\mu_1^{*}, \upsilon_1^{*}), (\mu_2^{*}, \upsilon_2^{*}), (\mu_3^{*}, \upsilon_3^{*})) = ((1,0),(1,0),(1,0))$$

$$A^{-} = ((\mu_1^{-}, \upsilon_1^{-}), (\mu_2^{-}, \upsilon_2^{-}), (\mu_3^{-}, \upsilon_3^{-})) = ((0,1),(0,1),(0,1))$$

Step 3. The weighted Hamming distance measures of each alternative from positive and negative ideal solutions are calculated as

$$d(A_1, A^{*}) = 0.3025, d(A_2, A^{*}) = 0.3590,$$
$$d(A_3, A^{*}) = 0.3825$$

$$d(A_1, A^{-}) = 0.8025, d(A_2, A^{-}) = 0.7225,$$
$$d(A_3, A^{-}) = 0.7750$$

Step 4. The relative closeness degrees are obtained as

$$C_1 = 0.7262, C_2 = 0.6681, C_3 = 0.6695$$

Step 5. Based on C_i values, the ranking result of the alternatives is $A_1 \succ A_3 \succ A_2$, and A_1 is the best desirable air-condition system. This reveals that the alternative is in agreement with the result in Li [31].

6. CONCLUSION

Intuitionistic fuzzy set is the extension of Zadeh's fuzzy set, and it can deal the decision with decision makers' hesitancy degree. This paper first construct a new intuitionistic fuzzy entropy measure and then puts forward a new decision making method for solving intuitionistic fuzzy MADM problem. Two weighting methods are proposed for the cases of weights are completely unknown and partially known. The new entropy measures can also be applied to the field of pattern recognition, image processing. The proposed intuitionistic fuzzy MADM method can also be applied to deal with other decision making problems, such as the project investment risk evaluation, material selection, robot selection and water quality evaluation.

ACKNOWLEDGEMENT

This study is partially supported by Natural Science Foundation of Hunan Province (No. 2015JJ3030 and No. 2016JJ4012) and Foundation of Hunan Educational Committee (No.15C0228). The author also gratefully acknowledges the helpful comments and suggestions of the reviewers, which have improved the presentation.

REFERENCES

1. Zadeh L. A., 1965. Fuzzy sets. Information and Control, 8(3): 338-353.
2. Dalman H., Güzel N., Sivri M., 2015. A fuzzy set-based approach to multi-objective multi-item solid transportation problem under uncertainty. International Journal of Fuzzy Systems, 18(4):716-729.
3. Lin P. C., Gu J. C., Yang M. T., 2014. Intelligent maintenance model for condition assessment of circuit breakers using fuzzy set theory and evidential reasoning. Iet Generation Transmission & Distribution, 8(7):1244-1253.
4. Ju H. C., 2014. Using the fuzzy set theory to developing an environmental impact assessment index for a thermal power plant. Quality & Quantity, 48(2):673-680.
5. Atanassov K. T., 1986. Intuitionistic fuzzy sets. Fuzzy Sets & Systems, 20(1):87-96.
6. Wang J. Q., Zhang H. Y., 2013. Multicriteria decision-making approach based on Atanassov's intuitionistic fuzzy sets with incomplete certain information on weights. IEEE Transactions on Fuzzy Systems, 21(3):510-515.
7. Gau W. L., Buehrer D. J., 1993. Vague sets. Systems IEEE Transactions on systems, Man & Cybernetics, 23(2):610-614.
8. Bustince H., Burillo P., 1996. Vague sets are intuitionistic fuzzy sets. Fuzzy Sets and Systems, 79(3): 403-405.
9. Xu Z., 2007. Some similarity measures of intuitionistic fuzzy sets and their applications to multiple attribute decision making. Fuzzy Optimization & Decision Making, 6(2):109-121.
10. Joshi D., Kumar S., 2014. Intuitionistic fuzzy entropy and distance measure based TOPSIS method for multi-criteria decision making. Egyptian Informatics Journal, 15(2):97-104.
11. Liang X., Wei C., 2013. An Atanassov's intuitionistic fuzzy multi-attribute group decision making method based on entropy and similarity measure. International Journal of Machine Learning & Cybernetics, 5(3):435-444.
12. Cheng H. D., Chen Y. H., Sun Y., 1999. A novel fuzzy entropy approach to image enhancement and thresholding. Signal Processing, 75(3):277-301.
13. Hsiao S. W., Chou J. R., 2006. A Gestalt-like perceptual measure for home page design using a fuzzy entropy approach. International Journal of Human-Computer Studies, 64(2):137-156.
14. Verma R., Sharma B. D., 2013. Exponential entropy on intuitionistic fuzzy sets. Kybernetika, 49(1):114-127.
15. Burillo P., Bustince H., 2001. Entropy on intuitionistic fuzzy sets and on interval-valued fuzzy sets. Fuzzy Sets and Systems, 118(3): 305-316.
16. Zhang Q. S., Jiang S. Y., 2008. A note on information entropy measure for vague sets. Information Sciences, 178, (21): 4184–4191.
17. Ye J., 2010.Two effective measures of intuitionistic fuzzy ertropy. Computing, 87(1/2): 55-62.
18. Verma R., Sharma B. D., 2013. Exponential entropy on intuitionistic fuzzy sets, Kybernetika, 49(1):114-127.

19. Wang J. Q., Wang P., 2012. Intuitionistic linguistic fuzzy multi-criteria decision-making method based on intuitionistic fuzzy entropy, Control and Decision, 27(11): 1694-1698.
20. Atanassov, K T., 1999. Intuitionistic Fuzzy Sets. New York: Physica-Verlag, Heidelberg.
21. Li D. F., 2012. Intuitionistic Fuzzy Set Decision and Game Analysis Methodologies, Beijing: National Defense Industry Press.
22. Xu Z., Zhang X., 2013. Hesitant fuzzy multi-attribute decision making based on TOPSIS with incomplete weight information. Knowledge-Based Systems, 52(6):53-64.
23. Biswas P., Pramanik S., Giri B. C., 2014. A new methodology for neutrosophic multi-attribute decision- making with unknown weight information. Neutrosophic Sets & Systems, 3(3):42-50.
24. Han Z. S., Liu P. D., 2011. A fuzzy multi-attribute decision-making method under risk with unknown attribute weights. Technological & Economic Development of Economy, 17(2):246-258.
25. Wei G. W., 2010. Extension of TOPSIS method for 2-tuple linguistic multiple attribute group decision making with incomplete weight information. Knowledge & Information Systems, 25(3):623-634.
26. Xu Z. S.,2007. Multi-person multi-attribute decision making models under intuitionistic fuzzy environment. Fuzzy Optimization Decision Making, 6(3):221-236.
27. Chen S. M., Tan J. M., 1994. Handing multicriteria fuzzy decision-making problems based vague set theory. Fuzzy Sets and Systems, 67(2):163-172.
28. Wu J. Z., Zhang Q.,2011. Multicriteria decision making method based on intuitionistic fuzzy weighted entropy. Expert Systems with Applications, 38(1), pp. 916-922.
29. Herrera F., Herrera-Viedma E., 2000. Linguistic decision analysis: steps for solving decision problems under linguistic information. Fuzzy Sets & Systems, 115(1):67-82.
30. Ye J., 2010. Fuzzy decision-making method based on the weighted correlation coefficient under intuitionistic fuzzy environment. European Journal of Operational Research, 205(1):202-204.
31. Li D. F., 2005. Multiattribute decision making models and methods using intuitionistic fuzzy sets. Journal of Computer and System Sciences, 70(1):73-85.

CHAPTER 10

New Methods for Comparing the Forecasts Accuracy

Bratu (Simionescu) Mihaela

Department of Statistics and Econometrics, Faculty of Cybernetics, Statistics and Economic Informatics, Bucharest, Romania

ABSTRACT

The main purpose of this research is to show the diversity of statistical methods that could be used to assess and compare forecasts accuracy. Some of the statistical approaches were not used before in literature to evaluate the forecasts accuracy. The different methods applied to compare the accuracy of the USA inflation forecasts on the horizon 1976-2012 started from the predictions provided by Survey of Professional Forecasters (SPF), Congressional Budget Office (CBO), Blue Chips (BC), and Administration, determining different results. According to U1 Theil's statistic, non-parametric tests and a new indicator proposed by us (RRSSE- ratio of radicals of sum of squared errors), the best forecasts were provided by Administration and the less accurate by SPF. The Spearman's and Kendall's coefficients of correlation and the ranks method gavea hierarchy of institutions performance regarding the accuracy that starts with BC and finished with SPF. The logistic regression computed by the author and the relative distance to the maximal performance method considered CBO as the best institution. Some methods of improving the forecasts accuracy were applied, getting more accurate predictions for the combined forecasts of BC and CBO using optimal scheme of combination. The smoothed predicted values based on Hodrick-Prescott filter outperformed all the initial predictions and the combined ones.

Keywords: forecasts, accuracy, logistic regression, combined forecasts, non-parametric tests, filters,multi-criteria ranking

1. INTRODUCTION

An auxiliary, but essential component of the forecasting process is the assessment of the accuracy, which reflects how closer the forecasted values of a variable are to its registered values. In USA there are more institutions that provide predictions for macroeconomic indicators. The main question is which of these institutions predicted the best an economic phenomenon. To answer this question we can use many methods. Some of the usual statistical methods were not applied in the forecasting context. It is important to analyze if more methods gave the same results. On the other hand, it is also essential to find out some empirical strategies to improve the forecasts accuracy.

In this study the accuracy is assessed in ex-post variant, resulting a mirror of the historical accuracy of institutions forecasts. Consequently, this analyze will be the best guide to choose the forecasts of a certain institution in the close future.

2. NEW STATISTICAL METHODS USED IN MAKING COMPARISONS BETWEEN PREDICTIONS

Some researchers were interested in evaluating the accuracy of macroeconomic forecasts made by some international institutions. However, they omitted to take into account the Administration anticipations.

Edge, Kiley and Laforte assessed the accuracy of predictions made by Federal Reserve staff and for those made starting from a DSGE model and a time-series model [7].

Abreu were interested in assessing the accuracy of predictions made by the Organization for Economic Co-operation and Development (OECD), International Monetary Fund (IMF), European Commission (EC) and two private institutions (Consensus Economics and The Economist) [1]. The directional accuracy and the ability of anticipating an economic crisis were deeply studied. The probability of USA recession was computed by Österholm (2012) starting from a BVAR model [9].

In general, the researchers used the classical measures of accuracy, like mean error, mean absolute errors, root mean squared error, U Theil's coefficient. Percentages errors, U statistics or mean absolute scaled errors are used to make comparisons in terms of accuracy.

One problem that we try to solve in this study is to bring an objective classification of forecasts. In practice, some of the accuracy measures recommend as better a certain forecast, while others show that other forecast is more accurate.

In order to solve this uncertain situation we applied the multi-criteria methods that take into account at the same time the values of all accuracy indicators.

The ranks method and the method of relative distance with respect to the maximal performance are applied to order the forecasts according to accuracy criterion.

For the forecasted variable X, the error is calculated as the difference between the real value and the predicted one. It is denoted by "e". Some of the measures of predictions accuracy are presented beWe selected 5 accuracy measures whose influence is taken into account at the same time.

If n is the length of the forecast horizon, then we computed:

1. The Mean error (ME)

$$ME = \frac{1}{n}\sum_{i=1}^{n} e_X \tag{1}$$

2. The Mean absolute error (MAE)

$$MAE = \frac{1}{n}\sum_{i=1}^{n} |e_X| \tag{2}$$

3. The Root Mean Squared Error (RMSE)

$$RMSE = \sqrt{\frac{1}{n}\sum_{i=1}^{n} e_X^2} \tag{3}$$

4. The U1 Theil's statistic

$$U_1 = \frac{\sqrt{\sum_{t=1}^{n}(r_t - f_t)^2}}{\sqrt{\sum_{t=1}^{n} r_t^2} + \sqrt{\sum_{t=1}^{n} f_t^2}} \quad (4)$$

r- the real values; f- the forecasted values

A higher accuracy is equivalent with a value closer to zero for U1 statistic.

5. U2 Theil's statistic

$$U_2 = \sqrt{\frac{\sum_{t=1}^{n-1}\left(\frac{f_{t+1} - r_{t+1}}{r_t}\right)^2}{\sum_{t=1}^{n-1}\left(\frac{r_{t+1} - r_t}{r_t}\right)^2}} \quad (5)$$

A value less than 1 for U2 confirms the superiority of the compared forecast, while a value greater than 1 shows a higher acccuracy for the benchmark forecast.

For comparisons with the naive forecasts a new indicator is introduced by us: ratio of radicals of sum of squared errors (RRSSE).

$$RRSSE = \frac{\sqrt[n]{\sum_{t=1}^{n} e_t^2}}{\sqrt[n-1]{\sum_{i=2}^{n}(X_i - X_{i-1})^2}}$$

In order to compare two predictions even for different variables, the values of this indicator are compared, a value closer to zero showing a better accuracy.

Ranks method has several steps:

1. Each accuracy measure receives a rank according to its value. (the value that indicates the a better degree accuracy has the rank 1);

The statistical units are represented by the number of institutions that provided forecasts. In our case study this number is 3. The rank corresponding to each institution is: $(r_{i\,ind_j})_{i=1,2,3,4}$ and ind_j-accuracy indicator j.

The ranks are sum up and institution with the lowest score receives the rank one:

$$S_i = \sum_{j=1}^{5} (r_{i\,ind_j}), \; i = 1,2,3,4 \tag{6}$$

2. The institution with the lowest score is the best one and it gets the final rank 1.

The method of relative distance related to the maximal performance supposes that for each accuracy measure the distance of each institution compared to the one with the best performance is calculated as:

$$d_{i\,ind_j} = \frac{ind_i^j}{\{\min abs(ind_i^j)\}_{i=1,\dots,4}}, \mathrm{i} = 1,2,3,4 \text{ and } \mathrm{j} = 1,2,..,5 \tag{7}$$

The relative distance is calculated as a ratio, where the denominator is the lowest value of the accuracy indicator for all institutions.

The geometric mean for the relative distances is calculated:

$$\overline{d_i} = \sqrt[5]{\Pi_{j=1}^{5} d_{i\,ind_j}}, i = 1,2,3,4 \tag{8}$$

The final ranks are assigned taking into account the values of average relative distances. The institution with the lowest average relative distance receives the rank 1. The location of each institution compared to the one with the best performance is a ratio: the average relative distance over the lowest average relative distance.

$$loc_i^{\%} = \frac{\overline{d_i}}{\min(d_i)_{i=\overline{1,4}}} \cdot 100 \tag{9}$$

Wilcoxon Signed Ranks test and Kruskall-Wallis test are nonparametric tests used when the series repartition is not known or non-normal. These tests are applied to check the differences between populations. In this case, the differences between the real values and the forecasted ones are checked using the two non-parametric tests. The null hypothesis refers to the lack of differences, while the alternative one shows that there are significant differences between the forecasts and the registered values. A p-value that is lower than 0.05 implies the rejection of null hypothesis. For small samples the chi-square approximation gives better results in most cases than Kruskal-Wallis test, according to Conover [6].

Comparisons between forecasts can be done using binary logistic regression when the dependent variable is a qualitative one. For this type of regression some assumptions are not considered (errors non-correlation, normality or homoscedasticity).Odd-ratios (OR) are computed to see how much the occurrence chances of an alternative of the dependent variable modify when the independent variable change with one unit. The coefficient of the exogenous variable from the regression model is denoted by b1.

If OR is higher than 1, an increase by one unit in the level of the exogenous variable implies a growth by $e^{\hat{b}_1}$ in the level of the dependent variable.

The absolute errors for each year in the forecasting horizon are computed. We test is each error differs significantly from a threshold fixed at 0.5%. We choose a threshold of 1%. The dependent variable has two alternatives:

$$\text{error\_significance} = \begin{cases} 1 & \text{higherror} \\ 0 & \text{lowerror} \end{cases}$$

For each forecasted value of the variable, the significance of the error is computed.

The Spearman's and Kendall's coefficients of correlation might be computed to see the associations between the real values and the predicted ones.

3. INFLATION FORECASTS COMPARISONS FOR USA

The forecasting horizon for USA inflation rate is 1978-2012 and the predictions are provided by Survey of Professional Forecasters (SPF), Congressional Budget Official (CBO), Blue Chips (BC) and Administration.

It is not recommend the use of a single measure of accuracy. In our study 5 accuracy indicators were selected.

According to U1 statistic, the best forecasts are provided by Administration, being followed by BCanticipations, CBO ones and finally the SPF forecasts. However, CBO predictions have the lowest mean error.

Table 1. The accuracy of inflation forecasts provided by SPF, CBO, BC and Administration (1978-2012)

Accuracy measure	SPF	CBO	BC	Administration
ME	2.9826	0.123	0.4527	0.6878
MAE	3.0597	1.8365	1.4673	1.4376
RMSE	4.1058	2.6445	2.5094	2.5279
U1	0.4525	0.3053	0.2985	0.2591
RRSSE	0.3467	0.3178	0.2856	0.2467
U2	0.8084	73.3077	152.8537	192.5686

Administration registered the lowest mean absolute error for their predictions. Only the SPF anticipations are better than the naïve forecasts. The indicator RRSSE introduced by us in literature gave the same results as U1.

The multi-criteria ranking solves the problem of contradictory measures of accuracy by considering their influence at the same time.

The ranks method recommends BC forecasts as the best and the SPF as the less accurate. CBO and Administration predictions have the same degree of accuracy.

Table 2. The ranks method for the comparison of USA inflation forecasts accuracy (1976-2012)

Criterion	SPF ranks	CBO ranks	BC ranks	Administration ranks
ME	4	1	2	3
MAE	4	3	2	1
RMSE	4	3	1	2
U1	4	3	1	2
U2	1	2	3	4
Sum of ranks	17	12	9	12
Final rank	4	2	1	3

According to the second method of multi-criteria ranking, the best forecasts on the horizon 1976-2012 were provided by CBO. The hierarchy of institutions is continued by: SPF, BC and Administration. So, there are differences regarding the hierarchy provided by the two methods. In general, the method of relative distance according to the best institution gives better results. However, CBO gave the best performance according to both methods.

Table 3. The method of relative distance related to the maximal performance for the comparison of USA inflation forecasts accuracy (1976-2012)

Criterion	SPF ranks	CBO ranks	BC ranks	Administration ranks
ME	24.2542	1.0000	3.6813	5.5934
MAE	2.1284	1.2775	1.0207	1.0000
RMSE	1.6362	1.0539	1.0000	1.0074
U1	1.7463	1.1783	1.1519	1.0000
U2	1.0000	90.6777	18839.741	2.39E+09
Average relative distance	2.7149	2.7013	9.600	106.0963
Final rank	2	1	3	4
Location (%)	1.0050	1.0000	3.5538	39.2753

The dependencies between the effective values and the forecasted ones are analyzed using non-parametric tests like Wilcoxon Sum Ranks and Kruskall-Wallis.

After the application of non-parametric tests we made the following conclusions with a probability of 95%:

The differences between CBO predictions and the registered values are not significant;

The differences between SPF predictions and the real values are not significant, but the Significance indicator is lower; this shows that CBO predictions are better than the SPF ones;

The differences are not significant between BC forecasts and the real values, but the p-valueis lower than that of the other two predictions; this implies that SPF and CBO expectations are better than BC ones;

There are significant differences between Administration forecasts and the effective values on inflation. The results of the tests applied in SAS are presented in **Appendix 1.**

So, the hierarchy given by the application of non-parametric tests is: Administration, CBO and Blue Chips. For all the SPF predictions the errors are significant and larger than de threshold of 1%.

The odds of having a low error for CBO forecasts grow with 36.6%, while the chances for Blue Chips increase with 25.3%. For Administration forecasts only few errors are not significant. So, the hierarchy provided by the analysis of binary logistic regression is: CBO, Blue Chips, Administration and SPF. The results of this procedure are displayed in **Appendix 2**.

The Spearman's and Kendall's coefficients of correlation are computed in **Appendix 3**. The strong correlation is between BC forecasts and real value, being followed by the one between Administration and the effective values and CBO and real values. The correlation between SPF expectation and the real values is not significant.

4. STRATEGIES OF IMPROVING THE FORECASTS ACCURACY

Bratu (2012) specified some strategies of improving the forecasts accuracy (application of filters and exponential smoothing techniques, combined forecasts, regressions models, historical errors method).

The most use approaches for combined forecasts are: optimal combination (OPT), inverse MSE weighting scheme (INV) and equal-weights-scheme (EW).

Bates and Granger) used two forecasts f1;t and f2;t, of the same variable Xt, derived h periods ago. For unbiased forecasts, the error is computed as: $e_{i,t} = X_{i,t} - f_{i,t}$. The errors follow a normal repartition of parameters 0 and σ_i^2. If ρ is the errors coefficient of correlation the covariance is $\sigma_{12} = \rho \cdot \sigma_1 \cdot \sigma_2$. The linear combination of the two predictions is: $c_t = m \cdot f_{1t} + (1-m) \cdot f_{2t}$. The error of the combined forecast is: $e_{c,t} = m \cdot e_{1t} + (1-m) \cdot e_{2t}$.The combined forecast mean is zero and the variance is: $\sigma_c^2 = m^2 \cdot \sigma_1^2 + (1-m)^2 \cdot \sigma_{2f}^2 + 2 \cdot m \cdot (1-m) \cdot \sigma_{12}$

The optimal value for m is determined by minimizing the error variance (m_{opt}) [2]:

$$m_{opt} = \frac{\sigma_2^2 - \sigma_{12}}{\sigma_1^2 + \sigma_2^2 - 2 \cdot \sigma_{12}} \tag{10}$$

The inverse weight (m_{inv}) is computed as:

$$m_{inv} = \frac{\sigma_2^2}{\sigma_1^2 + \sigma_2^2} \tag{11}$$

For equally weighted combined predictions (EW) the same weights are given to all models.

The U1 Theil's coefficient is computed for the combined forecasts based on the three schemes.

Table 4. The accuracy of USA inflation combined forecasts on the horizon 1976-2012

Combined forecasts	U1(OPT scheme)	U1 (INV scheme)	U1 (EW scheme)
SPF+CBO	0.6035	0.6219	0.3888
SPF+BC	0.6055	0.6139	0.4022
SPF+ADMINISTRATION	0.6299	0.6583	0.4318
CBO+BC	0.2493	0.2883	0.2890
CBO+ADMINISTRATION	0.2983	0.2999	0.3003
BC+ADMINISTRATION	0.3200	0.3011	0.3009

The combined forecasts of CBO and BC using OPT scheme improved the accuracy of all predictions. This type of combination gave better results than all the initial forecasts, excepting Administration ones. All the other combined forecasts excepting those where SPF anticipations are implied are better than SPF and CBO ones.

The application of filters to the initial forecasts and also the exponential smoothing techniques as Holts Winters are utilized for improving the forecasts accuracy [4].

The Hodrick–Prescott (HP) filter is used to extract the trend of the data series. Razzak explained that the Hodrick-Prescott filteris a true `filter' at the end of the sample and a `smoother' one over the sample [10]. The output gap from the true filter determines more accurate out-of-sample predictions of inflation. Christiano and Fitzgerald showed that Band-Pass filter is used to determine the component of the time series that is situated within a specific band of frequencies [5]. Christiano-Fitzegerald filter (CF filter) converges on long run to an optimal filter. It has a steep frequency response function at the band boundaries.

Holt-Winters (HW) Simple exponential smoothing technique is recommended for data set with linear trend and no seasonal variations. The filters and the HW technique are utilized to smooth the predictions provided by the four institutions. Then, the accuracy of the new forecasts is evaluated.

Table 5. The U1 values of USA inflation forecasts on the horizon 1976-2012

Technique of smoothing	SPF	CBO	BC	Administration
Hodrick-Prescott filter	0.6859	0.2474	0.2604	0.2784
Christiano-Fitzegerald filter	0.9656	0.9817	0.9639	0.9687
Baxter King filter	0.8976	0.8649	0.8535	0.8569
Holt-Winters technique	0.6651	0.2988	0.2878	0.3010

The Hodrick-Prescott technique and the Holt-Winters model improved the accuracy of BC and CBO forecasts. The great improvement was generated by Hodrick-Prescott filter for both types of predictions. For CBO anticipations, the accuracy is even better than that of the predictions provided by the other institutions or by the combined forecasts.

5. CONCLUSIONS

This research enriches the literature regarding the assessment and the improvement of forecasts accuracy.

According to U1 statistic and the new introduced RRSSE indicator and according to ranks method and to Spearman's coefficient of correlation the hierarchy of institutions that forecasted between the two-year inflation in 1982-2011 is: CBO, Administration and Blue Chips. The relative distance method with respect to the better institution, the logistic regression, the non-parametric tests provided the following ranking: Administration, CBO and Blue Chips.

The highest improvement in accuracy was brought by the combined forecasts of Blue Chips and Administration using inverse MSE scheme. The smoothed predicted values based on Holt-Winters technique, Hodrick-Prescott, Baxter King and Christiano-Fitzegerald filters din not improve the forecasts accuracy.

The novelty of this research consists in the application of some statistical approaches to compare the predictions accuracy, these methods never being mentioned in literature in this context. The results of the new approach are better than those provided by the U Theil's statistic, because more aspects of accuracy problem are taken into account.

REFERENCES

1. Abreu I., "International organizations' vs. private analysts' forecasts: an Evaluation", Banco de Portugal Papers, 29-34, 2011.
2. Bates, J., and C. W. J. Granger, "The Combination of Forecasts", Operations Research Quarterly, 20(4): 451-468, Jul. 1969.
3. Bratu, M. (2012). Strategies to Improve the Accuracy of Macroeconomic Forecasts in USA, LAP LAMBERT Academic Publishing, 2012, 6-28.
4. Bratu (Simionescu) M.,"Filters or Holt Winters technique to improve the forecasts for USA inflation rate ?", ActaUniversitatisDanubius. Œconomica, 9(1): 23-45, 2013.
5. Christiano, L. J. and Fitzgerald, T.J., "The Band Pass Filter",International Economic Review, 44(2): 435-465, 2003.
6. Conovor W.J., Practical nonparametric statistics, Wiley series in probability and statistics, New York: Wiley, 1999, 56-78.
7. Edge R.M., Kiley M.T. and Laforte J.-P., "A comparison of forecast performance between Federal Reserve Staff Forecasts simple reduced-form models and a DSGE model",Finance and Economics Discussion Series, 85-89, 2009.
8. Hodrick R. and Prescott, E.C., Postwar U.S. Business Cycles: An empirical investigation,Journal of Money, Credit and Banking, 1(16): 84-90, 2003.

9. Österholm, P., "The limited usefulness of macroeconomic Bayesian VARs when estimating the probability of a US recession", Journal of Macroeconomics, Elsevier, vol. 34(1): 76-86, 2012.
10. RazzakW.,The Hodrick-Prescott technique: A smoother versus a filter: An application to New Zealand GDP, Economics Letters, 57(2), 163-168, 1997.
11. http://www.cbo.gov/sites/default/files/cbofiles/ftpdocs/115xx/doc11553/forecastinga ccuracy.pdf.

CHAPTER 11

Diffusion Process Modeling by using Fractional-Order Models

Dominik Sierociuk [a], Tomas Skovranek [b,c], Michal Macias [a], Igor Podlubny [b], Ivo Petras [b], Andrzej Dzielinski [a], Pawel Ziubinski [a]

[a]*Institute of Control and Industrial Electronics, Warsaw University of Technology, Koszykowa 75, 00-662 Warsaw, Poland*

[b]*Institute of Control and Informatization of Production Processes, BERG Faculty, Technical University of Košice, B. Němcovej 3, 042 00 Košice, Slovakia*

[c]*Biomedical Imaging Group, EPFL STI IMT LIB, Swiss Federal Institute of Technology Lausanne, CH-1015 Lausanne, Switzerland*

ABSTRACT

This paper deals with a concept and description of a RC network as an electro-analog model of diffusion process which enables to simulate heat dissipation under different initial and boundary conditions. It is based on well-known analogy between heat and electrical conduction. In the paper are compared analytical solution together with numerical solution and experimentally measured data. For the first time a fractional order model of diffusion process and its modeling via lumped RC network has been used. Simple examples of simulations, measurements and their comparison are shown.

Keywords: Diffusion process, RC network, Fractional calculus, Matlab

1. INTRODUCTION

Fractional calculus and fractional differential equations have been used in various areas of the applications, for instance bioengineering [4], physics [2,23], chaos theory [11], viscoelasticity [5], control system engineering [1,7,13,22], fractional signal processing techniques [18] and many others areas (see e.g. [8,10,17]). One of the another important area of applications is the electrical engineering.

This article describes how to model diffusion using a RC network as an electro-analog model of diffusion process. Paper is based on well-known analogy between heat and electrical conduction. Such approach was already used for modeling the diffusion process described by integer-order differential equation [3]. In this article a fractional-order model of the diffusion process and its modeling via RC network is used. Experimentally obtained results are compared with analytical and numerical solution of the fractional order diffusion equation. For numerical computation of the fractional-order partial differential equation a matrix approach has been used [15,16].

2. FRACTIONAL CALCULUS

Fractional order differential calculus is a generalization of integer order integral and derivative to real or even complex order. This idea has first emerged at the end of 17th century, and has been developed in the area of mathematics throughout 18th and 19th century in the works of e.g. Liouville, Riemann, Cauchy, Abel, Grünwald and many others. More recently, by the end of 20th century, it turned out that some physical phenomena are modeled more accurately when fractional calculus is in use. There exist two (in fact three) main definitions of the fractional order integrals, derivatives and differences: Riemann–Liouville, Caputo, and Grünwald–Letnikov [6,8,10,13]. Some other, are also present in literature, but less commonly used in the applications. To be precise, the Riemann–Liouville and Caputo definitions concern both fractional derivative and integral.

To define the fractional order differ-integral, the definition of the $\Gamma(x)$ function is needed. The $\Gamma(x)$ function is given in the following way [13]:

$$\Gamma(x) = \int_0^\infty e^{-t} t^{x-1} dt, \tag{1}$$

where R(x)>0.

In this article we will consider the Caputo's definition, which can be written as [13]:

$$ {}_0D_t^{\alpha} f(t) = \frac{1}{\Gamma(n-\alpha)} \int_0^t \frac{f^{(n)}(\tau)}{(t-\tau)^{\alpha-n+1}} d\tau, \tag{2} $$

for n-1<α<n. The initial conditions for the fractional order differential equations with the Caputo derivatives are in the same form as for the integer-order differential equations. For the Caputo partial fractional derivative of order α of a function f(t,λ)with respect to variable t we will use the notation of the form ∂αf(t,λ)/∂tα, which is often used in the related literature.

The Laplace transform method is used often for solving engineering problems. The formula for the Laplace transform of the Caputo fractional derivative (2) has the form [13]:

$$ \int_0^{\infty} e^{-st} {}_0D_t^{\alpha} f(t) dt = s^{\alpha} F(s) - \sum_{k=0}^{n-1} s^{\alpha-k-1} f^{(k)}(0) \tag{3} $$

for n-1<α⩽n, where s∈C denotes the Laplace operator, and C denotes a set of complex numbers.

3. DESCRIPTION OF DIFFUSION PROCESS

One of a good example of diffusion process is a heat transfer action, when a heat (energy, temperature) is transported through a material. The description of such a processes can be meet in Sierociuk et al. [19]. Let us assume the heat transfer process in semi-infinite beam, with respect to the temperature (we will be able to control temperature at the beginning of the beam and observe also temperature at each point of the beam).

The analog equivalent of this process is a transmission line presented in Fig. 1. The voltage in each stage of RC-line represents a temperature in heat transfer process, and similarly the current represents a heat flux.

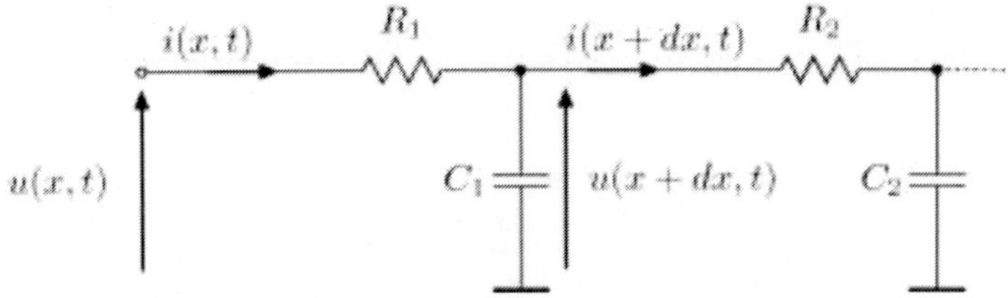

Figure. 1. Scheme of transmission line.

From the analog model the diffusion equation can be obtained very shortly. The voltage between two stages is a voltage on the resistor, and is proportional to the current in this stage. This can be written as follows:

$$u(x,t) - u(x+dx,t) = Ri(x,t) \tag{4}$$

what, in limit for dx→0, can be rewritten as

$$\frac{\partial}{\partial x}u(x,t) = Ri(x,t) \tag{5}$$

Similar, the voltage of the capacitor in each stage can be described

$$i(x,t) - i(x+dx,t) = C\frac{\partial}{\partial t}u(x,t) \tag{6}$$

what, in limit for dx→0, can be rewritten as

$$\frac{\partial}{\partial x}i(x,t) = C\frac{\partial}{\partial t}u(x,t) \tag{7}$$

This two equations (for resistor and capacitor) can be written together in the form of diffusion equation:

$$\frac{\partial^2}{\partial x^2}u(x,t) = RC\frac{\partial}{\partial t}u(x,t) \tag{8}$$

3.1. Evaluation of diffusion process equation

For a case of the ideal heating process, without energy loss, of a semi-infinite beam that equation can be written in the following form

$$\frac{\partial}{\partial t}T(t,\lambda)=\frac{1}{a^2}\frac{\partial^2 T(t,\lambda)}{\partial \lambda^2}, \tag{9}$$

with the following boundary conditions:

$$T(0,\lambda)=0, \quad T(t,0)=u(t), \tag{10}$$

where T(t,λ) is a temperature of the beam at time instant *t* and space coordinate (distance) λ, and 1a2 is a beam material conductivity.

As it was presented in Podlubny and Sierociuk et al. [13,19] the following relation between the heat flux and the temperature at the desired point holds (it can be obtained during solving the Eq. (9) in Laplace domain):

$$H(t,\lambda)=\frac{1}{a}\frac{\partial^{0.5}}{\partial t^{0.5}}T(t,\lambda). \tag{11}$$

Analytical solution for T(s,0)=c1(t) and α=1 is

$$T(t,\lambda_1)=c\,\mathrm{erfc}\left(\frac{\lambda_1 a}{2\sqrt{t}}\right) \tag{12}$$

3.2. Evaluation of diffusion process equation for loss case

Let as assume the situation that in each point of diffusive material length a part of the heat flux is dissipated. Moreover, the value of this dissipation flux is proportional to the temperature of this point. In order to achieve it, let us expand the Eq. (11), in the same way as in Sierociuk et al. [19], into the following form:

$$\frac{\partial}{\partial \lambda}H(t,\lambda)=\frac{1}{a}\frac{\partial^{0.5}}{\partial t^{0.5}}H(t,\lambda)+bH(t,\lambda), \tag{13}$$

where a is the same beam material conductivity as for ideal diffusion case and, b is a coefficient of heat flux dissipation.

This can be expressed as the analog model presented in Fig. 2, where losing of heat flux is represented by additional resistors connected parallel to capacitors.

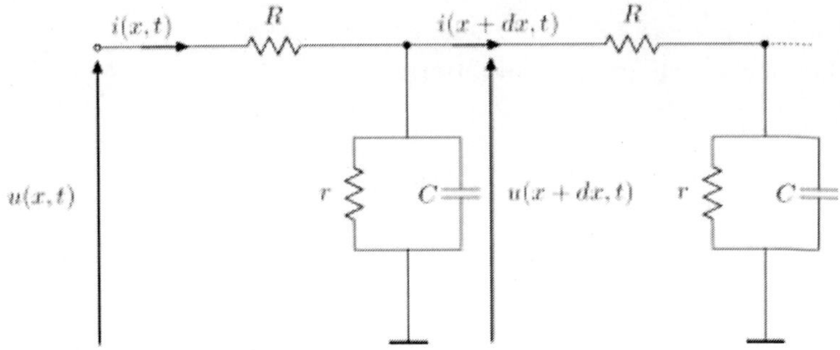

Figure. 2. Scheme of transmission line with heat dissipation.

The diffusion equation, in this case, has the following form

$$\frac{\partial^2}{\partial \lambda^2}T(t,\lambda) = a^2\frac{\partial}{\partial t}T(t,\lambda) + 2a^3b\frac{\partial^{0.5}}{\partial t^{0.5}}T(t,\lambda) + a^4b^2T(t,\lambda) \tag{14}$$

Solution of this equation in Laplace domain has the following form

$$T(s,\lambda_1) = e^{-\lambda_1\left(as^{0.5}+a^2b\right)}T(s,0). \tag{15}$$

what can be rewritten as

$$T(s,\lambda_1) = e^{-\lambda_1 a^2 b}e^{-\lambda_1 a s^{0.5}}T(s,0). \tag{16}$$

Analytical solution for T(s,0)=c1(t) and α=1 is

$$T(t,\lambda_1) = ce^{-\lambda_1 a^2 b}\mathrm{erfc}\left(\frac{\lambda_1 a}{2\sqrt{t}}\right) \tag{17}$$

4. MATRIX APPROACH FOR SOLVING FRACTIONAL ORDER DIFFUSION EQUATIONS

This approach is based on the fact that operation of the fractional calculus can be expressed by matrix [16]. It follows from Podlubny [14], that the left-sided Riemann–Liouville or Caputo fractional derivative v(α)(t)=0Dtαv(t) can be approximated in all nodes of the equidistant discretization net t=jτ (j=0,1,...,n) simultaneously with the help of the upper triangular strip matrix Bn(α) as:

$$\left[v_n^{(\alpha)}\ v_{n-1}^{(\alpha)}\ \dots\ v_1^{(\alpha)}\ v_0^{(\alpha)}\right]^T = B_n^{(\alpha)}\left[v_n\ v_{n-1}\ \dots\ v_1\ v_0\right]^T, \tag{18}$$

where

$$B_n^{(\alpha)} = \frac{1}{\tau^\alpha}\begin{bmatrix} \omega_0^{(\alpha)} & \omega_1^{(\alpha)} & \ddots & \ddots & \omega_{n-1}^{(\alpha)} & \omega_n^{(\alpha)} \\ 0 & \omega_0^{(\alpha)} & \omega_1^{(\alpha)} & \ddots & \ddots & \omega_{n-1}^{(\alpha)} \\ 0 & 0 & \omega_0^{(\alpha)} & \omega_1^{(\alpha)} & \ddots & \ddots \\ \dots & \dots & \dots & \ddots & \ddots & \ddots \\ 0 & \dots & 0 & 0 & \omega_0^{(\alpha)} & \omega_1^{(\alpha)} \\ 0 & 0 & \dots & 0 & 0 & \omega_0^{(\alpha)} \end{bmatrix} \tag{19}$$

$$\omega_j^{(\alpha)} = (-1)^j\binom{\alpha}{j}, \quad j = 0, 1, \dots, n. \tag{20}$$

Similarly, the right-sided Riemann–Liouville or Caputo fractional derivative v(α)(t)=tDbαv(t) can be approximated in all nodes of the equidistant discretization net t=jτ (j=0,1,...,n) simultaneously with the help of the lower triangular strip matrix Fn(α):

$$\left[v_n^{(\alpha)}\ v_{n-1}^{(\alpha)}\ \dots\ v_1^{(\alpha)}\ v_0^{(\alpha)}\right]^T = F_n^{(\alpha)}\left[v_n\ v_{n-1}\ \dots\ v_1\ v_0\right]^T \tag{21}$$

$$F_n^{(\alpha)} = \frac{1}{\tau^\alpha}\begin{bmatrix} \omega_0^{(\alpha)} & 0 & 0 & 0 & \dots & 0 \\ \omega_1^{(\alpha)} & \omega_0^{(\alpha)} & 0 & 0 & \dots & 0 \\ \omega_2^{(\alpha)} & \omega_1^{(\alpha)} & \omega_0^{(\alpha)} & 0 & \dots & 0 \\ \ddots & \ddots & \ddots & \ddots & \dots & \dots \\ \omega_{n-1}^{(\alpha)} & \ddots & \omega_2^{(\alpha)} & \omega_1^{(\alpha)} & \omega_0^{(\alpha)} & 0 \\ \omega_n^{(\alpha)} & \omega_{n-1}^{(\alpha)} & \ddots & \omega_2^{(\alpha)} & \omega_1^{(\alpha)} & \omega_0^{(\alpha)} \end{bmatrix} \tag{22}$$

The symmetric Riesz derivative of order β can be approximated based on its definition as a combination of the approximations (18) and (21) for the left and right-sided Riemann–Liouville derivatives, or using the centered fractional differences approximation of the symmetric Riesz derivative suggested recently by Ortigueira [9]. The general formula is the same:

$$\left[v_m^{(\beta)}\ v_{m-1}^{(\beta)}\ \dots\ v_1^{(\beta)}\ v_0^{(\beta)}\right]^T = R_m^{(\beta)}\left[v_m\ v_{m-1}\ \dots\ v_1\ v_0\right]^T$$

$$\left[v_m^{(\beta)}\ v_{m-1}^{(\beta)}\ \dots\ v_1^{(\beta)}\ v_0^{(\beta)}\right]^T = R_m^{(\beta)}\left[v_m\ v_{m-1}\ \dots\ v_1\ v_0\right]^T \tag{23}$$

In the first case, the approximation for the left-sided Caputo derivative is taken one step ahead, and the approximation for the right-sided Caputo derivative is taken one step back. This leads to the matrix

$$R_m^{(\beta)} = \frac{h^{-\alpha}}{2}\left[{}_{-1}U_m + {}_{+1}U_m\right] \tag{24}$$

In the second case (Ortigueira's definition) we have the following symmetric matrix:

$$R_m^{(\beta)} = h^{-\beta}\begin{bmatrix} \omega_0^{(\beta)} & \omega_1^{(\beta)} & \omega_2^{(\beta)} & \omega_3^{(\beta)} & \dots & \omega_m^{(\beta)} \\ \omega_1^{(\beta)} & \omega_0^{(\beta)} & \omega_1^{(\beta)} & \omega_2^{(\beta)} & \dots & \omega_{m-1}^{(\beta)} \\ \omega_2^{(\beta)} & \omega_1^{(\beta)} & \omega_0^{(\beta)} & \omega_1^{(\beta)} & \dots & \omega_{m-2}^{(\beta)} \\ \ddots & \ddots & \ddots & \ddots & \dots & \dots \\ \omega_{m-1}^{(\beta)} & \ddots & \omega_2^{(\beta)} & \omega_1^{(\beta)} & \omega_0^{(\beta)} & \omega_1^{(\beta)} \\ \omega_m^{(\beta)} & \omega_{m-1}^{(\beta)} & \ddots & \omega_2^{(\beta)} & \omega_1^{(\beta)} & \omega_0^{(\beta)} \end{bmatrix} \tag{25}$$

$$\omega_k^{(\beta)} = \frac{(-1)^k \Gamma(\beta+1)\cos(\beta\pi/2)}{\Gamma(\beta/2-k+1)\Gamma(\beta/2+k+1)}, \quad k=0,1,\ldots,m. \tag{26}$$

Both these approximations of symmetric Riesz derivatives give practically the same numerical results and in case of numerical solution of partial fractional differential equations lead to a well-posed matrix of the resulting algebraic system.

Similarly, if in addition to fractional-order time derivative we also consider symmetric fractional-order spatial derivatives, then we have to use all nodes at the considered time layer from the leftmost to the rightmost spatial discretization node.

Let us consider the nodes (ih,jτ),j=0,1,2,...,n, corresponding to all time layers at *i*th spatial discretization node. It has been shown in Podlubny [14] that all values of *α*th order time derivative of u(x,t) at these nodes are approximated using the discrete analogue of differentiation of arbitrary order:

$$\left[u_{i,n}^{(\alpha)}\ u_{i,n-1}^{(\alpha)}\ \ldots\ u_{i,2}^{(\alpha)}\ u_{i,1}^{(\alpha)}\ u_{i,0}^{(\alpha)}\right] = B_n^{(\alpha)}\left[u_{i,n}\ u_{i,n-1}\ \ldots\ u_{i,2}\ u_{i,1}\ u_{i,0}\right]^T. \tag{27}$$

In order to obtain a simultaneous approximation of *α*th order time derivative of u(x,t)in all nodes, we need to arrange all function values uij at the discretization nodes to the form of a column vector unm=[...]T, which has the structure described in Podlubny et al. [16].

In visual terms, we first take the nodes of *n*th time layer, then the nodes of (n-1)th time layer, and so forth, and put them in this order in a vertical column stack.

The matrix that transforms the vector Unm to the vector Ut(α) of the partial fractional derivative of order *α* with respect to time variable can be obtained as a Kronecker product of the matrix Bn(α), which corresponds to the fractional ordinary derivative of order *α* (recall that *n* is the number of time steps), and the unit matrix Em (recall that *m* is the number of spatial discretization nodes):

$$T_{mn}^{(\alpha)} = B_n^{(\alpha)} \otimes E_m \tag{28}$$

Similarly, the matrix that transforms the vector *U* to the vector Ux(β) of the partial fractional derivative of order *β* with respect to spatial variable can be obtained as a Kronecker product of the unit matrix En (recall that *m* is the number of spatial discretization nodes), and the matrix Rm(β), which corresponds to a symmetric Riesz ordinary derivative of order β[9], (recall that m is the number of time steps):

$$S_{mn}^{(\alpha)} = E_n \otimes R_m^{(\beta)}. \tag{29}$$

Having these approximations for partial fractional derivatives with respect to both variables, we can immediately discretize the general form of the fractional diffusion equation by simply replacing the derivatives with their discrete analogs. Namely, the equation

$${}_0^C D_t^{\alpha} u - \chi \frac{\partial^{\beta} u}{\partial |x|^{\beta}} = f(x,t) \tag{30}$$

is discretized as

$$\left\{ B_n^{(\alpha)} \otimes E_m - \chi E_n \otimes R_m^{(\beta)} \right\} u_{nm} = f_{nm}. \tag{31}$$

In the case of diffusion process with heat flux dissipation, the Eq. (14) can be rewritten into general discrete form using the Podlubny's matrix approach as

$$\left\{ \chi_1 B_n^{(\alpha)} \otimes E_m + \chi_2 B_n^{(\beta)} \otimes E_m + \chi_3 I_n \otimes E_m - E_n \otimes R_m^{(\gamma)} \right\} u_{nm} = f_{nm}. \tag{32}$$

Initial and boundary conditions must be equal to zero. If it is not so, then an auxiliary unknown function must be introduced, which satisfies the zero initial and boundary conditions.

In this way, the non-zero initial and boundary conditions moves to the right-hand side of the equation for the new unknown function.

5. RESULTS OF DIFFUSION PROCESS MODELING USING THREE DIFFERENT REPRESENTATIONS

In this section the data originating from physical system for diffusion process with dissipation are compared with the results obtained using analytical and numerical models.

Experimental setup is based on analog model of diffusion process with heat dissipation shown in Fig. 3, the analytical solution is presented in Section 3 and the numerical algorithm is presented in Section 4.

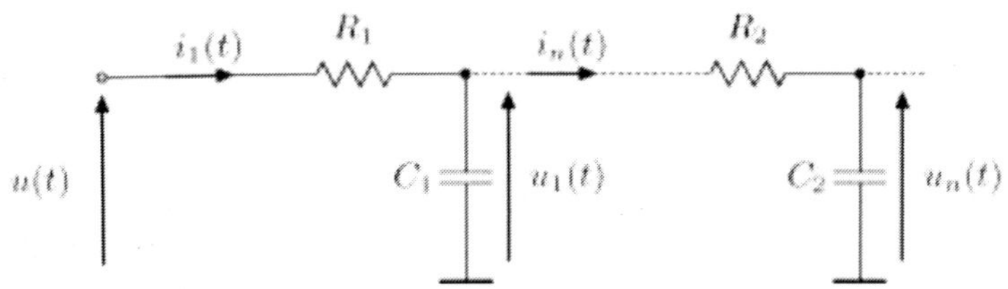

Figure. 3. Analog model of diffusion process with heat dissipation.

5.1. Experimental setup of the electrical circuit

In general the analog model of diffusion process is based on half order impedance implementation studied in Refs. [12,20,21]. It contains 200 elements and was designed and made according to the scheme presented in Fig. 3 with the following values of the passive elements: R1=2.4 kΩ, R2=8.2 kΩ, C1=330 nF and C2=220 nF. Due to decrease a noise ratio of the dSPACE A/D converters, which improve the accuracy of dSPACE card analog inputs, their input resistance was slightly reduced and comparable with resistors used in analog diffusion model. In this fact, when the voltages u1,...,un from Fig. 3 are directly connected to cards' inputs the setup stays a diffusion process with dissipation.

The experimental setup contains:

1) dSPACE DS1104 PPC card mounted in PC,
2) electrical circuit – the analog model of the diffusion process.

The experimental data were gathered by sensors connected to every 8th node of the analog model. To clarify the first voltage the sensors were placed into the input circuit terminals, the next measurement were gathered by second sensor (ADC channel) connected to the 8th node of domino ladder structure, and next sensors were connected to each next 8th node (e.g. 16,24,...).

The overview of the circuit boards for diffusion process with dissipation is presented in Fig. 4.

Figure. 4. A circuit board of the diffusion process with heat dissipation.

5.2. Comparison of the results of analog, analytical and numerical representations of diffusion process with dissipation

Results of analog modeling for diffusion process with dissipation were collected based on the experimental system presented in Figs. 3 (scheme) and 4 (real view). Based on numerical minimization of square error between experimental data and analytical solution, given by Eq. (17), the following parameters were obtained:

$$a = 0.281073708524755; \quad b = 1.674579293537321; \tag{33}$$

Due to used numerical algorithm, the identified parameters were obtained for minimization error for 3th–8th sensors.

Experimental data for diffusion process with dissipation, for sensors 1–11 and its comparison to the analytical solution given by Eq. (17), using identified parameters *a*and *b*, are presented in Fig. 5 and Table 1. The modeling error for analog and analytical results is presented in Fig. 6.

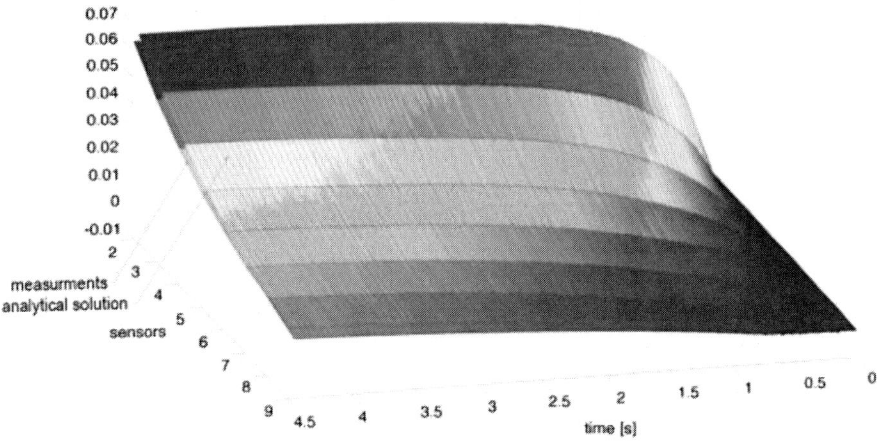

Figure 5. Comparison of analog modeling and analytical solution.

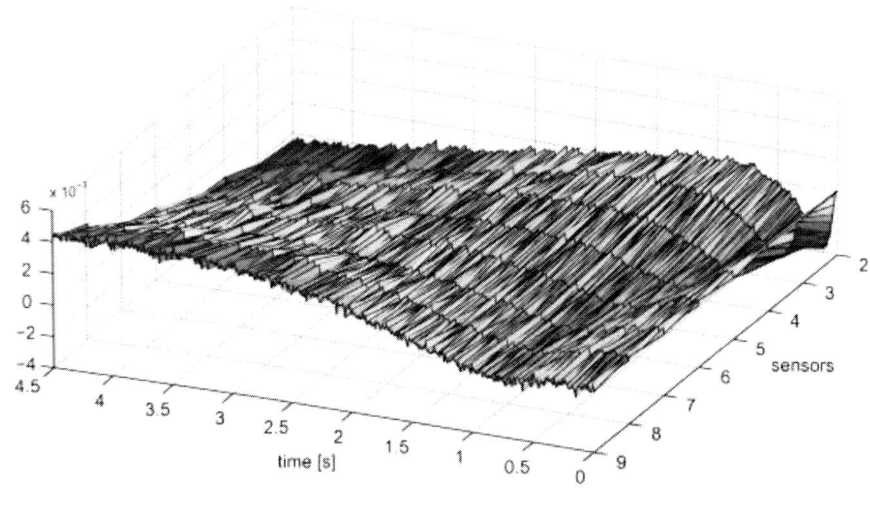

Figure. 6. Error between analog modeling and analytical solution.

Table 1. The squared error for diffusion process with heat flux dissipation. The bold font in the Table stays for the sum of errors for all sensors.

	Analog vs. analytical	Analog vs. numerical	Analytical vs. numerical
Sensor 2	1.0524e − 04	1.1270e − 04	16.0752e − 07
Sensor 3	2.6270e − 04	2.6622e − 04	6.3378e − 07
Sensor 4	4.9037e − 04	4.8290e − 04	15.6129e − 07
Sensor 5	7.0636e − 04	6.9736e − 04	10.7388e − 07
Sensor 6	9.1195e − 04	9.1108e − 04	9.8623e − 07
Sensor 7	13.4889e − 04	13.7164e − 04	13.6162e − 07
Sensor 8	17.3842e − 04	18.0666e − 04	24.9588e − 07
Sensor 9	22.3321e − 04	23.8260e − 04	51.8428e − 07
Sensor 10	28.4296e − 04	31.2802e − 04	110.4091e − 07
Sensor 11	39.1501e − 04	44.4508e − 04	232.4328e − 07
Sensor 12	7.9172e − 04	9.2346e − 04	479.2257e − 07
∑error	**153.4682e − 04**	**165.2770e − 04**	**.9711e − 04**

For numerical solution the model Eq. (14) was evaluated, and the fractional derivatives were calculated using the Podlubny's matrix approach. It was written in the discrete form Eq. (32), where the used parameters were α=1,β=1/2,χ1=a2,χ2=2a3b,χ3=a4b2. The numerical results were compared to the analog samples, shown in Fig. 7 (the smooth surface in the front is representing the numerical solution and the measured samples are represented by the grid surface behind), as well as to the analytical model Fig. 9 (numerical solution in the front, analytical solution behind). The error surfaces for all the comparisons of different approaches for describing heat transfer can be found in Figs. 6, 8, 10, respectively.

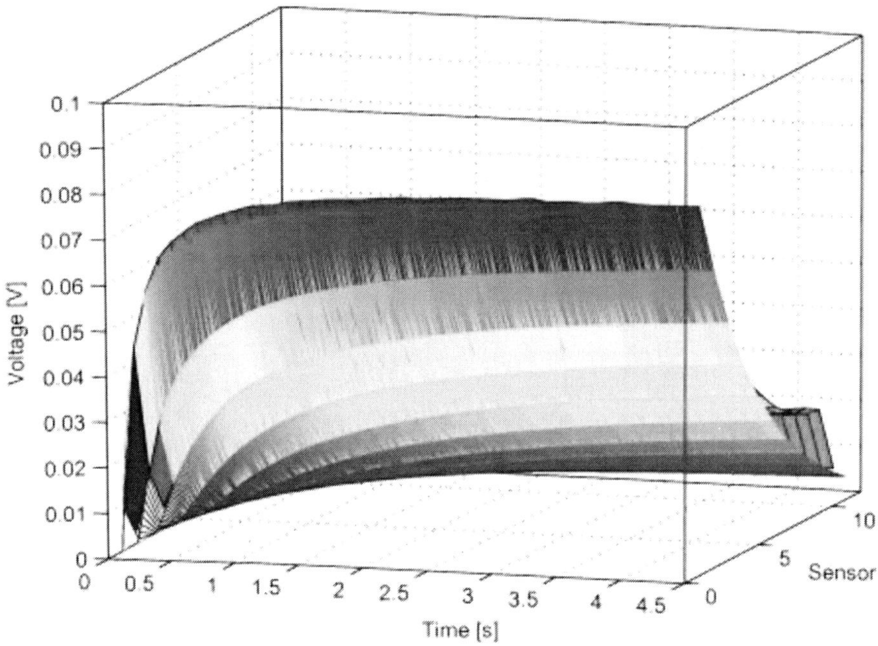

Figure. 7. Comparison of analog modeling and numerical solution.

The full list of least squared errors between the measured analog samples, analytical solution and numerical solution of the diffusion process with heat flux dissipation can be found in Table 1, where again the error for each sensor as well as the total error is given.

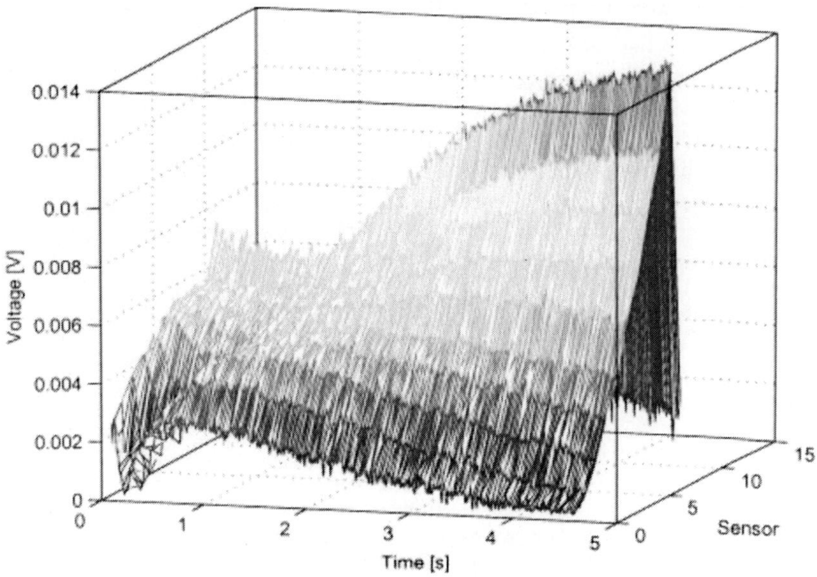

Figure. 8. Error between analog modeling and numerical solution.

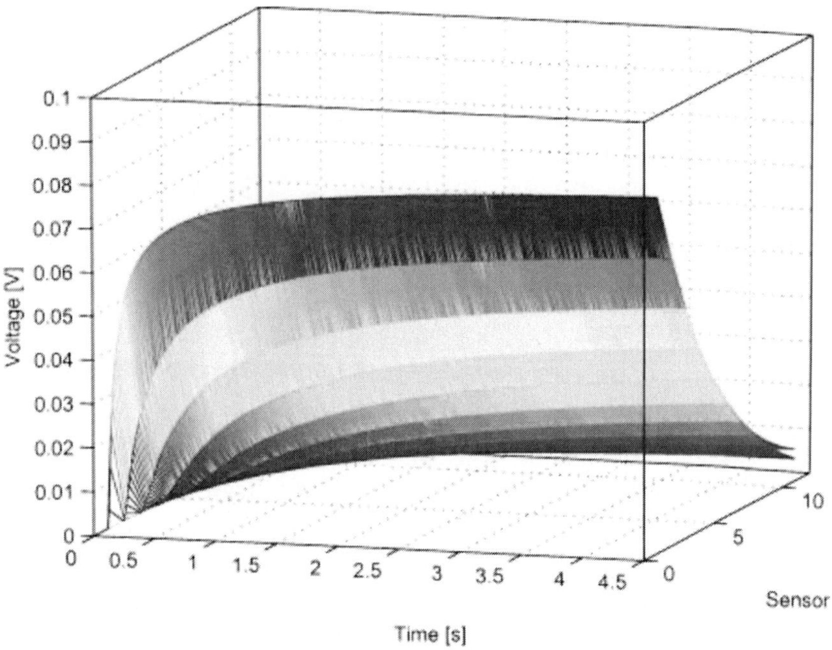

Figure. 9. Comparison of analytical and numerical solutions.

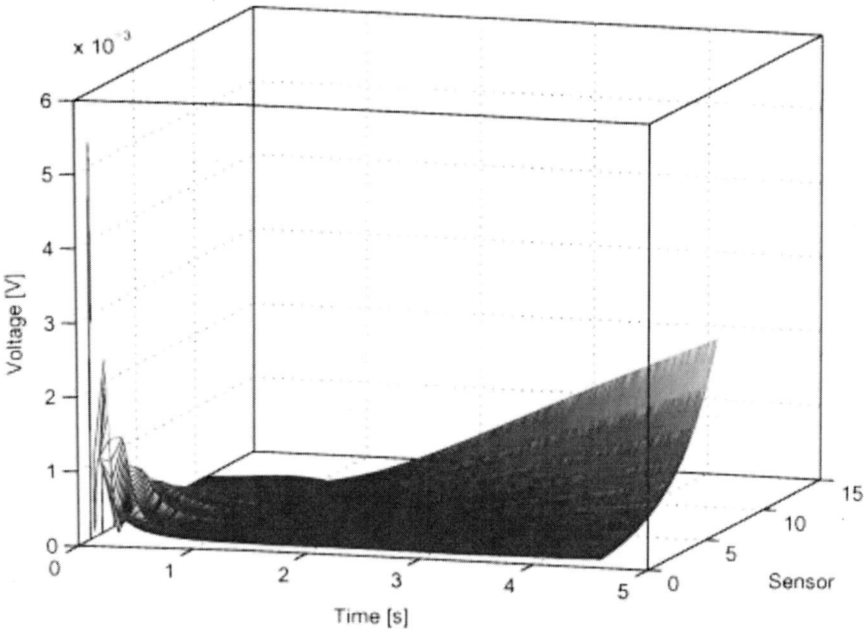

Figure. 10. Error between analytical and numerical solutions.

As it can be seen the error between analog and analytical results at the beginning of the plots is very limited and increase with the time and number of sensors. It is caused by the fact, that analytical solution is obtained for an assumption that the diffusion media (e.g. the heating beam) is infinite. The analog realization has only finite length, what has an effect in the dynamics of the system, the finite length domino ladder can be charged faster that infinite one. This can be observed at the point that the voltage value start to be constant when the analytical model still increase this value.

For the case of comparison between numerical and analytical solutions we can recognize that both solutions are very close. It confirms correctness of used matrix approach for solving such a type of fractional order partial differential equations.

6. CONCLUSIONS

The paper presents detailed formulation of diffusion process with dissipation and its description in the form of fractional order partial differential equation. For heat transfer case the diffusion with dissipation occurs when a part of the heat flux dissipate to environment in the whole length of the beam, what can be met when the beam is not ideally insulated from the environment. The realization of such a process was given also in the form of analog circuit and numerical algorithm. The analog model of diffusion equation was build based on modified domino ladder approximation of half order impedance. Numerical solution was obtained based on matrix approach for fractional order partial differential equations. The experimentally obtained results were compared with analytical and numerical solution, both based on fractional calculus and numerical solution. Comparison of results confirm high accuracy of used methods and important differences between used approach. The analytical and numerical solution are very close and can describe an infinite length case, while analog realization could describe only finite length case. The results presented in this paper allows to better understand the diffusion process with (heat flux, current) dissipation and its analog and numerical realizations.

ACKNOWLEDGMENTS

This work was supported in part by the Slovak Grant Agency for Science under Grants VEGA: 1/0552/14, 1/0497/11, and by the Slovak Research and Development Agency under Grants: SK-PL-0066-12, APVV-0482-11 and APVV-0040-07. The second author is a fellow of the Sciex-NMS*ch* programme.

REFERENCES

1. R. Caponetto, G. Dongola, L. Fortuna, I. Petras**FRACTIONAL ORDER SYSTEMS: MODELING AND CONTROL APPLICATIONS** World Scientific, Singapore (2010)
2. R. Hilfer**APPLICATIONS OF FRACTIONAL CALCULUS IN PHYSICS** World Scientific, Singapore (2000)
3. Kufudakis**NETWORK WITH LUMPED RC PARAMETERS AS AN ELECTRO-ANALOG MODEL OF DIFFUSION PROCESSES I. SIMULATION OF DIFFUSION THROUGH MEMBRANES** Czech. J. Phys. B, 21 (11) (1971), pp. 1163-1173
4. R.L. Magin**FRACTIONAL CALCULUS IN BIOENGINEERING** Begell House Publishers (2006)
5. F. Mainardi**FRACTIONAL CALCULUS AND WAVES IN LINEAR VISCOELASTICITY: AN INTRODUCTION TO MATHEMATICAL MODELS** Imperial College Press, Singapore (2010)
6. K.S. Miller, B. Ross**AN INTRODUCTION TO THE FRACTIONAL CALCULUS AND FRACTIONAL DIFFERENTIAL EQUATIONS** John Wiley and Sons Inc., New York (1993)
7. C.A. Monje, Y.Q. Chen, B.M. Vinagre, D. Xue, V. Feliu**FRACTIONAL-ORDER SYSTEMS AND CONTROLS, ADVANCES IN INDUSTRIAL CONTROL** Springer (2010)
8. K.B. Oldham, J. Spanier**THE FRACTIONAL CALCULUS** Academic Press, New York (1974)
9. M.D. Ortigueira**RIESZ POTENTIAL OPERATORS AND INVERSES VIA FRACTIONAL CENTERED DERIVATIVES** Int. J. Math. Math. Sci. (2006), pp. 1-12 (Article ID 48391)

10. Oustaloup**LA DERIVATION NON ENTIERE: THEORIE, SYNTHESE ET APPLICATIONS** Hermes, Paris (1995)
11. Petras**FRACTIONAL-ORDER NONLINEAR SYSTEMS** Springer, New York (2011)
12. Petras, D. Sierociuk, I. Podlubny**IDENTIFICATION OF PARAMETERS OF A HALF-ORDER SYSTEM** IEEE Trans. Signal Process., 60 (10) (2012), pp. 5561-5566
13. Podlubny**FRACTIONAL DIFFERENTIAL EQUATIONS** Academic Press, San Diego (1999)
14. Podlubny**MATRIX APPROACH TO DISCRETE FRACTIONAL CALCULUS** Fract. Calculus Appl. Anal., 3 (2000), pp. 359-386
15. Podlubny, T. Skovranek, B.M. Vinagre, Matrix approach to discretization of ODEs and PDEs of arbitrary real order, MathWorks, Inc., Matlab Central File Exchange, <www.mathworks.com/matlabcentral/fileexchange/22071>, 2008.
16. Podlubny, A. Chechkin, T. Skovranek, Y.Q. Chen, B.M.J. Vinagre**MATRIX APPROACH TO DISCRETE FRACTIONAL CALCULUS II: PARTIAL FRACTIONAL DIFFERENTIAL EQUATIONS** J. Comput. Phys., 228 (2009), pp. 3137-3153 ArticlePDF (3MB)
17. J. Sabatier, O.P. Agrawal, J.A. Tenreiro Machado (Eds.), Advances in Fractional Calculus: Theoretical Developments and Applications in Physics and Engineering, Springer (2007)
18. H. Sheng, Y.Q. Chen, T.S. Qiu**FRACTIONAL PROCESSES AND FRACTIONAL-ORDER SIGNAL PROCESSING** Springer, London (2012)
19. D. Sierociuk, A. Dzielinski, G. Sarwas, I. Petras, I. Podlubny, T. Skovranek**MODELLING HEAT TRANSFER IN HETEROGENEOUS MEDIA USING FRACTIONAL CALCULUS** Philos. Trans. R. Soc. A Math. Phys. Eng. Sci., vol. 371 (1990) (2013)
20. D. Sierociuk, A. Dzielinski, New method of fractional order integrator analog modeling for orders 0.5 and 0.25, Proc. of the 16th International Conference on Methods and Models in Automation and Robotics (MMAR), 2011, Miedzyzdroje, Poland, pp. 137–141, 2011.
21. D. Sierociuk, I. Podlubny, I. Petras**EXPERIMENTAL EVIDENCE OF VARIABLE-ORDER BEHAVIOR OF LADDERS AND NESTED LADDERS** IEEE Trans. Control Syst. Tech., 21 (2) (2013), pp. 459-466
22. D. Valério, J. Costa**AN INTRODUCTION TO FRACTIONAL CONTROL** IET, London (2013)
23. B. West, M. Bologna, P. Grigolini**PHYSICS OF FRACTAL OPERATORS** Springer, New York (2003)

CHAPTER 12

The Role of High Precision Arithmetic in Calculating Numerical Laplace and Inverse Laplace Transforms

Zinovi Krougly, Matt Davison, Sid Aiyar

Department of Applied Mathematics, Western University, London, Ontario, Canada

ABSTRACT

In order to find stable, accurate, and computationally efficient methods for performing the inverse Laplace transform, a new double transformation approach is proposed. To validate and improve the inversion solution obtained using the Gaver-Stehfest algorithm, direct Laplace transforms are taken of the numerically inverted transforms to compare with the original function. The numerical direct Laplace transform is implemented with a composite Simpson's rule. Challenging numerical examples involving periodic and oscillatory functions, are investigated. The numerical examples illustrate the computational accuracy and efficiency of the direct Laplace transform and its inverse due to increasing the precision level and the number of terms included in the expansion. It is found that the number of expansion terms and the precision level selected must be in a harmonious balance in order for correct and stable results to be obtained.

Keywords: Numerical Laplace Transform, Numerical Laplace Transform Inversion, Composite Simpson's Rule, Gaver-Stehfest Algorithm, High Precision Computation

1. INTRODUCTION

Laplace transforms play a key role in many applications of mathematics to the fields of engineering, physics, and finance, whenever probability density functions, or linear differential equations or integral equations are involved. Laplace transform techniques may simplify the task of solving systems of differential equations [1] , [2] , [3] , and can be considered in terms of typical applications [4] , [5] .

Numerical inversion of Laplace transform is crucial for many applications. Unfortunately, when considering interesting examples, it is often difficult to find an analytical expression for the inverse Laplace transform. Inverting the Laplace transform is a challenging task. This challenge faced in many application areas including the finding of various performance measures in queueing and related probability models [6] , [7] , [8] , [9] , in solving partial differential equations [10] , and in the pricing and hedging of financial derivatives [11] , [12] , [13] .

This paper investigate the complicated and very interesting relationship between numerical precision and the number of terms in one particular Laplace transform inversion algorithm, the Gaver-Stehfest algorithm, and illustrates this relationship using several carefully chosen numerical examples.

For numerous practical situations the inverse of Laplace transform is complicated and either doesn't have a closed form, or has a solution which cannot be represented by any simple formula, performed even in symbolic software (Maple or Mathematica). An alternative is to use a numerical technique for inversion. One way to choose among various alternative methods is to provide a large set of test problems, and to demonstrate how a specific algorithm works on each of them.

Several algorithms have been proposed for numerical Laplace transforms inversion, see for instance the surveys in [4] and [14] . The Gaver-Stehfest algorithm [15] is one of the most powerful algorithms for this purpose. The convergence of this algorithm has been examined in [16] . Unfortunately despite its theoretical advantages, in many practical applications, numerical inversion often encounters numerical accuracy problems [14] [17] [18] [19] [20]. As such, small rounding errors in computation in standard double arithmetic may signifi- cantly corrupt the results, rendering these algorithms impractical to apply. Ex- tended precision allows to add additional significant figures, and produce results that converge to the solution. Laplace and inverse transforms for the test functions used in numerical calculations are presented in Table 1. These complicated functions are used to test the accuracy of the numerical Laplace transform and its inverse.

In general, lowercase letters used to denote the function $f(t)$ to be transformed, and the uppercase letter $C(s)$ to denote its Laplace transform, for example, $\mathcal{L}\{f(t)\} = C(s)$. If the closed form of $C(s)$ inversion is unknown, the original $C(s)$ compared with numerical solution $\tilde{\tilde{C}}(s)$ after double transformation. The results are illustrated in the plots and error estimations.

Table 1. Laplace and inverse transforms for the test functions used in numerical calcu- lations.

Function	$C(s)$	$f(t)$
1	$\frac{1}{s+2}$	e^{-2t}
2	$\left((s+0.2)^2+1\right)^{-1}$	$e^{-0.2t}\sin(t)$
3	$\left(s^2+1\right)^{-1}$	$\sin(t)$
4	$\left(s^2-1\right)\left(s^2+1\right)^{-2}$	$t\cos(t)$
5	$\tan^{-1}\left(s^{-1}\right)$	$t^{-1}\sin(t)$
6	$1/\sqrt{1+s^2}$	$J_0(t)=\sum_{n=0}^{\infty}\frac{(-1)^n t^{2n}}{2^{2n}(n!)^n}$
7	$\frac{e^{-1/s}}{s^{3/2}}$	$\frac{\sin\left(2\sqrt{t}\right)}{\sqrt{\pi}}$

With the help of the arprec library [21] , C++ and MATLAB numerical class library [22] , [23] applied, to investigate the role that extended precision can play in accuracy of Laplace transform and inversions.

The remainder of the paper is organized as follows. In Section 2, a brief description of the underlying theory is given, to introduce numerical Laplace double transformation technique. Sections 3 and 4, apply the

Gaver-Stehfest algorithm to the test functions with various degrees numerical accuracy. In Sections 5 and 6, the stability and accuracy of the Laplace transform inversion and the role that the number of expansion terms and precision of the arithmetic play in the numerical results is described. Section 7 describes the algorithm and software implementation of the numerical direct Laplace transform. This section gives background material needed to provide the method, described in the next Section 8. In Section 8 the numerical double transformation technique to confirm agreement of the numerical inversion results is presented. In Section 9 compares the execution time for various arbitrary precision calculations. Concluding remarks are given in Section 10. The Appendix introduces C++ code used to implement numerical Laplace and inverse Laplace transform in arbitrary precision, and illustrates the corresponding graphical user interface with the help of several screenshots.

2. NUMERICAL LAPLACE TRANSFORMS AND THEIR INVERSES

2.1. Laplace Transform

Let $f(t)$ be a function defined for $t \geq 0$. Then the integral

$$\mathcal{L}\{f(t)\} = \int_0^\infty e^{-st} f(t)\,dt \tag{1}$$

is said to be the Laplace transform of $f(t)$, provided the integral converges. The symbol $\mathcal{L}$ is the Laplace transformation operator, which act on the function $f(t)$ and generates a new function, $C(s) = \mathcal{L}\{f(t)\}$.

2.2. Inverse Laplace Transform

If $C(s)$ represents the Laplace transform of a function $f(t)$, that is, $\mathcal{L}\{f(t)\} = C(s)$, then $f(t)$ is the inverse Laplace transform of $C(s)$ and $f(t) = \mathcal{L}^{-1}\{C(s)\}$. The inverse Laplace transform $\mathcal{L}^{-1}\{C(s)\}$ is uniquely determined in the sense that if $C(s) = G(s)$ and $f(t)$ and $g(t)$ are continuous functions, then $f(t) = g(t)$. This result is known as Lerch's theorem [4].

2.3. Numerical Laplace Transform and Inversion

The Laplace transform can be inverted either algebraically or numerically. The notation $\tilde{f}(t)$ used for the numerical approximation to $f(t)$ (numerical inversion of the Laplace transform $C(s)$), and $\tilde{C}(s)$ for the numerical Laplace transform of $f(t)$.

These results, together with the following well known properties which provide very useful numerical checks, may be applied to numerical algorithms and corresponding software.

$$\text{Integral property, } \int_0^\infty e^{-st} f(t)\,\mathrm{d}t = C(0) \tag{2}$$

$$\text{Initial value theorem, } \lim_{t\to 0} f(t) = \lim_{s\to\infty} sC(s) \tag{3}$$

$$\text{Final value theorem, } \lim_{t\to\infty} f(t) = \lim_{s\to 0} sC(s) \tag{4}$$

If X is the random variable with the probability density function f and the cumulative distribution function F, this gives

$$C(0) = \int_0^\infty e^{-st}\mathrm{d}F(t) = \int_0^\infty e^{-st} f(t)\,\mathrm{d}t = 1 \tag{5}$$

2.4. Numerical Laplace Double Transformation Technique

We define the following double transformation technique for the Laplace transform of the inversion:

$$\tilde{\tilde{C}}(s) = \mathcal{L}\{\mathcal{L}^{-1}\{C(s)\}\} \tag{6}$$

This definition will be used to estimate the accuracy of the Laplace transform inversion, when its closed form is unknown.

After applying the Laplace transform, the problem is said to be in the Laplace domain and it is denoted as a function of s not t. While calculations might be easier in the Laplace domain, leaving the solution in the Laplace domain is typically not useful. To transform the result back into the time-domain, inverse Laplace transforms are used. When the analytical answer is unknown, it is difficult to know whether or not the numerical inversion results are accurate. Moreover, it is hard to judge whether or not changes to the method improve or degrade the inversion estimate.

The following steps are used:

1) Begin with the Laplace domain function $C(s)$.

2) Compute the numerical inversion using some set of parameters. In this case, we will control the precision level and the number of terms in the approximation. Setting the precision level to N_1, we get

$$\hat{f}_{N_1}(t) = \mathcal{L}_{N_1}^{-1}\{C(s)\} \tag{7}$$

3) Take the Numerical Laplace Transform of $\hat{f}_{N_1}(t)$, resulting in

$$\mathcal{L}\{\hat{f}_{N_1}(t)\} = \tilde{\tilde{C}}_{N_1}(s) \tag{8}$$

4) Compare the functions $C(s)$ and $\tilde{\tilde{C}}_{N_1}(s)$, and define the error-function as:

$$\varepsilon_{N_1}(s) = \left|C(s) - \tilde{\tilde{C}}_{N_1}(s)\right| \tag{9}$$

5) Repeat the process with some other precision level N_2.

6) Compare $\varepsilon_{N_1}(s)$ and $\varepsilon_{N_2}(s)$. The precision level that provides lower errors is superior, and the difference between the error functions can provide a way of quantifying the accuracy improvement gained from increasing the precision level.

3. CHALLENGING EXAMPLES OF LAPLACE TRANSFORM AND THEIR INVERSES

This demonstration applies the Gaver-Stehfest algorithm [15] to determine the inverse Laplace transforms of the test functions to various degrees of numerical accuracy. The inverse functions and corresponding test functions are presented in Table 1.

The inverse Laplace transform of $C(s)$ is $f(t)$, defined such that the following must hold:

$$C(s)=\int_0^{\infty}e^{-st}f(t)\mathrm{d}t \tag{10}$$

Table 1 lists seven functions which have been used as tests of the numerical Laplace transform and inverse transform. Test 6 involves the Bessel function of order 0 [24] ,

$$J_0(t)=\sum_{n=0}^{\infty}\frac{(-1)^n t^{2n}}{2^{2n}(n!)^n},\ \text{and}\ J_0(s)=\frac{1}{\sqrt{1+s^2}} \tag{11}$$

4. GAVER-STEHFEST ALGORITHM OF INVERSE LAPLACE TRANSFORMS

The Gaver-Stehfest method [15] uses the summation:

$$f(t)\approx\frac{\ln(2)}{t}\sum_{n=1}^{L}K_nF\left(\frac{n\ln(2)}{t}\right), \tag{12}$$

where $F(\cdot)$ is the Laplace transform of $f(t)$. The coefficients K_n depends only on the (necessarily even) number of expansion terms, L, given by:

$$K_n=(-1)^{n+L/2}\sum_{k=\left\lfloor\frac{n+1}{2}\right\rfloor}^{\min(n,L/2)}\frac{k^{L/2}(2k)!}{(L/2-k)!k!(k-1)!(n-k)!(2k-n)!} \tag{13}$$

For each function an error E is calculated as the measure for the accuracy of the numerical solution. Let $f(t)$ be the analytical solution defined for $t = t_1, t_2, \cdots, t_m$. We define by $\tilde{f}(t) = \mathcal{L}^{-1}\{C(s)\}$ the numerical solution. Then E gives the root-mean-square deviation between the analytical and numerical solutions for the t values [19] :

$$E = \left(\sum_{i=1}^{m} \left(f(i/2) - \tilde{f}(i/2) \right)^2 / m \right)^{1/2}, i = 1, 2, \cdots m \tag{14}$$

The sum (12) doesn't provide convergence due to roundoff errors for the large number of terms L, usually if L exceeds the number of decimal digits of precision N (e.g. L greater than 16 for standard double precision arithmetics). The software implementation of the numerical Laplace transform and the Laplace transform inversion are given in the Appendix.

5. ACCURACY OF THE NUMERICAL LAPLACE TRANSFORM INVERSION AS A FUNCTION OF THE NUMBER OF EXPANSION TERMS AND PRECISION

Numerical inversion of the Laplace transform is an unstable process, so all algorithms are applied in arbitrary precision. In Table 2 and in Figures 2-8 numerical results related to the test functions presented in Table 1 are reported. The numerical solutions are given for different precisions: $N = 16$ (double precision), $N = 32$, $N = 64$, $N = 128$ and $N = 256$. First, as presented in Table 2, assigned $L = N$, and the number of terms L in the approximation equals to the digits of precision N. The measures (14) are used to calculate the errors E to the t values 0.5, 1, 1.5, ..., 35 for the functions 1 - 6, and to the t values 0.5, 1, 1.5, ..., 140 for the function 7. The reason to use this presentation is because t values require a large amount of space to realize the correct accuracy results.

Example 1. The first function explored is:

$$C(s) = \frac{1}{s+2}, \text{ with } f(t) = e^{-2t}, \tag{15}$$

where $f(t)=e^{-2t}$ is the exact solution of the inversion. The results correspond to finding numerical inverse Laplace transform are plotted in Figure 1. For this simple function standard double arithmetic algorithms work well. In Table 2 the calculations in double precision are at the order of 10^{-6}. In the plot, the inversion $\tilde{f}(t)=\mathcal{L}^{-1}\left\{\frac{1}{s+2}\right\}$ and the exact solution $f(t)=e^{-2t}$ visibly overlapping even for $N=16$. The precision from $N=32$ to $N=256$ gives the sequence of improvements. High accuracy is borne out by the errors at the order 10^{-32} and 10^{-69} respectively to the precision $N=128$ and $N=256$.

Table 2. Calculation errors of the inverse Laplace transform in arbitrary precision, Values 5.66e−6 $\equiv 5.66\times10^{-6}$.

Function	$C(s)$	$N=16$	$N=32$	$N=64$	$N=128$	$N=256$
1	$1/(s+2)$	5.66e−6	1.22e−9	6.52e−17	2.25e−32	7.02e−69
2	$((s+0.2)^2+1)^{-1}$	3.44e−2	7.33e−3	2.46e−4	1.54e−10	5.38e−36
3	$(s^2+1)^{-1}$	6.13e−1	5.04e−1	1.68e−1	2.44e−5	1.01e−27
4	$(s^2-1)(s^2+1)^{-2}$	1.43e+1	1.35e+1	6.34e0	1.33e−3	9.88e−26
5	$\tan^{-1}(s^{-1})$	3.51e−2	2.09e−2	4.80e−3	3.66e−7	3.84e−30
6	$1/\sqrt{1+s^2}$	1.12e−1	8.04e−2	2.26e−2	2.12e−6	2.78e−16
7	$e^{-1/s}/s^{3/2}$	2.59e−1	6.16e−3	3.83e−13	2.14e−41	2.84e−88

Example 2. The next test function inverted is

$$C(s)=\frac{1}{(s+0.2)^2+1}, \quad \text{with } f(t)=e^{-0.2t}\sin(t), \tag{16}$$

and the results correspond to numerical inverse Laplace transform are shown graphically in Figure 2. The accuracy, at double precision, in comparison with the exact solution is very poor. As can be seen from Table

2 the calculations in double precision are at the order of 10^{-2}. The precision level $N = 32$ and $N = 64$ show improvements in accuracy over double precision.

In the plot, the inversion for $N = 128$ and the exact solution are visibly overlapping. High accuracy is borne out in Table 2 by the errors at the order 10^{-10} and 10^{-35} respectively to the precision level $N = 128$ and $N = 256$.

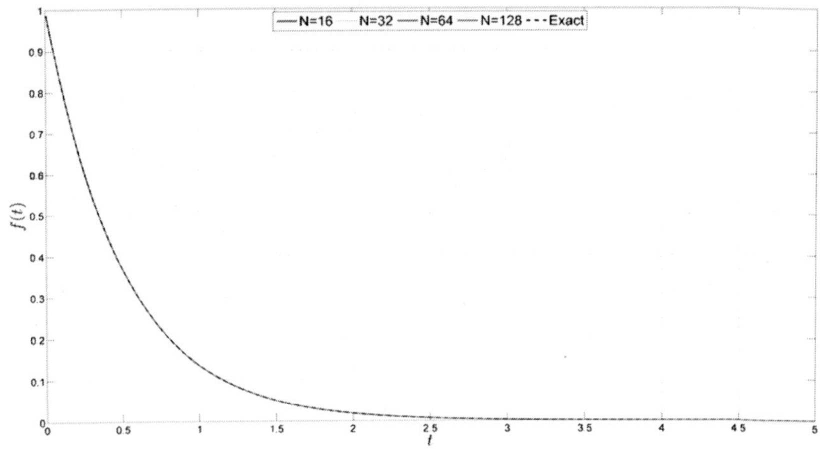

Figure 1. Given $C(s) = \frac{1}{s+2}$. Compute $\tilde{f}(t) = \mathcal{L}^{-1}\{C(s)\}$. The determining function is $f(t) = e^{-2t}$.

Example 3. The Figure 3 provides the inverse results for the function

$$C(s) = \frac{1}{s^2+1}, \text{ with } f(t) = \sin(t) \tag{17}$$

Steady improvement of the answer is observed through $N = 128$. By $N = 128$ the inversion is indistinguishable from the exact solution on the inter- val shown, but will eventually diverge from the exact solution for some higher values of t.

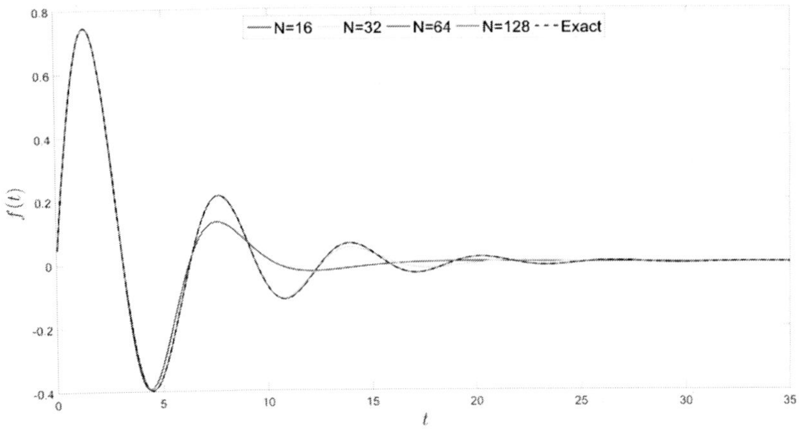

Figure 2. Given $C(s)=\frac{1}{(s+0.2)^2+1}$. Compute $\tilde{f}(t)=\mathcal{L}^{-1}\{C(s)\}$. The determining function is $f(t)=e^{-0.2t}\sin(t)$.

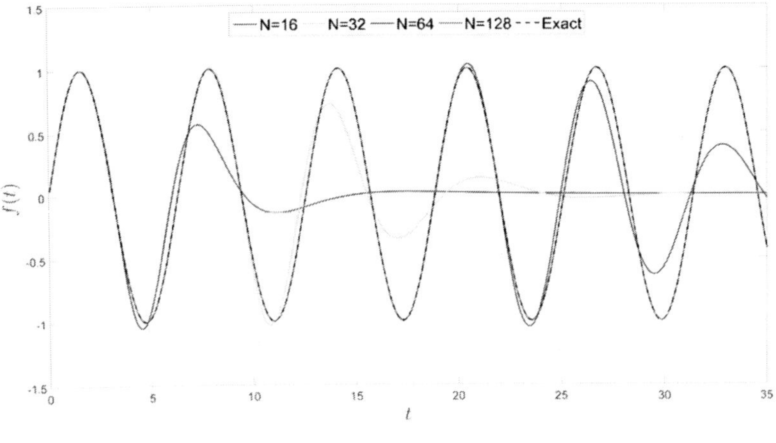

Figure 3. Given $C(s)=\frac{1}{s^2+1}$. Compute $\tilde{f}(t)=\mathcal{L}^{-1}\{C(s)\}$. The determining function is $f(t)=\sin(t)$.

Example 4. Figure 4 shows the inverse results to the function

$$C(s)=\frac{s^2-1}{\left(s^2+1\right)^2},\ \text{with}\ f(t)=t\cos(t) \tag{18}$$

Similar to the previous two test functions, as N increases, the interval on which the inversion is more accurate gets longer. Since the function is diverging and oscillating, the inaccuracies are more visible than in the previous two figures.

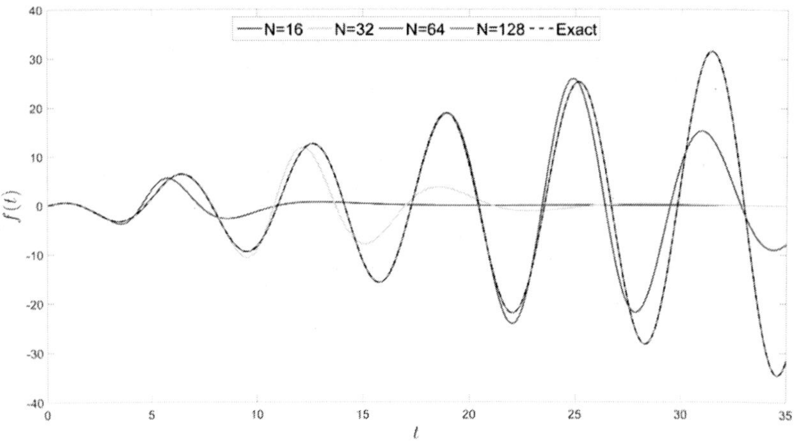

Figure 4. Given $C(s)=\frac{s^2-1}{\left(s^2+1\right)^2}$. Compute $\tilde{f}(t)=\mathcal{L}^{-1}\{C(s)\}$. The determining function is $f(t)=t\cos(t)$.

Example 5. Consider the inverse of the function

$$C(s)=\tan^{-1}\left(\frac{1}{s}\right),\ \text{with}\ f(t)=\frac{\sin(t)}{t} \tag{19}$$

In Figure 5 the precision level $N=32$ and $N=64$ show improvements in accuracy over double precision, but higher N do not result in a visibly better inversion. In the plot, the numerical inversion for $N\geq 64$ and the exact solution are visibly overlapping.

Example 6. The inverse results to the function

$$C(s)=\frac{1}{\sqrt{1+s^2}}, \quad \text{with} \quad f(t)=J_0(t) \tag{20}$$

is plotted in Figure 6. Varying precision levels $N=32$, $N=64$, $N=64$ and $N=128$ show successive improvements in accuracy over double precision. The numerical inversion for $N\geq 128$ and the exact solution are visibly overlapping.

Naively it would seem that continuing to increase the number of terms L would improve the accuracy of the approximation, since more terms seem to produce results closer to the exact solution. Consider the Bessel example with the precision level $N=32$, increasing the number of terms to $L=64$. The result is slightly better, close to 2.26×10^{-2}, that for $N=64$ and $L=64$. But using the precision level $N=32$ and the number of terms in the approximation $L=128$ (Figure 7), the error will be at the order of 10^{39} due to numerical error dominating the solutions that are obtained this way. The algorithm was unable to provide even an order of magnitude estimate.

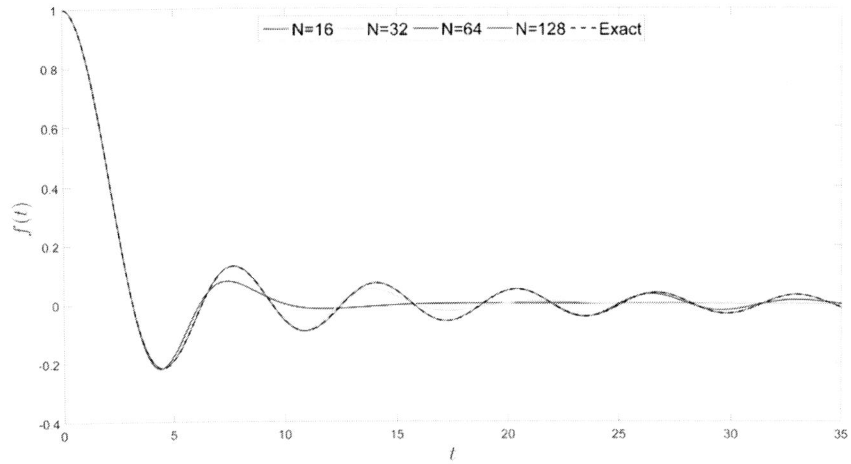

Figure 5. Given $C(s)=\tan^{-1}\left(\frac{1}{s}\right)$. Compute $\tilde{f}(t)=\mathcal{L}^{-1}\{C(s)\}$. The determining function is $f(t)=\frac{\sin(t)}{t}$.

Example 7. Consider the inverse of the function

$$C(s)=\frac{e^{-1/s}}{s^{3/2}},\quad \text{with } f(t)=\frac{\sin\left(2\sqrt{t}\right)}{\sqrt{\pi}} \tag{21}$$

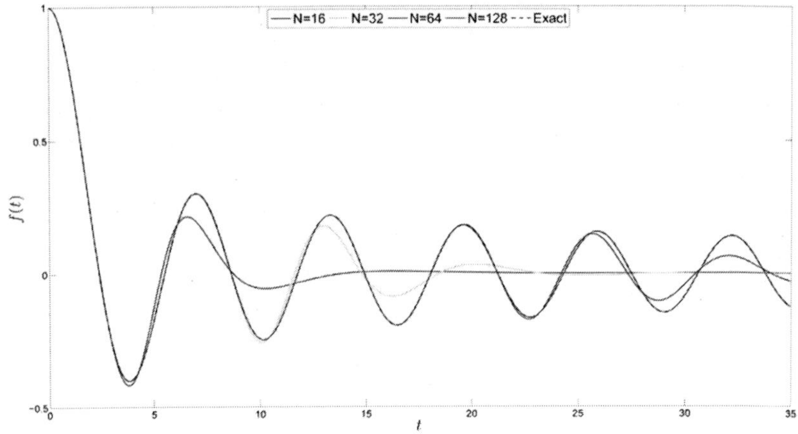

Figure 6. Given $C(s)=\frac{1}{\sqrt{1+s^2}}$. Compute $\tilde{f}(t)=\mathcal{L}^{-1}\{C(s)\}$. The determining function is $f(t)=J_0(t)$, Bessel function of order 0.

In Figure 8 steady improvement is observed through $N=64$ with the error at the order of 10^{-13} (13 decimal places accuracy, Table 2).

As can be seen from Table 2, all test functions have low level of accuracy for $N=16$, $N=32$ and for $N=64$ (except the last example).

For all functions the algorithm gave reasonable answer (at least 3 decimal places of accuracy), increasing the precision level up to 128. Finally all functions have high level of accuracy (at least 16 decimal place accuracy) increasing the precision level up to 256 digits.

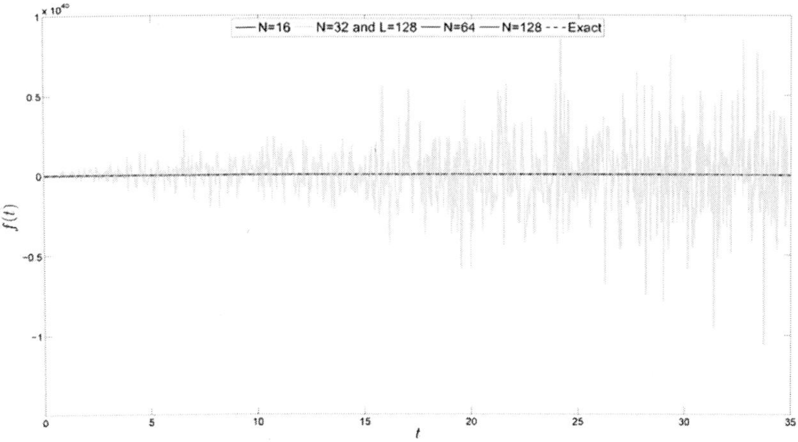

Figure 7. Given $C(s)=\frac{1}{\sqrt{1+s^2}}$. The precision level is $N=32$ and the number of expansion terms is $L=128$. Compute $\tilde{f}(t)=\mathcal{L}^{-1}\{C(s)\}$. The determining function is $f(t)=J_0(t)$. The algorithm was unable to provide even an order of magnitude estimate.

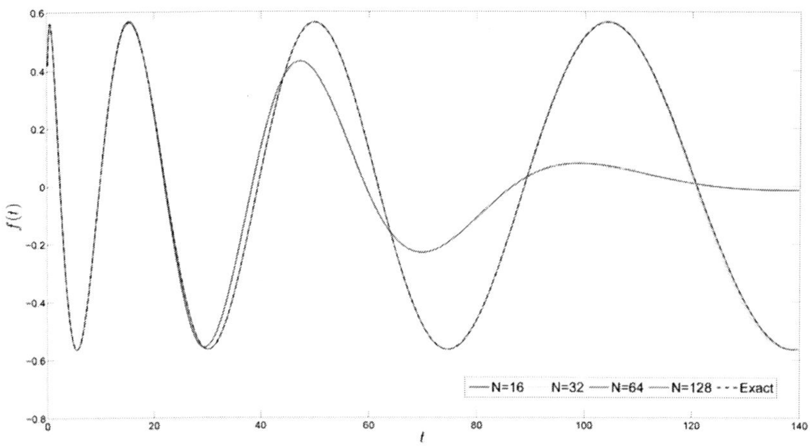

Figure 8. Given $C(s)=\frac{e^{-1/s}}{s^{3/2}}$. Compute $\tilde{f}(t)=\mathcal{L}^{-1}\{C(s)\}$. The determining function is $f(t)=\frac{\sin\left(2\sqrt{t}\right)}{\sqrt{\pi}}$.

6. THE ROLE OF THE NUMBER OF EXPANSION TERMS AND PRECISION IN THE NUMERICAL ACCURACY

In Table 2 the numerical solutions are given for the number of expansion terms equal to the precision, $L=N$. Now the accuracy of the numerical inversion is investigated, varying the number of expansion terms and precision. The Table 3 and Figure 9 give the error estimates of the numerical inverse using the Gaver- Stehfest implementation for the function $f(t)=e^{-0.2t}\sin(t)$ having the La- place transform $C(s)=\left((s+0.2)^2+1\right)^{-1}$. The solutions are observed while increasing the number of expansion terms L. However, there is a limitation to adding additional terms.

Let $N=16$. Increasing the number of terms in the computation to $L\geq 64$ it appears that the numerical inversion becomes unstable and functions are dominated by numerical errors. If $N=32$ and $N=64$, the numerical inversion becomes unstable for $L\geq 128$. If $N=128$, the numerical inversion becomes unstable for $L\geq 256$. On the other hand there is only slight improvement if the precision exceeds the number of terms, $N>L$.

Table 3. Numerical errors of the inverse Laplace transform $\tilde{f}(t)=\mathcal{L}^{-1}\left\{\left((s+0.2)^2+1\right)^{-1}\right\}$ as a function of the number of expansion terms L and precision N, Values 3.44e−2 $\equiv 3.44\times10^{-2}$.

$C(s)=\left((s+0.2)^2+1\right)^{-1}$	$N=16$	$N=32$	$N=64$	$N=128$	$N=256$
$L=16$	3.44e−2	3.44e−2	3.44e−2	3.44e−2	3.44e−2
$L=32$	7.33e−3	7.33e−3	7.33e−3	7.33e−3	7.33e−3
$L=64$	4.66e+10	2.66e−4	2.46e−4	2.46e−4	2.46e−4
$L=128$	2.78e+53	6.86e+38	1.01e+10	1.54e−10	1.54e−10
$L=256$	2.99e+139	5.47e+124	9.72e+95	4.57e+23	5.38e-36

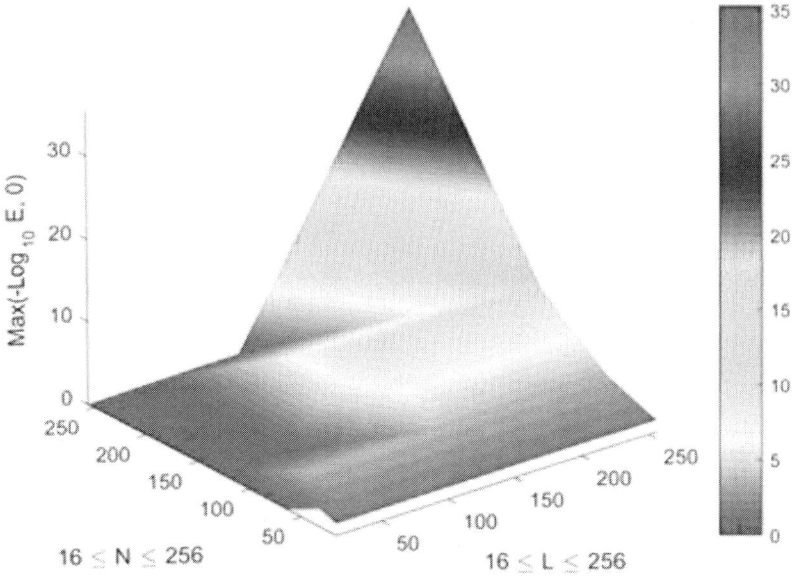

Figure 9. Numerical errors E of the Laplace transform inversion $\tilde{f}(t)=\mathcal{L}^{-1}\left\{\left((s+0.2)^2+1\right)^{-1}\right\}$ as a function of the number of expansion terms L and precision N. Estimated data presented in terms of $\mathrm{MAX}(-\mathrm{Log}_{10}E,0)$, here 0 denotes the algorithm has failed.

Table 3 clearly shows that there is no point in increasing the precision beyond a point warranted by the accuracy of the number of terms in Gaver-Stehfest algorithm. On the other hand, using a large number of terms will not be of benefit if the precision with which each term is calculated, is insufficient. For example, $L=256$, $N=128$ is a useless result while $N=256$ gives high accuracy. The example, $N=16$, $L=16$ gives the same accuracy as $N=256$, $L=16$.

Similar conclusions may be drawn from the results reported in Table 4 and in Figure 10.

The error estimates are given for the example

$$f(t)=\frac{\sin\left(2\sqrt{t}\right)}{\sqrt{\pi}},\quad \text{with}\quad C(s)=\frac{e^{-1/s}}{s^{3/2}} \tag{22}$$

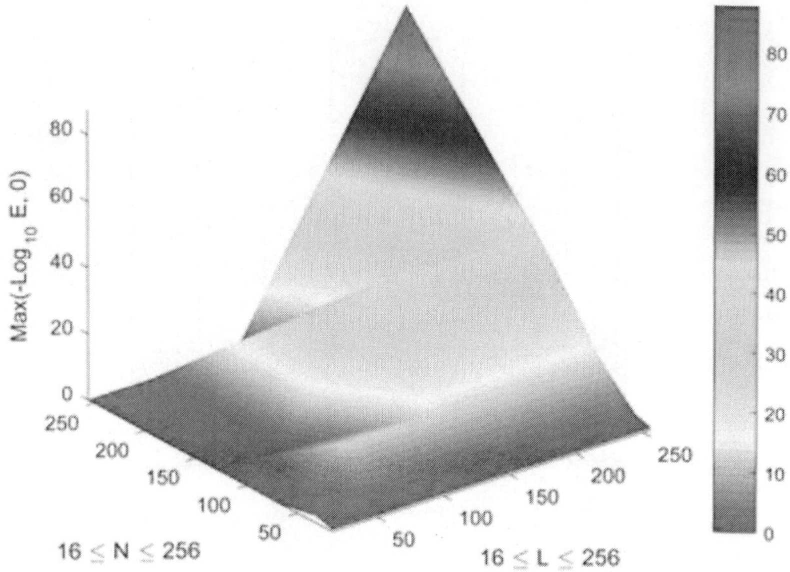

Figure 10. Numerical errors E of the Laplace transform inversion $\tilde{f}(t)=\mathcal{L}^{-1}\{(e^{-1/s}/s^{3/2}\}$ as a function of the number of expansion terms L and precision N Estimated data presented in terms of $\mathrm{MAX}(-\mathrm{Log}_{10}E,0)$, here 0 denotes the algorithm has failed.

Table 4. Numerical errors of the inverse Laplace transform $\tilde{f}(t)=\mathcal{L}^{-1}\{e^{-1/s}/s^{3/2}\}$.

$C(s)=e^{-1/s}/s^{3/2}$	$N=16$	$N=32$	$N=64$	$N=128$	$N=256$
$L=16$	2.59e−1	2.59e−1	2.59e−1	2.59e−1	2.59e−1
$L=32$	6.16e−3	6.16e-3	6.16e−3	6.16e−3	6.16e−3
$L=64$	3.33e+13	1.19e−1	3.83e−13	3.83e−13	3.83e−13
$L=128$	4.64e+56	2.00e+42	3.08e+13	2.14e−41	2.14e−41
$L=256$	1.60e+143	6.17e+128	1.14e+100	7.94e+27	2.84e−88

Let $N=16$. Increasing the number of terms in the computation to $L\geq 64$ we can see that the numerical inversion becomes unstable. If $N=32$ and $N=64$, the numerical inversion failed for $L\geq 128$. In case of $N=128$, the numerical inversion becomes unstable for $L\geq 256$. There is only a slight improvement if the precision exceeds the number of terms $N>L$. Again as can be seen from Table 3, the number of terms and precision must be in a harmonious balance for good results to be obtained.

Evidently the two plots in Figure 9 and Figure 10, are very much alike. They show similar tracking boundary movements and illustrate whether the algorithm has succeeded in obtaining high-order accuracy or fails due to numerical instability.

7. NUMERICAL COMPUTATION OF THE DIRECT LAPLACE TRANSFORM

The Laplace transform of a function $f(t)$ is defined by (1) on the interval $[0,\infty]$. The problem of an infinite upper limit of integration may be removed by the substitution $t=-\ln(u),\ \mathrm{d}t=u^{-1}\mathrm{d}u$ which replaces infinite by finite limits.

When $t=0$, $u=1$ and when $t\to\infty$, $u\to 0$. Then

$$\int_0^{\infty} e^{-st} f(t)\,\mathrm{d}t=\int_0^{\infty} e^{\ln(u^s)} f(-\ln(u))u^{-1}\mathrm{d}u=\int_0^{1} u^{s-1} f(-\ln(u))\,\mathrm{d}u \tag{23}$$

The behaviour of the function to be transformed must be considered at the new limits, and the exponential function inside the integral requires special examination in terms of high accuracy.

Compute the Direct Laplace Transform by Composite Simpson's Rule

For integration over the interval $[a,b]$, an even n is chosen such that the function is adequately smooth over each subinterval $[x_j,x_{j+1}]$ where $x_j=a+jh$ for all $j\in\{0,1,2,\cdots,n\}$ with $h=\dfrac{b-a}{n}$. In particular, $x_0=a$ and $x_n=b$. Then, the composite Simpson's Rule is given by [25] :

$$\int_a^b f(x)\,dx \approx \frac{h}{3}\left[f(x_0)+2\sum_{j=1}^{n/2-1} f(x_{2j})+4\sum_{j=1}^{n/2} f(x_{2j-1})+f(x_n)\right] \quad (24)$$

Applying this to the transformed integrand from the Equation (23) we get $u_j = jh$ for all $j \in \{0,1,2,\cdots,n\}$ with $h=\frac{1}{n}$. Therefore,

$$C(s) \approx \frac{1}{3n}\left[0^{s-1} f(-\ln(0))+2\sum_{j=1}^{n/2-1} u_{2j}^{s-1} f(-\ln(u_{2j})) + 4\sum_{j=1}^{n/2} u_{2j-1}^{s-1} f(-\ln(u_{2j-1}))+1^{s-1} f(-\ln(1))\right] \quad (25)$$

The basic Simpson's rule formula divides the interval $[a,b]$ of integration into two pieces. To apply the composite Simpson's rule, the interval $[a,b]$ must

be divided into an even number of subintervals $n = 2m$. Then $h=\frac{b-a}{n}=\frac{b-a}{2m}$

Let f be a function with four continuous derivatives. Then, the composite Simpson's rule converges to the true value of the integral with rapidity n^{-4} at worst. The error committed by the composite Simpson's rule is bounded in absolute value by [25] .

$$E_n = -\frac{(b-a)^5}{180n^4}\max\left|f^4(\xi)\right|, \quad a<\xi<b \quad (26)$$

For numerical Laplace transform in the obtaining approximate integral by this rule, some modifications must be made. First, the term $0^{s-1} f(-\ln(0))$ in (25) must be addressed. This term addresses the behaviour of the function at infinity. If the Laplace transform exists, the $\lim_{t\to\infty} e^{-st} f(t) \to 0$, meaning that the exponential dampening term outweighs the value of $f(t)$ at infinity. Therefore, $\lim_{t\to\infty} f(t)$ must either exist or oscillate between some finite bounds. As such, it may be concluded that the term $0^{s-1} f(-\ln(0))$ vanishes and may be dropped from the formulation.

Next, we need to examine the last term, $1^{s-1} \cdot f(-\ln(1)) = f(0)$. Evaluating this term should require that the function be defined at $t = 0$. This can however prove problematic since many applications of the Laplace transform result in t-domain functions that have singularities at $t = 0$. As such, we change the domain of integration to be $(0, 1-\epsilon)$ as opposed to $(0,1)$, where ϵ is the machine epsilon depending on the precision level used. For example, in double precision $(N = 16)$, $\epsilon \approx 2.22 \times 10^{-16}$.

Therefore we have:

$$C(s) \approx \frac{1}{3n}\left[2\sum_{j=1}^{n/2-1} u_{2j}^{s-1} f\left(-\ln\left(u_{2j}\right)\right) + 4\sum_{j=1}^{n/2} u_{2j-1}^{s-1} f\left(-\ln\left(u_{2j-1}\right)\right) + u_n^{s-1} f\left(-\ln\left(u_n\right)\right)\right], \quad (27)$$

where n is the number of subintervals, $h = \frac{1-\epsilon}{n}$, and ϵ is the machine epsilon at the precision level.

Our C++ software implementation of the numerical Laplace transform is based on Equation (27). The following improvements were made to speed up the calculations. Notice that only the powers of u depend on s in the Equation (27). As such, the function evaluations, $f(-\ln(u_j))$, need not be evaluated every time a new s value is calculated. This is especially useful in the double transformation calculations because each evaluation of the function $\hat{f}(t)$ is the numerical inversion of $C(s)$ at some point t. Depending on the precision level, this can be an extremely time consuming step. The implementation of the method is shown in Appendix.

Example 8. The numerical Laplace transform for the Bessel function $J_0(t)$ is plotted in Figure 11. We used a composite Simpson's Rule calculation with up to 25,000 subintervals to ensure high accuracy. Comparing Laplace transform $C(s) = \mathcal{L}\{J_0(t)\}$ and the exact solution $C(s) = \frac{1}{\sqrt{1+s^2}}$ for Bessel function $J_0(t)$ is shown in Table 5.

Example 9. Consider the following Laplace transform example

$$f(t) = \frac{\sin\left(2\sqrt{t}\right)}{\sqrt{\pi}}, \quad \text{and} \quad C(s) = \frac{e^{-1/s}}{s^{3/2}} \quad (28)$$

Numerical Laplace transform results are shown in Figure 12 and Table 6.

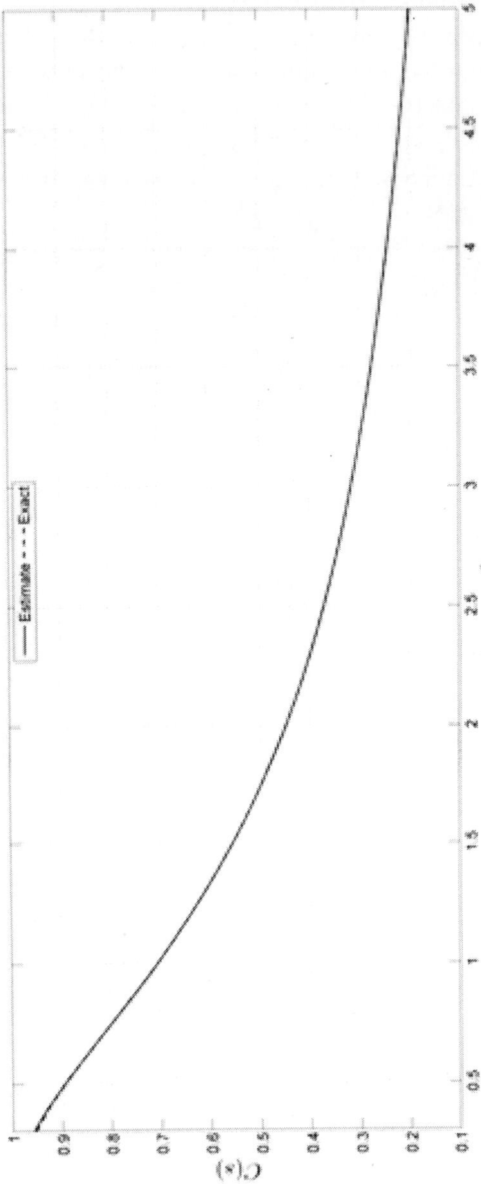

Figure 11. Given $f(t)=\sum_{n=0}^{\infty}\frac{(-1)^n t^{2n}}{2^{2n}(n!)^n}$. Compute $\tilde{C}(s)=\mathcal{L}\{f(t)\}$. The determining function is $C(s)=\frac{1}{\sqrt{1+s^2}}$.

8. VALIDATION OF THE NUMERICAL INVERSION USING DOUBLE TRANSFORMATION TECHNIQUE

Example 10. The first computation presented in this section is the theoretical error for Numerical Laplace transform of the function

$$f(t)=e^{-2t}, \quad \text{with} \quad C(s)=\frac{1}{s+2}, \tag{29}$$

where $C(s)$ is the exact Laplace transform solution. Let the number of subintervals $n=2m=256$. From the Equation (26) we have

$$E_n \le \frac{1}{180\times 256^4}\max\left|f^{(4)}(u)\right| \tag{30}$$

The integral $\int_0^1 u^{s-1} f(-\ln(u))\,\mathrm{d}u$ is used, where $u=e^{-t}$. Now $f(u)=u^2$. This yields

$$\int_0^1 u^{s-1} f(-\ln(u))\,\mathrm{d}u=\int_0^1 u^{s-1}u^2\,\mathrm{d}u=\int_0^1 u^{s+1}\,\mathrm{d}u \tag{31}$$

The integrand $f(u)=u^{s+1}$ has four continuous derivatives and

$$f^{(4)}(u)=s(s-1)(s+1)(s-2)u^{s-3}.$$

Because $0\le u\le 1$, then for example if $s=0.7$,

$$E_n \le \frac{1}{180\times 256^4}\max\left|f^{(4)}(u)\right|=\frac{1}{180\times 256^4}\times 0.464\approx 6\times 10^{-13}$$

Next we compare this theoretical error bound result with the numerical error obtaining by two step double transformation technique, on

$$\tilde{\tilde{C}}(s)=\mathcal{L}\left\{\mathcal{L}^{-1}\left\{\frac{1}{s+2}\right\}\right\} \tag{32}$$

Table 5. Comparison of the Laplace transform $\tilde{C}(s)=\mathcal{L}\{J_0(t)\}$ with the exact solution $C(s)=\frac{1}{\sqrt{1+s^2}}$ for Bessel function $J_0(t)$. Values 5.26e−2 $\equiv 5.26\times10^{-2}$.

s	$\mathcal{L}\{J_0(t)\}$	$\frac{1}{\sqrt{1+s^2}}$	Error
0.1	0.942376	0.995037	5.266103e−2
0.2	0.964656	0.980581	1.592439e−2
0.3	0.953056	0.957826	4.770068e−3
1	0.707105	0.707107	1.438876e−06
2	0.447214	0.447214	3.906561e−11
3	0.316228	0.316228	1.001639e−13
4	0.242536	0.242536	9.999995e−14
5	0.196116	0.196116	9.999975e−14

Table 6. Comparison of the Laplace transform $\tilde{C}(s)=\mathcal{L}\left(\frac{\sin(2\sqrt{t})}{\sqrt{\pi}}\right)$ with the exact solution $C(s)=\frac{e^{-1/s}}{s^{3/2}}$. Values 5.62E−6 $\equiv 5.62\times10^{-6}$.

s	$\mathcal{L}\left(\frac{\sin(2\sqrt{t})}{\sqrt{\pi}}\right)$	$\frac{e^{-1/s}}{s^{3/2}}$	Error
1	3.678744e−1	3.678794e−1	5.625486e−06
2	2.144409e−1	2.144410e−1	2.318805e−08
10	2.861345e−2	2.861347e−2	2.317171e−08
30	5.886267e−3	5.886290e−3	2.317492e−08
50	2.772397e−3	2.772421e−3	2.317815e−08
100	1.719596e−3	1.719619e−3	2.318121e−08
120	7.543895e−4	7.544127e−4	2.318945e−08
140	5.993618e−4	5.993850e−4	2.319269e−08

Figure 13 displays the numerical error $\varepsilon_N(s)$ in varying precision levels for some range of s beginning at very low values. From Figure 13 it is clear that increasing the decimal precision greatly increases the accuracy of the estimates for low values of t.

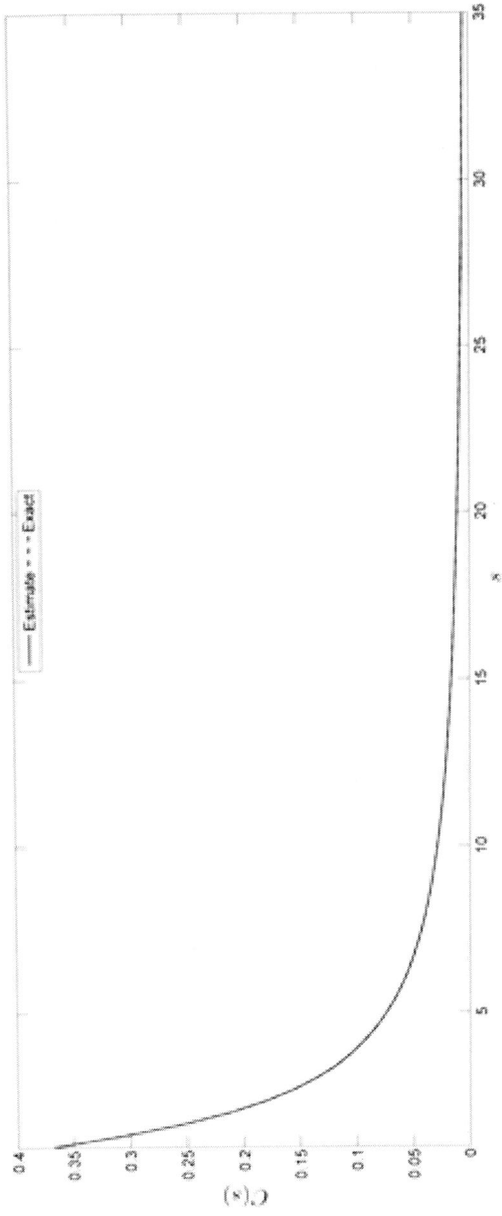

Figure 12. Given $f(t)=\frac{\sin\left(2\sqrt{t}\right)}{\sqrt{\pi}}$. Compute $\tilde{C}(s)=\mathcal{L}\left(\frac{\sin\left(2\sqrt{t}\right)}{\sqrt{\pi}}\right)$. The determining function is $C(s)=\frac{e^{-1/s}}{s^{3/2}}$.

Each time N is doubled, $\varepsilon_N(1)$ and $\varepsilon_N(2)$ are nearly halved. Examining lower values of s reveals that the difference between the various $\varepsilon_N(s)$ is not as pronounced as for $s=1$ or $s=2$. This suggests that increasing N does not have a very significant impact on the double transfor- mation accuracy for higher values of t. As such, depending on the needs of the application, one might stop increasing precision past $N=64$ since the double transformation errors are not showing a worthwhile improvement for the large increase in computational cost.

The error at $s \approx 0.7(\log_2 s = -0.5)$ for $N=250$ subintervals. For the precision level 64 the error is at the order of 10^{-17}, and for the precision level 128 the accuracy is much better, at the order 10^{-30}.

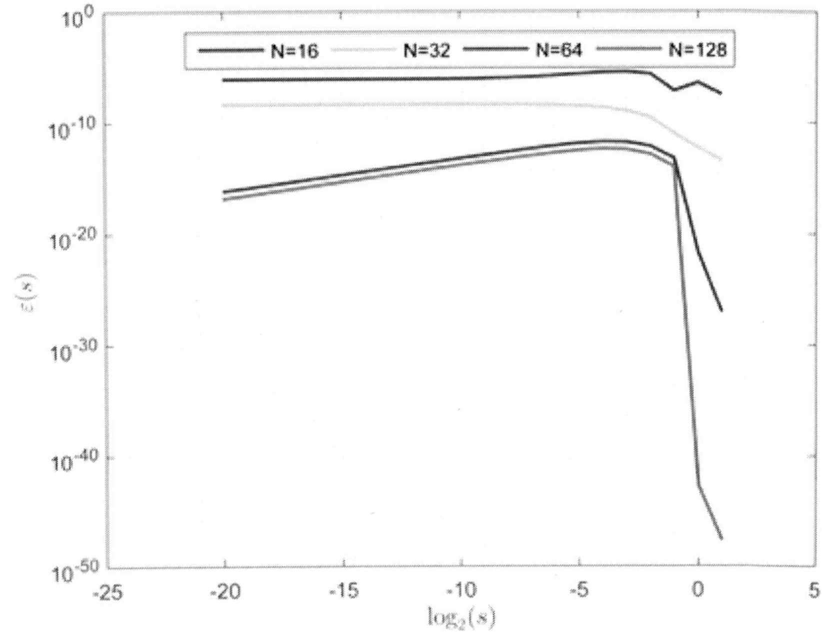

Figure 13. Numerical error $\varepsilon_N(s)=\left|C(s)-\tilde{\tilde{C}}_N(s)\right|$ for double transformation $\tilde{\tilde{C}}_N(s)=\mathcal{L}\left\{\mathcal{L}^{-1}\left\{\frac{1}{s+2}\right\}\right\}$ in varying precision levels.

For the precision levels 16 and 32 the accuracy are at the order 10^{-7} and 10^{-12} respectively. For the precision levels 16 and 32, the accuracy of the answer has deteriorated due to roundoff error. Note that this error of the inverse Laplace transform is ignored in the first step of the calculation as it is much smaller than suggested by the composite Simpson's rule in the second step of the double transformation algorithm.

Example 11. This example illustrates two steps of the numerical double transformation calculation

$$\tilde{\tilde{C}}(s)=\mathcal{L}\left\{\mathcal{L}^{-1}\left\{\frac{e^{-1/s}}{s^{3/2}}\right\}\right\} \tag{33}$$

Figure 14 depicts $\varepsilon_N(s)$ in varying precision levels for some range of s from the very low values.

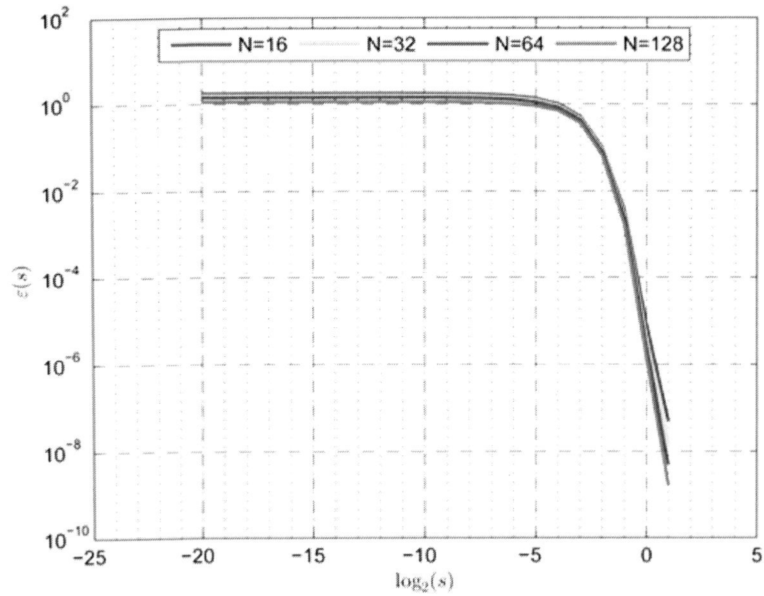

Figure 14. Numerical error $\varepsilon_N(s)=\left|C(s)-\tilde{\tilde{C}}_N(s)\right|$ for double transformation $\tilde{\tilde{C}}_N(s)=\mathcal{L}\left\{\mathcal{L}^{-1}\left\{\frac{e^{-1/s}}{s^{3/2}}\right\}\right\}$ in varying precision levels.

Example 12. Numerical double transformation for Bessel function of order 0 is given in Figure 15and Table 7. The plots in Figure 11 and Figure 15 are indistinguishable. High accuracy is borne out by comparison of the approximation with the exact solution.

Table 7. Comparison of $\tilde{\tilde{C}}(s)=\mathcal{L}\{\mathcal{L}^{-1}\{C(s)\}\}$ for Bessel function in Table 1 with the exact solution $C(s)=\frac{1}{\sqrt{1+s^2}}$. Values 7.3e−3 $\equiv 7.3\times10^{-3}$.

s	$\mathcal{L}\left\{\mathcal{L}^{-1}\left\{\frac{1}{\sqrt{1+s^2}}\right\}\right\}$	$\frac{1}{\sqrt{1+s^2}}$	Error
0.1	0.987647	0.995037	7.390034e−3
0.2	0.979209	0.980581	1.371815e−3
0.3	0.957524	0.957826	3.018401e−4
1	0.707107	0.707107	3.477729e−07
2	0.447214	0.447214	5.855865e−07
3	0.316228	0.316228	1.241552e−07
4	0.242536	0.242536	4.256322e−08
5	0.196116	0.196116	1.135660e−08

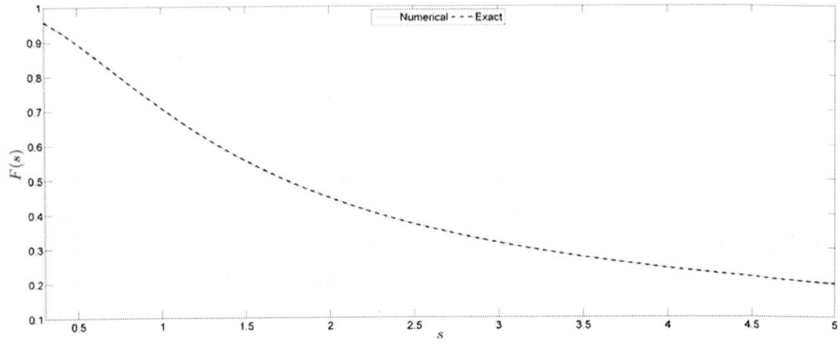

Figure 15. Given $C(s)=\frac{1}{\sqrt{1+s^2}}$. Compute $\tilde{\tilde{C}}=\mathcal{L}\{\mathcal{L}^{-1}\{C(s)\}\}$.

Example 13. Numerical double transformation results for the function $C(s)=\frac{e^{-1/s}}{s^{3/2}}$ are given in Figure 16 and in Table 8. The plots in Figure 12 and Consider again double transformation results for the function $C(s)=\frac{e^{-1/s}}{s^{3/2}}$.

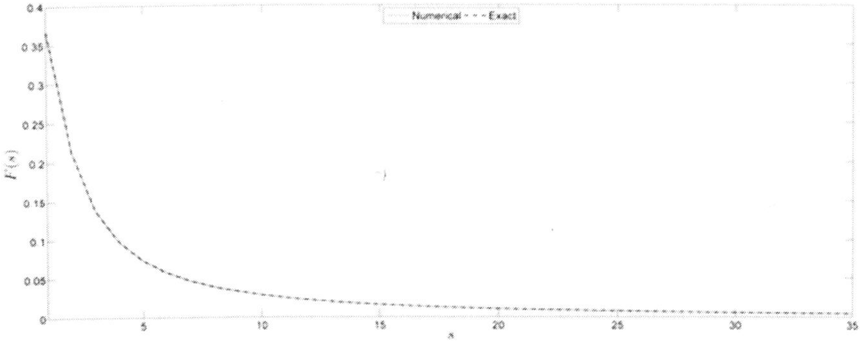

Figure 16. Given $C(s)=\frac{e^{-1/s}}{s^{3/2}}$. Compute $\tilde{\tilde{C}}=\mathcal{L}\{\mathcal{L}^{-1}\{C(s)\}\}$.

Figure 16 are nearly identical. As can be seen from Table 6 and Table 8, the accuracy is very high.

Let $s=2(\log_2 s=1)$ and the precision level N = 32. The results illustrated in the four plots (Figure 17) correspond to different number of subintervals n in composite Simpson's rule to approximate the Laplace transform. The error estimations are at the order $10^{-5}, 10^{-6}, 10^{-7}$ and 10^{-8} respectively with n = 32, 64, 128 and 512 subintervals.

9. COMPARISON OF RUNNING TIME IN ARBITRARY PRECISION CALCULATIONS

The execution time for the test functions is shown in Table 1. The direct Laplace transform and inverse Laplace transform are computed using different precision levels N. The number of subintervals used in the direct Laplace transform calculations is $L=512$. Table 9 gives the relative CPU time of the numerical solutions. All times are relative runs performing in double precision (16 digits). Computer configuration used: AMD 8350 8-Core processor 4 GHz, 8 GB RAM, 240 GB SDD, Windows 10 Pro.

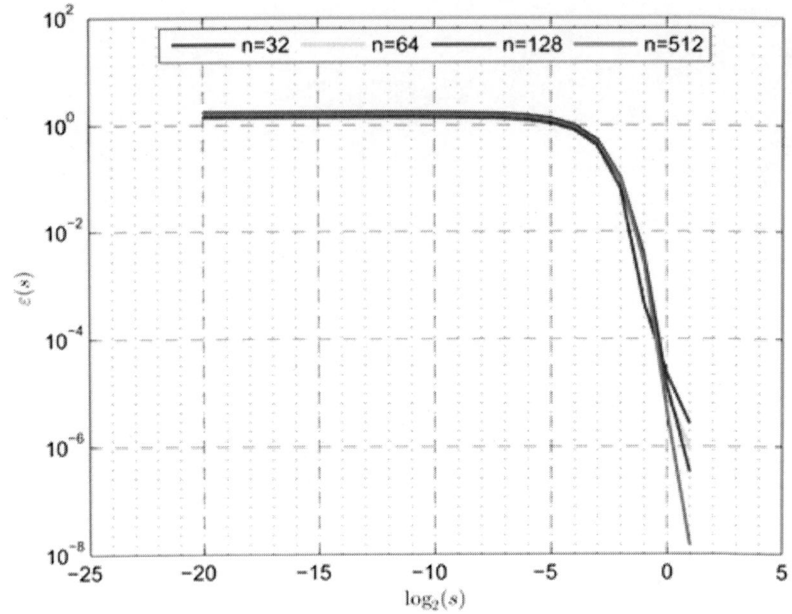

Figure 17. Given $C(s)=\frac{e^{-1/s}}{s^{3/2}}$. The precision level $N=32$. Compute the error estimations for different number of subintervals with n = 32, 64, 128 and 512 in composite Simpson's rule to approximate the Laplace transform.

Let N be the precision level. We approximate the relative CPU time $t(N)$ for inverse and direct Laplace transform algorithms by the polynomial $p(N)$ of degree 6. There are seven coefficients and the polynomial is:

$$p(N)=p_1*x^6+p_2)*x^5+p_3*x^4+p_4*x^3+p_5*x^2+p_6*x+p_7 \quad (34)$$

For inverse Laplace transform algorithm, the coefficients in the polynomial $p(N)$ of degree 6 are:

Table 8. Comparison of $\tilde{C}(s)=$ with original Laplace transform $C(s)=\frac{e^{-1/s}}{s^{3/2}}$. Values 4.3e−07 $\equiv 4.3\times 10^{-7}$.

s	$\mathcal{L}\left\{\mathcal{L}^{-1}\left\{\frac{e^{-1/s}}{s^{3/2}}\right\}\right\}$	$\frac{e^{-1/s}}{s^{3/2}}$	Error
1	3.678790e−1	3.678794e−1	4.315168e−07
2	2.144410e−1	2.144410e−1	8.855806e−11
10	2.861347e−2	2.861347e−2	9.927583e−11
30	5.886290e−3	5.886290e−3	1.452239e−10
50	2.772420e−3	2.772420e−3	1.461473e−10
100	9.900497e−4	9.900498e−4	1.436632e−10
120	7.544126e−4	7.544127e−4	1.431769e−10
140	5.993848e−4	5.993850e−4	1.4283671e−10

Table 9. Relative CPU time for inverse and direct Laplace transform algorithms as a function of the precision level. All times are relative to run in double precision (16 digits).

Precision Level (*N*)	16	32	64	128	256	512	750	1000
Inverse Laplace Transform	1	3	25	200	1.80e3	1.91e4	7.95e4	2.17e5
Direct Laplace Transform	1	1.65	2.04	4.52	10.25	22.07	37.33	70.51

$$\begin{aligned}p(6)&=(p_1,p_2,p_3,p_4,p_5,p_6,p_7)\\&=(-4.673630e-13, 9.235484e-10, -4.371448e-07,\\&\quad 2.108954e-04, -1.350226e-02, 5.725313e-01, -6.348896e+00).\end{aligned}$$

For the direct Laplace transform, the coefficients in the approximation polynomial $p(N)$ of degree 6 are:

$$\begin{aligned}p(6)&=(p_1,p_2,p_3,p_4,p_5,p_6,p_7)\\&=(-6.42651e-17, 2.43029e-14, 3.199151e-10,\\&\quad -4.079160e-07, 1.878244e-04, 9.7309e-03, 9.178160e-01).\end{aligned}$$

In Figure 18 we plotted relative CPU time for inverse Laplace transform and for direct Laplace transform as a function of the precision level N.

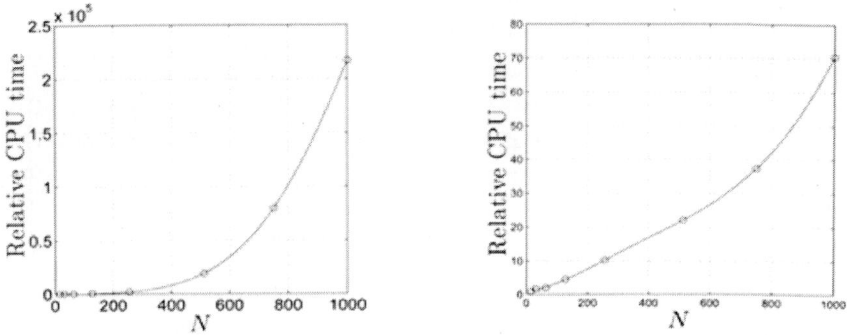

Figure 18. Relative CPU time for inverse Laplace transform (left) and for direct Laplace transform (right) as a function of the precision level N. All times are relative to run in double precision (16 digits).

10. CONCLUSION

Laplace Transform applications often require high accuracy beyond IEEE double precision. Common situations involve calculations that are numerically unstable, and even double-double precision is not sufficient to reach the necessary accuracy. Roundoff and underflow/overflow errors that occur during the computations can cause severe stability problems. There are several numerical inverse Laplace transform methods, each successful in some fields. The problem is to find methods successful in stability, accuracy and computational efficiency. Overall, the presented double transformation approach provides an effective way to compare the effectiveness of numerical inversion methods. In order to validate and improve the inversion solution, the numerical direct Laplace transform are used for this inversion, and compared it with the original Laplace transform. We implemented the composite Simpson's Rule for direct Laplace transform, and investigate the role that extended precision can play obtaining accurate transform inversions. The high precision approach seems to be an effective way to handle the challenging problems dealing with periodic functions. We demonstrated that the level of precision chosen must match algorithms properly. In the Gaver-Stehfest algorithm the balance is between the truncation error and roundoff error. High precision in an inaccurate algorithm yield little benefit, while a potentially highly accurate algorithm may be defeated by roundoff error if inadequate accuracy is used.

ACKNOWLEDGEMENTS

Matt Davison thanks to Natural Sciences and Engineering Research Council of Canada for research funding. We are also grateful to the anonymous referees for useful suggestions which improved the present work.

REFERENCES

1. Cengel, Y. and Palm, W. (2013) Differential Equations for Engineers and Scientists. McGraw-Hill, New York.
2. Edwards, C. and Penney, C. (2015) Differential Equations and Boundary Value Problems. 5th Edition, Pearson, London.
3. Polyanin, A.D. (2016) Handbook of Linear Partial Differential Equations for Engineers and Scientists. 2nd Edition, Chapman and Hall/CRC, UK.
4. Cohen, A.M. (2007) Numerical Methods for Laplace Transform Inversion. Springer, Berlin.
5. Kreyszig, E. (2011) Advanced Engineering Mathematics. 10th Edition, Wiley, Hoboken.
6. Abate, J., Chudhuru, G. and Whitt, W. (2000) An Introduction to Numerical Transform Inversion and Its Application to Probability Models. In: Grassmann, W., Ed., Computational Probability, Springer, Berlin, 257-323.
7. Badescu, A., Breuer, L., Da Silva Soares, A., Latouche, G., Remiche, M.A. and Stanford, D.A. (2005) Risk Processes Analyzed as Fluid Queues. Scandinavian Actuarial Journal, 2005, 127-141.
8. Kao, E. (1997) An Introduction to Stochastic Processes. Duxbury Press, Pacific Grove, CA.
9. Nadarajah, S. and Kotz, S. (2006) On the Laplace transform of the Pareto Distribution. Queueing Systems, 54, 243-244.
10. Davison, M. and Doeschl, A. (2004) A Hyperbolic PDE with Parabolic Behavior. SIAM Reviev, 46, 115-127.
11. Cruz-Baez, D.I. and Gonazalex-Rodriguez, J.M. (2008) A Different Approach for Pricing Asian Options. Applied Mathematics Letters, 21, 303-306.
12. Deakin, A. and Davison, M. (2010) An Analytic Solution for a Vasicek Interest Rate Convertible Bond Model. Journal of Applied Mathematics, 2010, Article ID: 263451.
13. Fusai, G. and Roncoroni, A. (2008) Implementing Models in Quantitative Finance: Methods and Cases. In: Avellaneda, M., Barone-Adesi, G., Broadie, M., Davis, M., Derman, E., Klüppelberg, C. and Schachermayer, W., Eds., Springer Finance, Springer-Verlag, Berlin, Heidelberg.
14. Kuhlman, K.L. (2013) Review of Inverse Laplace Transform Algorithms for Laplace-Space Numerical Approaches. Numerical Algorithms, 63, 339-355.
15. Stehfest, H. (1970) Algorithm 368: Numerical Inversion of Laplace Transform. Communications of the ACM, 13, 47-49.
16. Kuznetsov, A. (2013) On the Convergence of the Gaver—Stehfest Algorithm. SIAM Journal of Numerical Analysis, 51, 2984-2998.
17. Abate, A. and Valko, P. (2004) Multi-Precision Laplace Transform Inversion. International Journal for Numerical Methods in Engineering, 60, 979-993.
18. Duffy, D.G. (2015) Green's Functions with Applications. 2nd Edition, CRC Press, Boca Raton, FL.

19. Davies, B. and Martin, B. (1979) Numerical Inversion of the Laplace Transform, a Survey and Comparison of Methods. Journal of Computational Physics, 33, 1-32.
20. Murli, A. and Rizzardi, M. (1990) Algorithm 682: Talbot's Method for the Laplace Inversion Problem. ACM Transactions on Mathematical Software, 16, 158-168.
21. Bailey, D.H., Hida, Y., Li, X.S and Thompson, B. (2002) ARPREC: An Arbitrary Precision Computation Package. Tech. Rep. LBNL-53651, Lawrence Berkley National Lab.
22. Krougly, Z.L. and Jeffrey, D.J. (2009) Implementation and Application of Extended Precision in Matlab. Proceedings of the 11th WSEAS International Conference on Mathematical Methods and Computational Techniques in Electrical Engineering, Athens, 28-30 September 2009, 103-108.
23. Krougly, Z.L., Jeffrey, D.J. and Tsarapkina, D. (2013) Software Implementation of Numerical Algorithms in Arbitrary Precision. 15th International Symposium on Symbolic and Numeric Algorithms for Scientific Computing (SYNASC), Timisoara, 23-26 September 2013, 131-137.
24. Abramowitz, M. and Stegun, I.A. (1964) Handbook of Mathematical Functions. National Bureau of Standards.
25. Atkinson, K.E. (1989) An Introduction to Numerical Analysis. 2nd Edition, John Wiley & Sons, Hoboken.

CHAPTER 13

Approximate solutions for steady boundary layer MHD viscous flow and radiative heat transfer over an exponentially porous stretching sheet

Remus-Daniel Enea [a], Vasile Marinca [b,c]

[a]*University Politehnica Timişoara, Department of Mathematics, Timişoara 300006, Romania*

[b] *University Politehnica Timişoara, Department of Mechanics and Vibration, Timişoara 300222, Romania*

[c] *Department of Electromechanics and Vibration, Center for Advanced and Fundamental Technical Research, Romania Academy, Timişoara 300223, Romania*

ABSTRACT

In the present paper, the approximate solutions for steady boundary layer of the MHD viscous flow and radiative heat transfer over an exponentially porous stretching sheet are given. The nonlinear partial differential equations are reduced to an ordinary differential equations by the similarity transformations, taking into account velocity slip, thermal slip and the boundary conditions. These equations are solved approximately by means of the Optimal Homotopy Asymptotic Method (OHAM). This approach is highly efficient and it controls the convergence of the approximate solutions. OHAM is very efficient in practice, ensuring a very rapid convergence of the solutions after only one iteration. It does not need small or large parameters in the governing equations. Approximate solutions obtained through OHAM are compared with the results obtained by shooting method. It is found a very good agreement between these solutions.

Keywords: Optimal homotopy asymptotic method, Exponential stretching, Boundary layer flow, Velocity slip, Thermal slip, Stretching sheet

1. INTRODUCTION

It is well-known that in general the study of nonlinear phenomena is of great importance in many engineering processes, but it is very difficult to solve these nonlinear problems. The incompressible flow of viscous fluid and heat transfer of a viscous fluid over an exponentially stretching sheet have several engineering applications within, for instance polymer processing, the cooling process of metallic plate, drawing of plastic films and wires, glass fiber, manufacture of foods, crystal growing, etc. A great deal of works has been carried out on various aspects of fluid flow and heat transfer over a stretching sheet since the pioneering work of Sakiadis [1] and Crane [2]. Magyari and Keller [3] studied the steady boundary layers on an exponentially stretching continuous surface with an exponential temperature distribution. Wang [4] investigated the partial slip effects on the planar stretching flow. Raptis et al. [5] studied the effect of thermal radiation on the MHD flow of a viscous fluid past a semi-infinite stationary plate. The contributions of the viscous dissipation and elastic deformation in a visco-elastic boundary layer fluid flow over an exponentially stretching sheet using confluent hypergeometric solutions are investigated by Sanjayanand and Khan [6]. Sahoo [7] investigated the effects of partial slip in the MHD flow and mass transfer of an electrically conducting second grade fluid past an axisymmetric stretching sheet. Another important aspect is the investigation of multiple solutions for hydrodynamic flow of a second grade fluid over a stretching or shrinking sheet by Van Gorder and Vajravelu [8]. Fang et al. [9]considered the effects of second order slip on the flow of a shrinking sheet. The effect of radiation on the boundary layer flow and heat transfer of a viscous fluid over an exponentially stretching sheet is studied by Sajid and Hayat [10]. The exact analytical solutions for the flow and heat transfer over permeable stretching/shrinking surface taking into account a second order slip model is proposed by Turkilmazoglu [11]. In [12] Mukhopadhyay studied the boundary layer flow and heat transfer towards a porous exponential stretching sheet in presence of magnetic field and taking into account the velocity slip and thermal slip by means of numerical solutions obtained by shooting method.

The objective of the present paper is to propose an accurate procedure to nonlinear differential equations for the steady boundary layer MHD viscous flow and radiative heat transfer over an exponentially porous stretching sheet, using OHAM. A version of the OHAM is applied in this study to derive highly accurate analytical expressions of the solutions. Our procedure does not depend upon any small or large parameters, contradistinguishing from other known methods in literature. The main advantage of this approach is the control of the convergence of approximate solutions in a very rigorous way. For the first time, we present an effective analytical approximate solution with a new concepts as: linear operator, auxiliary functions $H_i(C_i) \neq C_i$ and especially in the selection of the first approximations f_1 and θ_1 in an original forms. A very good agreement was found between our approximate solutions and numerical solutions obtained by means the shooting method in combination with the fourth-order Runge Kutta explicit method, which proves that our method is very efficient and accurate.

2. EQUATIONS OF MOTION

In what follows, we consider the two-dimensional laminar flow of an incompressible viscous electrically conducting fluid past a stretching sheet coinciding with the plane y=0, the flow being confined to $y > 0$ [10–12]. The x-axis is taken along the stretching surface in the direction of the motion while y-axis is perpendicular to it. Two equal and opposite forces are applied along the x-axis so that the wall is stretched keeping the origin fixed (Fig. 1). The flow is assumed to be generated by stretching of the elastic boundary sheet from a slit with a large force such that the velocity of the boundary sheet is an exponential order of the flow directional coordinate x. A variable magnetic field B(x)=B0ex2L is applied normal to the sheet, B_0 being a constant depending on the nature of material. If u and v are components of velocity respectively in the x and y directions, ν=μρ is the kinematic viscosity, ρ is the fluid constant density, μ is the coefficient of fluid viscosity, σ is the electrical conductivity, $c_p$is the specific heat at constant pressure, k is the thermal conductivity of the fluid and q_r is the radiative heat flux, then the continuity, momentum and energy equations governing are written in the form

$$\frac{\partial u}{\partial x} + \frac{\partial v}{\partial y} = 0 \tag{1}$$

$$u\frac{\partial u}{\partial x}+v\frac{\partial u}{\partial y}=\nu\frac{\partial^2 u}{\partial y^2}-\frac{\sigma B^2}{\rho}u \tag{2}$$

$$\rho c_p\left(u\frac{\partial T}{\partial x}+v\frac{\partial T}{\partial y}\right)=k\frac{\partial^2 T}{\partial y^2}-\frac{\partial q_r}{\partial y} \tag{3}$$

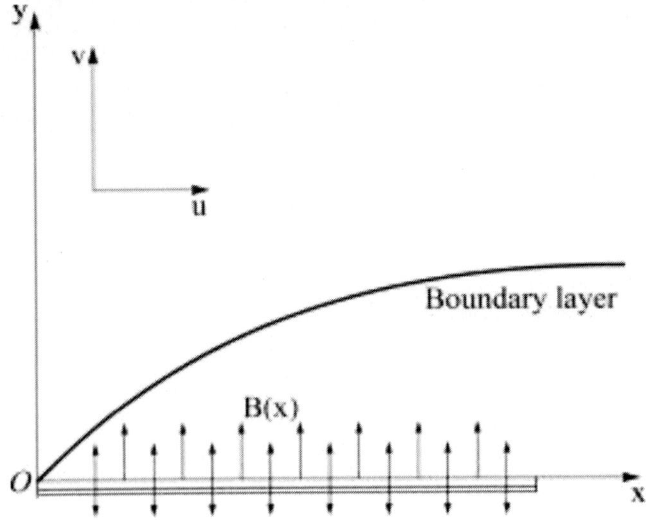

Figure. 1. Sketch of physical model.

The initial/boundary conditions for this problem are given as

$$\begin{aligned} u&=U_0e^{\frac{x}{L}}+N_0e^{-\frac{x}{2L}}\nu\frac{\partial u}{\partial y},\quad v=-V_0e^{-\frac{x}{2L}}, \\ T&=T_\infty+T_0e^{\frac{x}{2L}}+D_0e^{-\frac{x}{2L}}\frac{\partial T}{\partial y} \end{aligned} \quad \text{at } y=0$$

$$u\to 0,\ \ T\to T_\infty \quad \text{as } y\to\infty \tag{4}$$

$$q_r=-\frac{4\delta^*}{3k^*}\frac{\partial T^4}{\partial y} \tag{5}$$

in which U_0 is the reference velocity, N_0 is the initial value of velocity slip factor, V_0 is the initial strength of suction ($V_0 > 0$) or blowing ($V_0 < 0$), T_0 and T_∞ are respectively the temperatures at far and away from the plate and L is a constant, D_0 is the initial value of thermal slip factor.

By means of Rosseland approximation for radiation one has:

$$q_r = -\frac{4\delta^*}{3k^*}\frac{\partial T^4}{\partial y} \tag{6}$$

where δ^* is the Stefan Boltzman constant and k^* is the absorption coefficient. Assuming that T^4 is a linear function of temperature, then

$$T^4 = 4T_\infty{}^3T - 3T_\infty{}^4 \tag{7}$$

From Eqs. (7) and (3) it holds that

$$\rho c_p\left(u\frac{\partial T}{\partial x} + v\frac{\partial T}{\partial y}\right) = \left(k + \frac{16\delta^* T_\infty{}^3}{3k^*}\right)\frac{\partial^2 T}{\partial y^2} \tag{8}$$

$$\eta = \sqrt{\frac{U_0}{2\nu L}}e^{\frac{x}{2L}}y, \quad u = U_0 e^{\frac{x}{L}} f'(\eta)$$
$$v = -\sqrt{\frac{\nu U_0}{2L}}e^{\frac{x}{2L}}[f(\eta) + f'(\eta)], \quad \theta(\eta) = \frac{T - T_\infty}{T_0}e^{-\frac{x}{2L}} \tag{9}$$

Eq. (1) is automatically satisfied and Eqs. (2) and (8) reduce to

$$f'''(\eta) + f(\eta)f''(\eta) - 2[f'(\eta)]^2 - M^2 f'(\eta) = 0 \tag{10}$$

$$\left(1 + \frac{4}{3}R\right)\theta''(\eta) + Pr[f(\eta)\theta'(\eta) - f'(\eta)\theta(\eta)] = 0 \tag{11}$$

with the initial/boundary conditions

$$f(0) = S, \;\; f'(0) = 1 + \lambda f''(0), \quad f'(\infty) = 0 \tag{12}$$

$$\theta(0) = 1 + \delta\theta'(0), \quad \theta(\infty) = 0 \tag{13}$$

where the prime denotes differentiation with respect to η, $M = \sqrt{\frac{2\sigma B_0^2 L}{\rho U_0}}$ is the magnetic parameter, $\lambda = N_0\sqrt{\frac{\nu U_0}{2L}}$ is the velocity slip parameter, $\delta = D_0\sqrt{\frac{U_0}{2\nu L}}$ is the thermal slip parameter, $S = V_0\sqrt{\frac{2L}{U_0}}$ is suction ($V_0 > 0$) or blowing ($V_0 < 0$) parameter, $R = \frac{4\delta^* T_\infty^3}{kk^*}$ is the radiation parameter, Pr=μcpk is the Prandl number.

In the following, the nonlinear differential Eqs. (10) and (11) with initial/boundary conditions (12) and (13) can be solved analytically using OHAM.

3. BASIC IDEAS OF THE OPTIMAL HOMOTOPY ASYMPTOTIC METHOD

Eqs. (10) or (11) with initial/boundary conditions (12) and (13) can be written in a more general form

$$N[f(\eta)] = 0 \tag{14}$$

where N is a given nonlinear differential operator depending on the unknown function $f(\eta)$, subjected to the initial and boundary conditions

$$B\left(f(\eta), \frac{df(\eta)}{d\eta}\right) = 0. \tag{15}$$

Let $f_0(\eta)$ be an initial approximation of $f(\eta)$ and L an arbitrary linear operator such as

$$L[f_0(\eta)] = 0, \quad B\left(f_0(\eta), \frac{df_0(\eta)}{d\eta}\right) = 0. \tag{16}$$

It should be emphasized that this linear operator L is not unique.

If $p \in [0,1]$ denotes an embedding parameter and F is an analytic function, then we propose to construct a homotopy [13–16]:

$$\mathcal{H}[L(F(\eta,p)),\ H(\eta,C_i),\ N(F(\eta,p))],\quad i=1,2,\ldots,s \tag{17}$$

with the properties

$$\mathcal{H}[L(F(\eta,0)),\ H(\eta,C_i),\ N(F(\eta,0))] = L(F(\eta,0)) = L(f_0(\eta)) = 0 \tag{18}$$

$$\mathcal{H}[L(F(\eta,1)),\ H(\eta,C_i),\ N(F(\eta,1))] = H(\eta,C_i)N(F(\eta),1) = 0,\quad i=1,2,\ldots,s \tag{19}$$

where $H(\eta, C_i) \neq 0$ is an arbitrary auxiliary function and C_i, i=1,2,...,s are unknown parameters at this moment.

Let we consider the function F in the form

$$F(\eta,p) = f_0(\eta) + pf_1(\eta,C_i). \tag{20}$$

By substituting Eq. (20) into equation obtained by means of homotopy (17)

$$\mathcal{H}[L(F(\eta,p)),\ H(\eta,C_i),\ N(F(\eta,p))] = 0,\ \ i=1,\ldots,s \tag{21}$$

and then equating the coefficients of p^0 and p^1, we obtain:

$$\mathcal{H}[L(F(\eta,p)),\ H(\eta,C_i),\ N(F(\eta,p))] = L(f_0(\eta)) + p[L(f_1(\eta,C_i)) - L(f_0(\eta)) - H(\eta,C_i)N(f_0(\eta))] = 0,\quad i=1,2,\ldots,s. \tag{22}$$

From Eq. (22) we obtain the governing equation of $f_0(\eta)$ given by Eq. (16) and the governing equation of $f_1(\eta)$, i.e.

$$L(f_1(\eta,C_i)) = H(\eta,C_i)N(f_0(\eta)),\quad B\left(f_1(\eta,C_i),\ \frac{df_1(\eta,C_i)}{d\eta}\right) = 0,\quad i=1,\ldots,s \tag{23}$$

in which we find the following expression for the nonlinear operator:

$$N(f_0(\eta)) = \sum_{i=1}^{m} h_i(\eta) g_i(\eta) \tag{24}$$

where the functions $h_i(\eta)$ and $g_i(\eta)$, i=1,...,m are known and depend on the function $f_0(\eta)$ and also on the nonlinear operator, m being a known integer number.

In this way, taking into account Eq. (20) for p=1, we obtain the first-order approximate solution which becomes

$$\overline{f}(\eta, C_i) = f_0(\eta) + f_1(\eta, C_i), \quad i = 1, \ldots, s \tag{25}$$

It should be emphasized that $f_0(\eta)$ and $f_1(\eta, C_i)$ are governed by the linear Eqs. (16)and (23) respectively with initial/boundary conditions that come from the original problem. It is known that the general solution of nonhomogeneous linear Eq. (23) is equal to the sum of general solution of the corresponding homogeneous equation and of some particular solutions of the nonhomogeneous equation. However, the particular solutions are readily selected only in the exceptional cases.

In what follows we do not solve Eq. (23), but from the theory of differential equations, taking into considerations the method of variation of parameters, Cauchy method, method of influence function, the operator method [28] and so on, is more convenient to consider the unknown function $f_1(\eta)$, in the form

$$f_1(\eta, C_j) = \sum_{i=1}^{n} H_i(\eta, h_j(\eta), C_j) g_i(\eta), \quad j = 1, \ldots, s$$
$$B\left(f_1(\eta, C_i), \frac{df_1(\eta, C_i)}{d\eta} \right) = 0 \tag{26}$$

where within expression of $H_i(\eta, h_j(\eta), C_j)$ appear linear combinations of some functions h_j, some terms which are given by the corresponding homogeneous equation and the unknown parameters $C_j, j = 1, \ldots, s$. In the sum $\sum_{i=1}^{n} H_i g_i$ Higiappear an arbitrary number of n the such terms. We have a large freedom to choose the value of n. We cannot demand $f_1(\eta, C_i)$ to be solutions of Eq. (23) but f¯(η,Ci)given by Eq. (25) with $f_1(\eta, C_i)$ given by Eq. (26), are the solutions of Eq. (14). This is underlying idea of our method. The convergence of the approximate solution f¯(η,Ci) given by Eq. (25) depends upon the auxiliary functions $H_i(\eta, h_i, C_j)$, j=1,...,s. There are many possibilities to choose these functions H_i. We try to choose H_i so that within Eq. (26) the term $\sum_{i=1}^{n} H_i(\eta, h_j(\eta), C_j)g_i(\eta)$ be of the same shape with the term $\sum_{i=1}^{m} h_i(\eta)g_i(\eta)$ given by Eq. (24).

The first-order approximate solution f˘(η,Ci) also depend on the parameters C_j, j=1,...,s. The values of these parameters can be optimally identified via various methods, such as: the least-square method, the Galerkin method the collocation method, the Ritz method, and so on. The first option should be minimizing the square residual error:

$$J(C_1, C_2, \ldots, C_s) = \int_{(D)} R^2(\eta, C_1, C_2, \ldots, C_s)\, d\eta \tag{27}$$

where the residual R is given by

$$R(\eta, C_1, C_2, \ldots, C_s) = N(\overline{f}(\eta, C_i)). \tag{28}$$

The unknown parameters C1,C2,...,Cs can be identified from the conditions:

$$\frac{\partial J}{\partial C_1} = \frac{\partial J}{\partial C_2} = \ldots = \frac{\partial J}{\partial C_s} = 0. \tag{29}$$

With these parameters known (called optimal convergence-control parameters), the first-order approximate solution given by Eq. (25) is well-determined.

It should be emphasized that our procedure contains the auxiliary functions $H_i(\eta, f_i, C_j)$, i=1,...,n,j=1,...,s which provides us with a simple way to adjust and control the convergence of the approximate solutions. It is very important to properly choose these functions $H_i(\eta, f_i, C_j)$ which appear in the construction in the first-order approximation.

Other interesting alternatives to homotopy method are given in [17–27].

4. APPLICATION OF THE OHAM TO THE STEADY BOUNDARY LAYER MHD VISCOUS FLOW AND RADIATIVE HEAT TRANSFER

In what follows we apply our procedure to obtain approximate solution of Eqs. (10), (11), (12) and (13). For this purpose, we choose the linear operator for Eqs. (10) and (12) of the form:

$$L[f(\eta)] = f'''(\eta) - K^2 f'(\eta) \tag{30}$$

where K is an unknown positive parameter and will be determined later.

We mention that the linear operator is not unique. Also, we have freedom to choose

$$L[f(\eta)] = f'''(\eta) + K f'(\eta). \tag{31}$$

$$L[f(\eta)] = f'''(\eta) + \frac{3K}{K\eta + 1} f''(\eta). \tag{32}$$

$$L[f(\eta)] = f'''(\eta) - \frac{6K^2}{(K\eta + 1)^2} f'(\eta). \tag{33}$$

The initial approximation $f_0(\eta)$ can be obtained from Eq. (16) with initial/boundary conditions:

$$f_0(0) = S, \quad f_0'(0) = 1 + \lambda f_0''(0), \quad f_0'(\infty) = 0. \tag{34}$$

Eq. (16) with the linear operator (30) has the solution

$$f_0(\eta) = S + \frac{1 - e^{-K\eta}}{K(K\eta + 1)} \tag{35}$$

The nonlinear operator corresponding to nonlinear differential Eq. (10) and linear operator given by Eq. (30) is defined by

$$N[f(\eta)] = (K^2 - M^2)f'(\eta) + f(\eta)f''(\eta) - 2[f'(\eta)]^2 \tag{36}$$

By substituting Eq. (35) into Eq. (36) it holds that

$$N[f_0(\eta)] = \frac{(K^2 - M^2 - KS)(K\lambda + 1) - 1}{(K\lambda + 1)^2} e^{-K\eta} - \frac{1}{(K\lambda + 1)^2} e^{-2K\eta} \tag{37}$$

Comparing Eqs. (24) and (37), one can get

$$\begin{aligned} h_1(\eta) &= \frac{(K^2 - M^2 - KS)(K\lambda + 1) - 1}{(K\lambda + 1)^2}, \quad g_1(\eta) = e^{-K\eta}, \\ h_2(\eta) &= -\frac{1}{(K\lambda + 1)^2}, \quad g_2(\eta) = e^{-2K\eta}, \quad . \end{aligned} \tag{38}$$

The first approximation $f_1(\eta)$ given by Eq. (26) becomes

$$\begin{aligned} & f_1(\eta, C_j) = H_1(\eta, C_j)e^{-K\eta} + H_2(\eta, C_j)e^{-2K\eta} \\ & f_1(0, C_j) = 0, \quad f_1'(0, C_j) = \lambda f_1''(0, C_j), \quad f_1'(\infty, C_j) = 0. \end{aligned} \tag{39}$$

where we have freedom to choose a lot of possibilities for the unknown functions H_1 and H_2. If we choose

$$H_1(\eta, C_j) = C_1\eta + \frac{2\lambda K + 1}{2\lambda} C_1\eta^2 + C_2\eta^3 + C_3\eta^4 \tag{40}$$

$$H_2(\eta, C_j) = C_4\eta + \frac{4\lambda K + 1}{2\lambda} C_4\eta^2 + C_5\eta^3 \tag{41}$$

then the first approximation becomes

$$f_1(\eta, C_j) = \left(C_1\eta + \frac{2\lambda K + 1}{2\lambda} C_1\eta^2 + C_2\eta^3 + C_3\eta^4\right)e^{-K\eta} + \left(C_4\eta + \frac{4\lambda K + 1}{2\lambda} C_4\eta^2 + C_5\eta^3\right)e^{-2K\eta} \tag{42}$$

The first-order approximate solution of Eqs. (10) and (12) given by Eq. (25) is obtained from Eqs. (35) and (42):

$$\bar{f}(\eta, C_j) = S + \frac{1}{K(K\lambda+1)} + \left(-\frac{1}{K(K\lambda+1)} + C_1\eta + \frac{2\lambda K+1}{2\lambda}C_1\eta^2 + C_2\eta^3 + C_3\eta^4\right)e^{-K\eta} + \left(C_4\eta + \frac{4\lambda K+1}{2\lambda}C_4\eta^2 + C_5\eta^3\right)e^{-2K\eta} \tag{43}$$

where C_j, j=1,2,3,4 and $K > 0$ are unknown parameters.

Now, for Eqs. (11) and (13), the linear operator can be written as

$$L[\theta(\eta)] = \theta''(\eta) + K^*\theta'(\eta) \tag{44}$$

where K^* is an unknown positive parameter.

The initial approximation $\theta_0(\eta)$ can be obtained from equations

$$\theta_0''(\eta) + K^*\theta_0'(\eta) = 0, \quad \theta_0(0) = 1 + \delta\theta_0'(0), \quad \theta_0(\infty) = 0 \tag{45}$$

which has the solution

$$\theta_0(\eta) = \frac{1}{1+\delta K^*}e^{-K^*\eta} \tag{46}$$

The nonlinear operator for Eq. (11) with linear operator given by Eq. (44) becomes

$$N[\theta(\eta)] = -K^*\left(1 + \frac{4}{3}R\right)\theta'(\eta) + Pr\left[f(\eta)\theta'(\eta) - f'(\eta)\theta(\eta)\right] \tag{47}$$

By substituting Eqs. (46) and (45) into Eq. (47) it holds that

$$N[\theta_0(\eta)] = \left[\frac{K^{*2} - 3K^*S + 4RK^{*2}}{3(1+\delta K^*)} - \frac{K^*}{K(K\lambda+1)(1+K^*\delta)}\right]e^{-K^*\eta} + \frac{K^* - K}{K(K\lambda+1)(1+K^*\delta)}e^{-(K+K^*)\eta} \tag{48}$$

Comparing Eqs. (24) and (48) one can get

$$h^*{}_1(\eta) = \frac{K^{*2} - 3K^*S + 4RK^{*2}}{3(1+\delta K^*)} - \frac{K^*}{K(K\lambda+1)(1+K^*\delta)},$$
$$g^*{}_1(\eta) = e^{-K^*\eta},$$
$$h^*{}_2(\eta) = \frac{K^* - K}{K(K\lambda+1)(1+K^*\delta)}, \quad g^*{}_2(\eta) = e^{-(K+K^*)\eta} \tag{49}$$

The first approximation $\theta_1(\eta, D_j)$ given by Eq. (26) becomes

$$\theta_1(\eta, D_j) = H_3(\eta, D_j)e^{-K^*\eta} + H_4(\eta, D_j)e^{-(K+K^*)\eta},$$
$$\theta_1(0) = \delta\theta_1'(0), \quad \theta_1(\infty) = 0 \tag{50}$$

We choose the auxiliary functions H_3 and H_4 in the form

$$H_3(\eta, D_j) = D_1\eta + D_2\eta^2 + D_3\eta^3 + D_4\eta^4 \tag{51}$$

$$H_4(\eta, D_j) = -D_1\eta + D_5\eta^2 + D_6\eta^3 \tag{52}$$

such that the first-order approximate solution of Eqs. (11) and (13) is obtained from Eqs. (25), (46), (50) and (51):

$$\overline{\theta}(\eta, D_j) = \left(\frac{1}{1+\delta K^*} + D_1\eta + D_2\eta^2 + D_3\eta^3 + D_4\eta^4\right)e^{-K^*\eta} + (-D_1\eta + D_5\eta^2 + D_6\eta^3)e^{-(K+K^*)\eta} \tag{53}$$

where D_j, j=1,2,...,6 and $K^* > 0$ are unknown parameters.

In this way, we can obtain many other solutions.

5. NUMERICAL RESULTS

We illustrate the accuracy of our procedure for different values of the coefficients *S, M, λ, R, Pr* and *δ*. We represent graphically the behavior of the functions $\bar{f}$ and $\bar{\theta}$,and we compare results obtained through our procedure with numerical results. Also, in some cases we represent graphically the behavior of the residual $R\bar{f}$ and $R\bar{\theta}$.

CASE 5.1. First, we consider S=0.1,M=0,λ=0.1,R=0.1,Pr=0.7,δ=0.1

The optimal convergence-control parameters C_j and D_j in all cases are determined by means of the least-square method and are:

$$C_1 = -0.0088492634, \quad C_2 = 0.0056210808, \quad C_3 = -0.0004350582,$$
$$C_4 = -0.0029961570, \quad C_5 = -0.0038069657, \quad K = 1.1076523227,$$

The first-order approximate solution obtained by means of OHAM becomes

$$\begin{aligned} &D_1 = 0.0514240617, \quad D_2 = 0.0140292628, \quad D_3 = -0.0004662431, \\ &D_4 = 0.0000315621, \quad D_5 = -0.0119793890, \quad D_6 = -0.0044331549, \\ &K^* = 0.6816198449 \end{aligned} \tag{53}$$

$$\begin{aligned} \overline{f}(\eta) = {} & 0.9127823420 + (-0.8127823420 - 0.0088492634\eta \\ & - 0.0540482243\eta^2 + 0.0056210808\eta^3 - 0.0004350582\eta^4)e^{-1.1076523227\eta} \\ & + (-0.0029961570\eta - 0.0216181859\eta^2 - 0.0038069657\eta^3)e^{-2.2153046455\eta} \end{aligned} \tag{54}$$

$$\begin{aligned} \overline{\theta}(\eta) = {} & (0.9361875956 + 0.0514240617\eta + 0.0140292628\eta^2 - 0.0004662431\eta^3 + 0.0000315621\eta^4)e^{-0.6816198449\eta} \\ & + (-0.0514240617\eta - 0.0119793890\eta^2 - 0.0044331549\eta^3)e^{-1.7892721677\eta} \end{aligned} \tag{55}$$

CASE 5.2. ForS=0.5,M=0.1,λ=0.1,R=0.1,Pr=0.7,δ=0.1, the optimal convergence-control parameters are

$$\begin{aligned} &C_1 = -0.0071875319, \quad C_2 = 0.0065323642, \quad C_3 = -0.0005043266, \\ &C_4 = -0.0025346928, \quad C_5 = -0.0054136041, \quad K = 1.3355505019, \\ &D_1 = 0.0457401659, \quad D_2 = 0.0119544326, \quad D_3 = -0.0006295169, \\ &D_4 = 0.0000318274, \quad D_5 = -0.0141812138, \quad D_6 = -0.0052909996, \\ &K^* = 0.8142449950 \end{aligned}$$

The first-order approximate solution obtained with OHAM can be written as

$$\begin{aligned} \overline{f}(\eta) = {} & 1.1605368772 + (-0.6605368772 - 0.0071875319\eta \\ & - 0.0455369718\eta^2 + 0.0065323642\eta^3 - 0.0005043266\eta^4)e^{-1.3355505019\eta} \\ & + (-0.0025346928\eta - 0.0194438849\eta^2 - 0.0054136041\eta^3)e^{-2.6711010038\eta} \end{aligned} \tag{56}$$

$$\begin{aligned} \overline{\theta}(\eta) = {} & (0.9247062559 + 0.0457401659\eta + 0.0119544326\eta^2 - 0.0006295169\eta^3 + 0.0000318274\eta^4)e^{-0.8142449950\eta} \\ & + (-0.0457401659\eta - 0.0141812138\eta^2 - 0.0052909996\eta^3)e^{-2.1497954970\eta} \end{aligned} \tag{57}$$

CASE 5.3. In the caseS=0.3,M=0.1,λ=0.1,R=0.1,Pr=0.7,δ=0.1, we give only the first-order approximate solution:

$$\begin{aligned}\overline{f}(\eta) = {} & 1.0306379756 + (-0.7306379756 - 0.0079946059\eta \\ & -0.0497253361\eta^2 + 0.0062090225\eta^3 - 0.0004743630\eta^4)e^{-1.2198608360\eta} \\ & +(-0.0027287522\eta - 0.0203011572\eta^2 - 0.0044690471\eta^3)e^{-2.4397216720\eta}\end{aligned} \tag{58}$$

$$\begin{aligned}\overline{\theta}(\eta) = {} & (0.9307222412 + 0.0490143654\eta + 0.0130216563\eta^2 - 0.0005497484\eta^3 + 0.0000309591\eta^4)e^{-0.7443440767\eta} \\ & +(-0.0490143654\eta - 0.0134044621\eta^2 - 0.0049285032\eta^3)e^{-1.9642049127\eta}\end{aligned} \tag{59}$$

CASE 5.4. ForS=0.1,M=0.1,λ=0.1,R=0.1,Pr=0.7,δ=0.1, the first-order approximate solution can be written in the form:

$$\begin{aligned}\overline{f}(\eta) = {} & 0.9086599850 + (-0.8086599850 - 0.0088108438\eta \\ & -0.0538587912\eta^2 + 0.0057414582\eta^3 - 0.0004443943\eta^4)e^{-1.1127847009\eta} \\ & +(-0.0029367177\eta - 0.0212194580\eta^2 - 0.0035688107\eta^3)e^{-2.2255694019\eta}\end{aligned} \tag{60}$$

$$\begin{aligned}\overline{\theta}(\eta) = {} & (0.9362951489 + 0.0517012567\eta + 0.0143527176\eta^2 - 0.0004817767\eta^3 + 0.0000329395\eta^4)e^{-0.6803928341\eta} \\ & +(-0.0517012567\eta - 0.0122130857\eta^2 - 0.0044706180\eta^3)e^{-1.7931775351\eta}\end{aligned} \tag{61}$$

CASE 5.5. ForS=0,M=0.1,λ=0.1,R=0.1,Pr=0.7,δ=0.1, the first-order approximate solution is:

$$\begin{aligned}\overline{f}(\eta) = {} & 0.8507603570 + (-0.8507603570 - 0.0092222223\eta \\ & -0.0559099398\eta^2 + 0.0054863768\eta^3 - 0.0004348803\eta^4)e^{-1.0625235105\eta} + \\ & +(-0.0030376081\eta - 0.0216431007\eta^2 - 0.0031008859\eta^3)e^{-2.1250470210\eta}\end{aligned} \tag{62}$$

$$\begin{aligned}\overline{\theta}(\eta) = {} & (0.9389047069 + 0.0525544702\eta + 0.0151864803\eta^2 - 0.0004622337\eta^3 + 0.0000356287\eta^4)e^{-0.6507081350\eta} \\ & +(-0.0525544702\eta - 0.0113572926\eta^2 - 0.0041778981\eta^3)e^{-1.7132316455\eta}\end{aligned} \tag{63}$$

CASE 5.6. IfS=−0.3,M=0.1,λ=0.1,R=0.1,Pr=0.7,δ=0.1, we obtain:

$$\begin{aligned}\overline{f}(\eta) = {} & 0.6897905868 + (-0.9897905868 - 0.0104598318\eta \\ & -0.0619723165\eta^2 + 0.0047564051\eta^3 - 0.0004390001\eta^4)e^{-0.9247908996\eta} \\ & +(-0.0032915263\eta - 0.0225455791\eta^2 - 0.0015636266\eta^3)e^{-1.8495817992\eta}\end{aligned} \tag{64}$$

$$\overline{\theta}(\eta) = (0.9459958284 + 0.050101643760\eta + 0.0192015824\eta^2 \\ - 0.000571852564\eta^3 + 0.0000557533\eta^4)e^{-0.5708711389\eta} \\ + (-0.0501016437\eta - 0.0065704464\eta^2 - 0.0027775363\eta^3)e^{-1.4956620385\eta} \tag{65}$$

CASE 5.7. In the caseS=−0.5,M=0.1,λ=0.1,R=0.1,Pr=0.7,δ=0.1, the first-order approximate solution of Eqs. (10), (11), (12) and (13):

$$\overline{f}(\eta) = 0.5932102075 + (-1.0932102075 - 0.0111087816\eta \\ - 0.0649150014\eta^2 + 0.0040974729\eta^3 - 0.0004500438\eta^4)e^{-0.8435752407\eta} \\ + (-0.0034922697\eta - 0.0233533332\eta^2 - 0.0016696977\eta^3)e^{-1.6871504815\eta} \tag{66}$$

$$\overline{\theta}(\eta) = (0.9501040463 + 0.0376050133\eta + 0.0249758984\eta^2 - 0.0009924090\eta^3 + 0.0000871197\eta^4)e^{-0.5251630471\eta} \\ + (-0.0376050133\eta + 0.0010097896\eta^2 - 0.0007813015\eta^3)e^{-1.3687382879\eta} \tag{67}$$

CASE 5.8. ForS=0.1,M=0.6,λ=0.1,R=0.1,Pr=0.7,δ=0.1, we have:

$$\overline{f}(\eta) = 0.7965964439 + (-0.6965964439 - 0.0070302206\eta \\ - 0.0441033674\eta^2 + 0.0066673174\eta^3 - 0.0005509236\eta^4)e^{-1.2733973248\eta} \\ + (-0.00226190097\eta - 0.0170701022\eta^2 - 0.0036403974\eta^3)e^{-2.5467946497\eta} \tag{68}$$

$$\overline{\theta}(\eta) = (0.9397127073 + 0.0617627386\eta + 0.0219888320\eta^2 - 0.0008183675\eta^3 + 0.0000731113\eta^4)e^{-0.6415502542\eta} \\ + (-0.0617627386\eta - 0.0206043665\eta^2 - 0.0067231497\eta^3)e^{-1.9149475790\eta} \tag{69}$$

CASE 5.9. In the last case, we considerS= 0.1,M=0.3, λ=0.1, R=0.1,Pr=0.7,δ=0.1, and therefore, the first-order approximate solution is written as:

$$\overline{f}(\eta) = 0.8780236617 + (-0.7780236617 - 0.0082359519\eta \\ - 0.0506715762\eta^2 + 0.0058647919\eta^3 - 0.0004474304\eta^4)e^{-1.1524856382\eta} \\ + (-0.0028031861\eta - 0.0204771942\eta^2 - 0.0040416806\eta^3)e^{-2.3049712764\eta} \tag{70}$$

$$\overline{\theta}(\eta) = (0.9371335972 + 0.0542997895\eta + 0.0165468521\eta^2 - 0.0005734010\eta^3 + 0.0000427240\eta^4)e^{-0.6708371460\eta} \\ + (-0.0542997895\eta - 0.0143185202\eta^2 - 0.0049761140\eta^3)e^{-1.8233227842\eta} \tag{71}$$

In Tables 1–5 we present a comparison between the derivative of the first-order approximate solutions (54)–(63) respectively with numerical results for some values of variable η and the relative error

$$\varepsilon_{f'}(\eta) = |\overline{f}'(\eta) - f'_{\text{numerical}}(\eta)|,\ \varepsilon_{\theta'}(\eta) = |\overline{\theta}'(\eta) - \theta'_{\text{numerical}}(\eta)|.$$

Table 1. Comparison between the derivative of the first-order approximate solution ($\bar{f}'$,$\bar{\theta}'$) obtained from Eqs. (54) and (55) respectively obtained by OHAM with numerical results.

η	$\bar{f}'(\eta)$ Eq. (54)	$f'_{\text{numerical}}(\eta)$	$\varepsilon_{f'}(\eta)$	$\bar{\theta}'(\eta)$ Eq. (55)	$\theta'_{\text{numerical}}(\eta)$	$\varepsilon_{\theta'}(\eta)$
0	0.88843482	0.88843481	9.99×10^{-9}	−0.63812404	−0.63812414	9.99×10^{-8}
1	0.28545290	0.28541390	3.90×10^{-5}	−0.30015583	−0.30015412	1.71×10^{-6}
2	0.10483255	0.10485346	2.09×10^{-5}	−0.15902631	−0.15902833	2.01×10^{-6}
3	0.04062500	0.04062055	4.45×10^{-6}	−0.08801940	−0.08801839	1.01×10^{-6}
4	0.01608714	0.01607463	1.25×10^{-5}	−0.04953981	−0.04953894	8.73×10^{-7}
5	0.00641367	0.00641565	1.97×10^{-6}	−0.02806573	−0.02806663	9.06×10^{-7}
6	0.00256161	0.00256937	7.75×10^{-6}	−0.01594297	−0.01594346	4.96×10^{-7}
7	0.00102562	0.00103041	4.78×10^{-6}	−0.00906679	−0.00906639	3.96×10^{-7}
8	0.00041282	0.00041346	6.33×10^{-7}	−0.00515840	−0.00515785	5.43×10^{-7}
9	0.00016750	0.00016593	1.57×10^{-6}	−0.00293496	−0.00293478	1.80×10^{-7}
10	0.00006860	0.00006660	1.99×10^{-6}	−0.00166979	−0.00166998	1.84×10^{-7}

Table 2. Comparison between the derivative of the first-order approximate solution ($\bar{f}'$,$\bar{\theta}'$) obtained from Eqs. (56) and (57) respectively obtained by OHAM with numerical results.

η	$\bar{f}'(\eta)$ Eq. (56)	$f'_{\text{numerical}}(\eta)$	$\varepsilon_{f'}(\eta)$	$\bar{\theta}'(\eta)$ Eq. (57)	$\theta'_{\text{numerical}}(\eta)$	$\varepsilon_{\theta'}(\eta)$
0	0.87245813	0.87245812	1.00×10^{-8}	−0.75293744	−0.75293745	9.99×10^{-9}
1	0.22828030	0.22825759	2.27×10^{-5}	−0.31820241	−0.31820124	1.16×10^{-6}
2	0.06722555	0.06723918	1.36×10^{-5}	−0.14889548	−0.14889559	1.08×10^{-7}
3	0.02055908	0.02054601	1.31×10^{-5}	−0.07176848	−0.07176748	9.99×10^{-7}
4	0.00635117	0.00634994	1.22×10^{-6}	−0.03490253	−0.03490395	1.41×10^{-6}
5	0.00196270	0.00196944	6.73×10^{-6}	−0.01702253	−0.01702275	2.21×10^{-7}
6	0.00060749	0.00061147	3.97×10^{-6}	−0.00831015	−0.00830922	9.24×10^{-7}
7	0.00018965	0.00018989	2.39×10^{-7}	−0.00405759	−0.00405701	5.79×10^{-7}
8	0.00006020	0.00005895	1.24×10^{-6}	−0.00198084	−0.00198101	1.66×10^{-7}
9	0.00001953	0.00001828	1.25×10^{-6}	−0.00096681	−0.00096734	5.27×10^{-7}
10	6.48×10^{-6}	5.65×10^{-6}	8.34×10^{-7}	−0.00047188	−0.00047236	4.75×10^{-7}

Table 3. Comparison between the derivative of the first-order approximate solution ($\bar{f}'$,$\bar{\theta}'$) obtained from Eqs. (58) and (59) respectively obtained by OHAM with numerical results.

η	$\bar{f}'(\eta)$ Eq. (58)	$f'_{\text{numerical}}(\eta)$	$\varepsilon_{f'}(\eta)$	$\bar{\theta}'(\eta)$ Eq. (59)	$\theta'_{\text{numerical}}(\eta)$	$\varepsilon_{\theta'}(\eta)$
0	0.88055329	0.88055328	1.00×10^{-8}	−0.69277758	−0.69277768	1.00×10^{-7}
1	0.25600972	0.25598142	2.82×10^{-5}	−0.31007655	−0.31007448	2.07×10^{-6}
2	0.08444370	0.08446263	1.89×10^{-5}	−0.15514672	−0.15514822	1.50×10^{-6}
3	0.02914513	0.02913259	1.25×10^{-5}	−0.08052316	−0.08052139	1.77×10^{-6}
4	0.01021339	0.01020682	6.57×10^{-6}	−0.04231594	−0.04231659	6.52×10^{-7}
5	0.00358957	0.00359586	6.28×10^{-6}	−0.02233555	−0.02233653	9.81×10^{-7}
6	0.00126294	0.00126930	6.35×10^{-6}	−0.01180890	−0.01180848	4.12×10^{-7}
7	0.00044654	0.00044836	1.82×10^{-6}	−0.00624694	−0.00624610	8.43×10^{-7}
8	0.00015953	0.00015841	1.11×10^{-6}	−0.00330482	−0.00330450	3.18×10^{-7}
9	0.00005783	0.00005597	1.85×10^{-6}	−0.00174809	−0.00174836	2.73×10^{-7}
10	0.00002130	0.00001978	1.52×10^{-6}	−0.00092454	−0.00092505	5.07×10^{-7}

Table 4. Comparison between the derivative of the first-order approximate solution $(\bar{f}', \bar{\theta}')$ obtained from Eqs. (60) and (61) respectively obtained by OHAM with numerical results.

η	$\bar{f}'(\eta)$ Eq. (60)	$f'_{\text{numerical}}(\eta)$	$\varepsilon_{f'}(\eta)$	$\bar{\theta}'(\eta)$ Eq. (61)	$\theta'_{\text{numerical}}(\eta)$	$\varepsilon_{\theta'}(\eta)$
0	0.88843482	0.88843481	9.99×10^{-9}	−0.63812404	−0.63812414	9.99×10^{-8}
1	0.28545290	0.28541390	3.90×10^{-5}	−0.30015583	−0.30015412	1.71×10^{-6}
2	0.10483255	0.10485346	2.09×10^{-5}	−0.15902631	−0.15902833	2.01×10^{-6}
3	0.04062500	0.04062055	4.45×10^{-6}	−0.08801940	−0.08801839	1.01×10^{-6}
4	0.01608714	0.01607463	1.25×10^{-5}	−0.04953981	−0.04953894	8.73×10^{-7}
5	0.00641367	0.00641565	1.97×10^{-6}	−0.02806573	−0.02806663	9.06×10^{-7}
6	0.00256161	0.00256937	7.75×10^{-6}	−0.01594297	−0.01594346	4.96×10^{-7}
7	0.00102562	0.00103041	4.78×10^{-6}	−0.01594297	−0.01594346	4.96×10^{-7}
8	0.00041282	0.00041346	6.33×10^{-7}	−0.00515840	−0.00515785	5.43×10^{-7}
9	0.00016750	0.00016593	1.57×10^{-6}	−0.00293496	−0.00293478	1.80×10^{-7}
10	0.00006860	0.00006660	1.99×10^{-6}	−0.00166979	−0.00166998	1.84×10^{-7}

Table 5. Comparison between the derivative of the first-order approximate solution $(\bar{f}', \bar{\theta}')$ obtained from Eqs. (62) and (63) respectively obtained by OHAM with numerical results.

η	$\bar{f}'(\eta)$ Eq. (62)	$f'_{\text{numerical}}(\eta)$	$\varepsilon_{f'}(\eta)$	$\bar{\theta}'(\eta)$ Eq. (63)	$\theta'_{\text{numerical}}(\eta)$	$\varepsilon_{\theta'}(\eta)$
0	0.89169305	0.89169304	1.00×10^{-8}	−0.61095293	−0.61095303	1.00×10^{-7}
1	0.29811553	0.29807936	3.61×10^{-5}	−0.29375661	−0.29375586	7.54×10^{-7}
2	0.11413100	0.11415378	2.27×10^{-5}	−0.16002100	−0.16002252	1.51×10^{-6}
3	0.04625779	0.04625340	4.38×10^{-6}	−0.09149886	−0.09149890	3.83×10^{-8}
4	0.01920647	0.01919124	1.52×10^{-5}	−0.05335199	−0.05335060	1.39×10^{-6}
5	0.00804207	0.00804286	7.86×10^{-7}	−0.03135997	−0.03136038	4.03×10^{-7}
6	0.00337607	0.00338496	8.89×10^{-6}	−0.01849636	−0.01849706	6.99×10^{-7}
7	0.00142071	0.00142715	6.44×10^{-6}	−0.01092565	−0.01092564	1.24×10^{-8}
8	0.00060059	0.00060216	1.56×10^{-6}	−0.00645769	−0.00645731	3.78×10^{-7}
9	0.00025561	0.00025415	1.46×10^{-6}	−0.00381762	−0.00381737	2.45×10^{-7}
10	0.00010964	0.00010728	2.35×10^{-6}	−0.00225693	−0.00225695	2.11×10^{-8}

Numerical solutions of the Eqs. (10)–(13) are obtained by shooting method in combination with the fourth-order Runge Kutta method, using Wolfram Mathematica 6.0 software. It is very important for this method the values of the skin-friction coefficient $|\bar{f}''(0)|$ and the heat transfer coefficient $\bar{\theta}'(0)$ (Nusselt number).

In Table 6 are presented comparisons of our results with some available results of Ishak [29] and Mukhopadhyay [12] for different values of Prandtl number *Pr* and radiation number *R* and for $\lambda=0, M=0, \delta=0$. The results are found to be very accurate.

Table 6. Values of $\bar{\theta}'(0)$ for several values of Prandtl number *Pr* and radiation parameter *R* for $\lambda=0, M=0, \delta=0$.

Pr	*R*	Ishak [29]	Mukhopadhyay [12]	present study
1	0	−0.9548	−0.9547	−0.9547013278
1	1	−0.5312	−0.5311	−0.5311135791
2	0.5		−1.0734	−1.0739984172
2	1		−0.8626	−0.8626001271

The residuals for Eqs. (10) and (11) are respectively

$$R_f(\eta) = \bar{f}'''(\eta) + \bar{f}(\eta)\bar{f}''(\eta) - 2[\bar{f}'(\eta)]^2 - M^2\bar{f}'(\eta) \tag{72}$$

$$R_\theta(\eta) = \left(1 + \frac{4}{3}R\right)\bar{\theta}''(\eta) + Pr[\bar{f}(\eta)\bar{\theta}'(\eta) - \bar{f}'(\eta)\bar{\theta}(\eta)] \tag{73}$$

It can be seen from the above tables and figures that the solutions obtained by the proposed procedure are nearly identical with the numerical solution obtained using the shooting method in combination with the fourth-order Runge Kutta method.

In order to analyze the results obtained by OHAM and the results obtained by numerical integration, in Figs. 2–6 are shown variation of velocity profiles f' and of temperature θ for different values of the parameters.

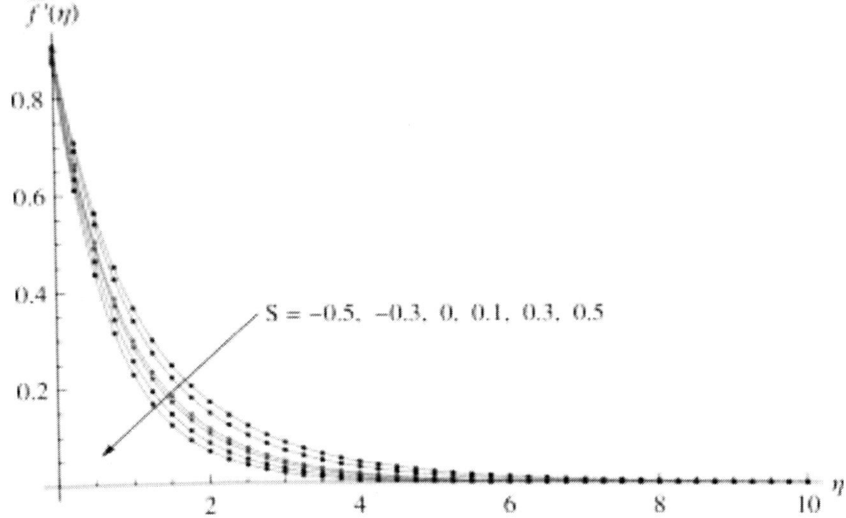

Figure. 2. Variation of horizontal velocity $f'(\eta)$ with η for M=0.1,λ=0.1 and for several values of suction/blowing parameter S: — fnumerical′; .... f⁻′.

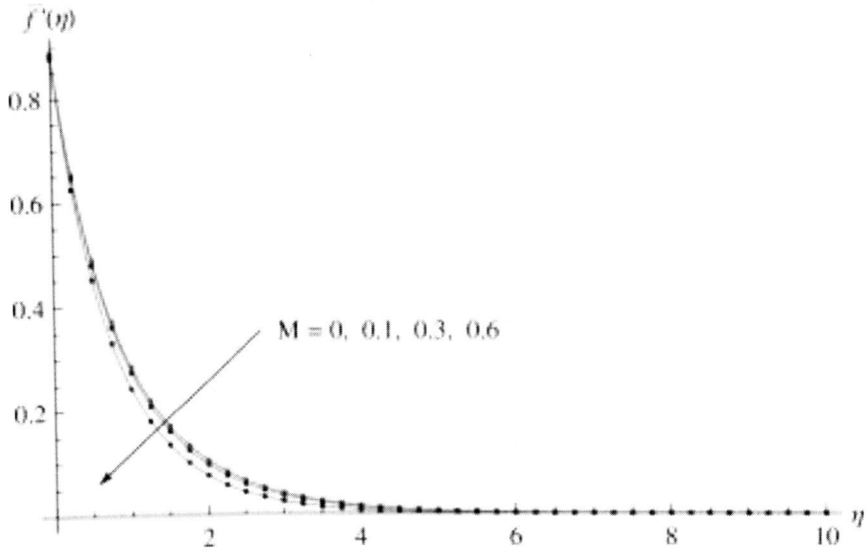

Figure. 3. Variation of horizontal velocity $f'(\eta)$ with η for S=0.1,λ=0.1 and for several values of magnetic parameter M: — fnumerical′; .... f⁻′.

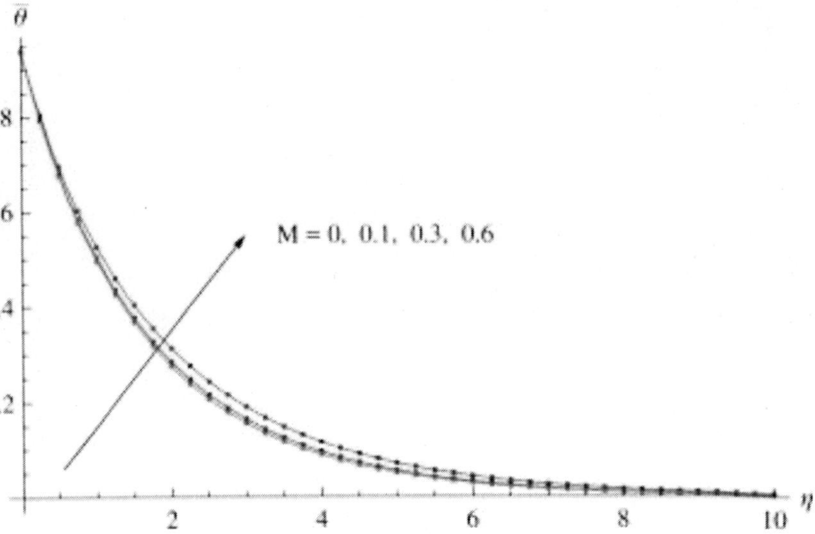

Figure. 4. Variation of temperature $\theta(\eta)$ with η for several values of magnetic parameter M and for R=0.1,Pr=0.7,δ=0.1,S=0.1,λ=0.1: — $\theta_{numerical}$; .... θ¯.

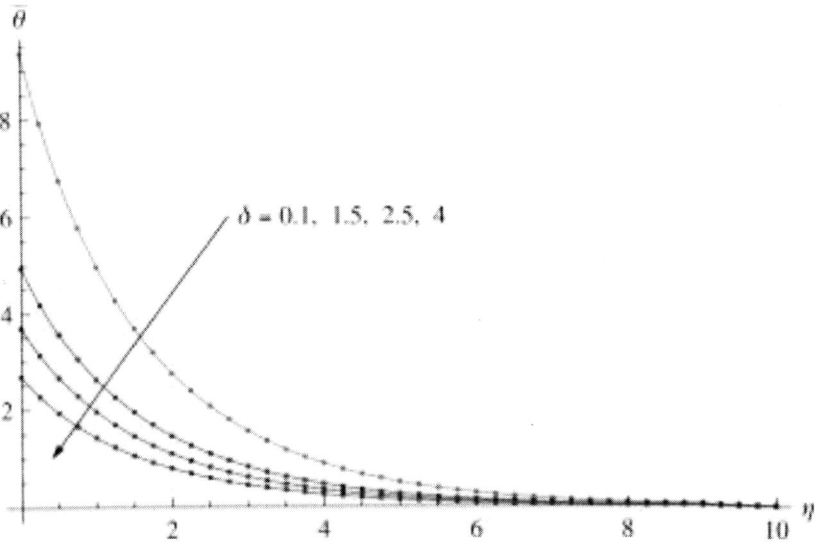

Figure. 5. Variation of temperature $\theta(\eta)$ with η for several values of thermal slip parameter δ and for R=0.1,Pr=0.7,S=0.1,M=0.1,λ=0.1: — $\theta_{numerical}$; .... θ¯.

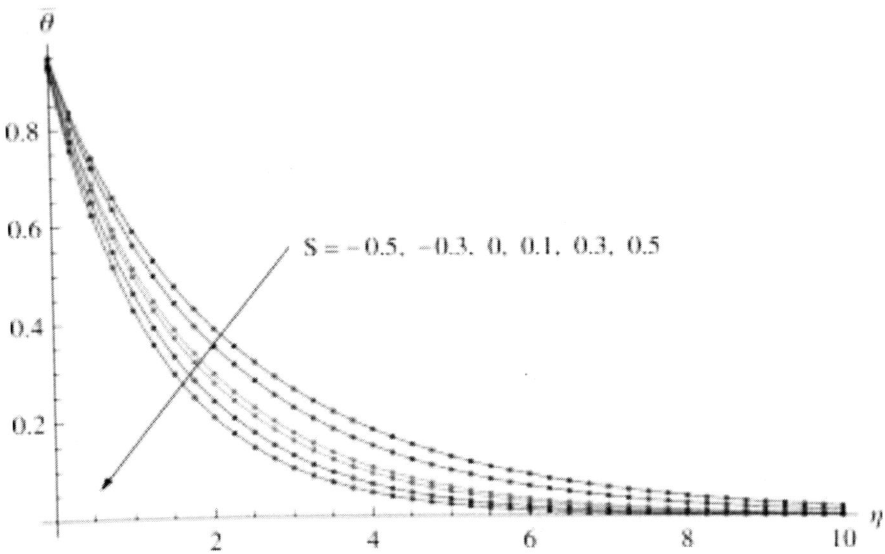

Figure. 6. Variation of temperature $\theta(\eta)$ with η for several values of suction/blowing parameter S and for R=0.1,Pr=0.7,δ=0.1,S=0.1,λ=0.1: — $\theta_{\text{numerical}}$; .... θ^{-}.

The velocity profiles for different values of the suction or blowing parameter *S* and different values of the magnetic parametric *M* presented in Figs. 2 and 3, respectively. It is clear that the rate of transport is considerably reduced with the increase of *S* or *M*. Also, the velocity profile decreases with the increase of *S* or *M*. In these cases the velocity vanishes at relative large distance from the sheet (at η=6).

Temperature profiles are presented in Figs. 4–6 for the variation of magnetic parameter *M*, thermal slip parameter δ and the suction/blowing parameters *S*, respectively. With the increasing *M*, the temperature is found to increase (Fig. 4).

From Eqs. (5) and (6) it is seen that the temperature decreases with increasing slip thermal parameter and with increasing suction/blowing parameter, respectively.

Fig. 7 depicts the nature of skin-friction coefficient $|f''(0)|$ for λ=0.1 and S=0.1. It is clear that skin-friction coefficient increases with increasing magnetic parameter *M*.

Fig. 8 exhibits the nature of skin-friction coefficient for λ=0.1 and M=0.1. It is found that skin-friction coefficient also increases with increasing suction parameter *S*.

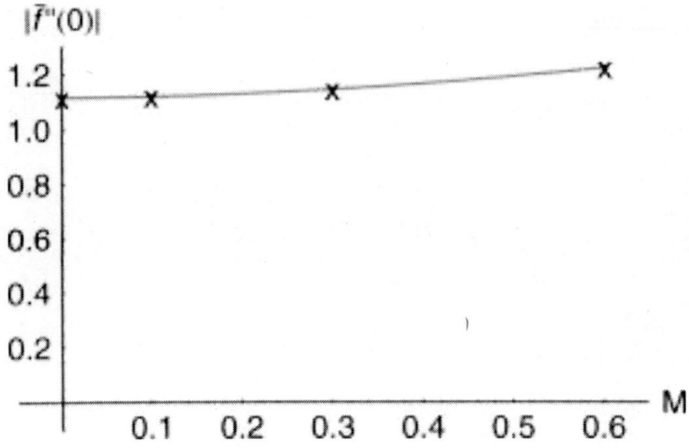

Figure. 7. Skin-friction coefficient $|\bar{f}''(0)|$ against magnetic parameter *M* for S=0.1.

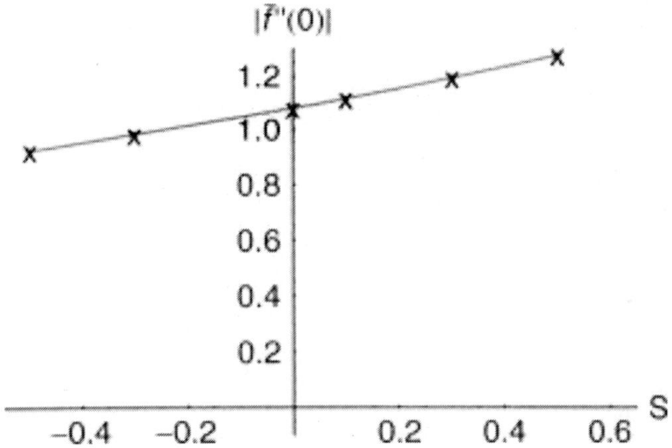

Figure. 8. Skin-friction coefficient $|\bar{f}''(0)|$ against suction (blowing) parameter *S* for M=0.1.

Fig. 9 present the effect of thermal slip parameter δ on Nusselt number for M=0.1,λ=0.1,R=0.1,Pr=0.7. It is observed that the Nusselt number decreases with increasing values of thermal slip parameters. Nusselt number decreases with suction (blowing) parameter *S* (Fig. 10).

Estimation of skin-friction and Nusselt number are very important from the industrial application point of view. Also the radiative effects have importance in physics, technology and engineering. Some applications in this field may occur in the polishing of artificial heart valves and internal cavity or in a variety of stretching sheet problems or yet in polymer fiber coating, film cooling, prevent corrosion and so on.

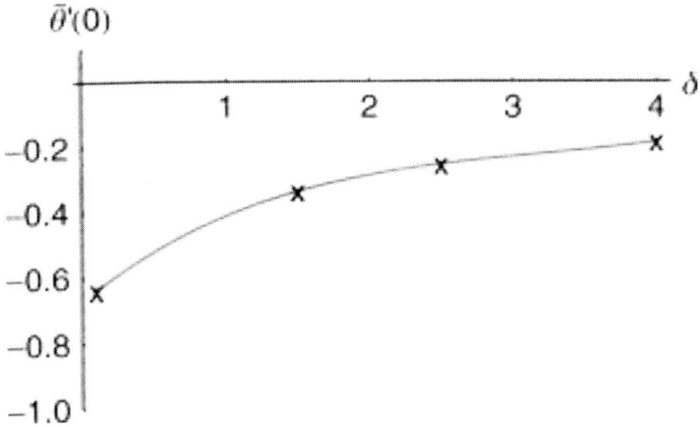

Figure. 9. Nusselt number $\theta^{-\prime}(0)$ against thermal slip parameter δ for S=0.1,M=0.1,λ=0.1,R=0.1.

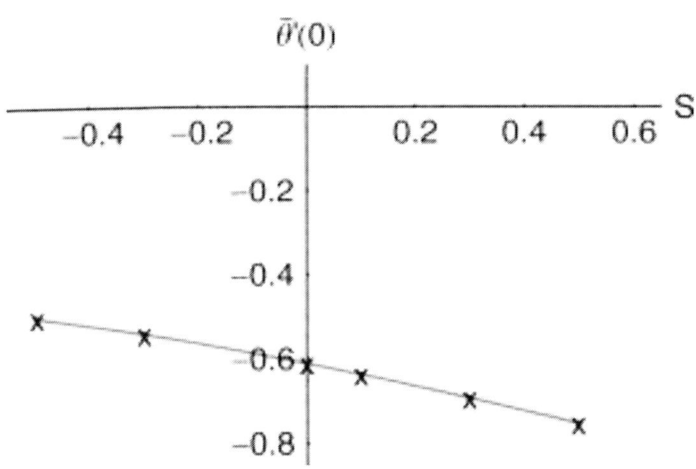

Figure. 10. Nusselt number $\theta^{-\prime}(0)$ against suction (blowing) parameter S for M=0.1,λ=0.1,R=0.1,δ=0.1.

6. CONCLUSIONS

The Optimal Homotopy Asymptotic Method is employed to propose new analytic approximate solutions for steady boundary layer MHD viscous flow and radiative heat transfer over an exponentially porous stretching sheet. Our procedure is valid even if the nonlinear differential equations does not contain any small or large parameters. In construction of the homotopy appear some distinctive concepts as: optimal auxiliary functions H1(η,Ci)−H4(η,Ci), the linear operators $L(f)$ and $L(\theta)$ and several optimal convergence-control parameters C_1, C_2,... which ensure a fast convergence of all the solutions. The examples presented in this work, lead to the conclusion that the obtained results are of the exceptional accuracy using only one iteration. The OHAM provides us with a simple and rigorous way to control and adjust the convergence of the solutions through the auxiliary functions $H_i(\eta, C_i)$ involving several parameters C_1, C_2,... which are optimally determined. Actually, the capital strength of OHAM is its fast convergence after only one iteration which proves that our procedure is very efficient in practice.

Based on the present investigations we conclude that the displacement increases when the suction/blowing parameter increases and the displacement decreases with the increasing of magnetic parameter. The velocity profiles decreases with the increase of suction/blowing parameter and magnetic parameters. The velocity vanishes at relative large distance from the sheet. Temperature profiles increases with the increasing of magnetic parameter, but temperature profiles decreases when slip thermal parameter or suction/blowing parameter increase.

REFERENCES

1. B.C. Sakiadis**BOUNDARY-LAYER BEHAVIOR ON CONTINUOUS SOLID SURFACE: I. BOUNDARY LAYER EQUATIONS FOR TWO-DIMENSIONAL AND AXISYMMETRIC FLOW** AlChE J., 7 (1961), pp. 26-28
2. L.J. Crane**FLOW PAST A STRETCHING PLATE** Z. Angew. Math. Phys., 21 (1970), pp. 645-647
3. E. Magyari, B. Keller**HEAT AND MASS TRANSFER IN THE BOUNDARY LAYERS ON AN EXPONENTIALLY STRETCHING CONTINUOUS SURFACE** J. Phys. D Appl. Phys., 32 (1999), pp. 577-585
4. C.Y. Wang**ANALYSIS OF VISCOUS FLOW DUE TO A STRETCHING SHEET WITH SURFACE SLIP AND SUCTION** Nonlinear Anal. Real World Appl., 10 (2009), pp. 375-380
5. Raptis, C. Perdikis, H.S. Takhar**EFFECTS OF THERMAL RADIATION ON MHD FLOW** Appl. Math. Comput., 153 (2004), pp. 645-649
6. E. Sanjayamand, S.K. Khan**ON HEAT AND MASS TRANSFER IN A VISCOELASTIC BOUNDARY LAYER FLOW OVER AN EXPONENTIALLY STRETCHING SHEET** Int. J. Therm. Sci., 45 (2006), pp. 819-828

7. B. Sahoo**EFFECT OF SLIP VISCOUS DISSIPATION AND JOULE HEATING ON MHD FLOW AND HEAT TRANSFER OF A SECOND GRADE FLUID PAST A RADIALLY STRETCHING SHEET** Appl. Math. Mech., Engl. Ed., 31 (2) (2010), pp. 159-173
8. R.A.V. Gorder, K. Vajravelu**MULTIPLE SOLUTIONS FOR HYDRODYNAMIC FLOW OF A SECOND GRADE FLUID OVER A STRETCHING OR SHRINKING SHEET** Quart. Appl. Math., 69 (2011), pp. 404-424
9. T.G. Fang, S. Yao, J. Zhang, A. Aziz**VISCOUS FLOW OVER A SHRINKING SHEET WITH A SECOND ORDER SLIP FLOW MODEL** Commun. Nonliner Sci. Numer. Simul., 15 (2010), pp. 1831-1842
10. M. Sajid, P. Hayat**INFLUENCE OF THERMAL RADIATION ON THE BOUNDARY LAYER FLOW DUE TO AN EXPONENTIALLY STRETCHING SHEET** Int. Commun. Heat and Mass Transf., 35 (2008), pp. 347-356
11. M. Turkyilmazoglu**HEAT AND MASS TRANSFER ON MHD SECOND ORDER SLIP FLOW** Comput. Fluids, 71 (2013), pp. 426-434
12. S. Mukhopadhyay**SLIP EFFECTS ON MHD BOUNDARY LAYER FLOW OVER AN EXPONENTIALLY STRETCHING SHEET WITH SUCTION/BLOWING AND THERMAL RADIATION** Ain Shams Eng. J., 4 (2013), pp. 485-491
13. V. Marinca, N.H. sanu, C. Bota, B. Marinca**AN OPTIMAL HOMOTOPY ASYMPTOTIC METHOD APPLIED TO THE STEADY FLOW OF A FOURTH GRADE FLUID PAST A POROUS PLATE** Appl. Math. Lett., 22 (2009), pp. 245-251
14. V. Marinca, N.H. sanu Nonlinear Dynamical Systems in Engineering – Some Approximate Approaches, Springer Verlag, Heidelberg (2011)
15. V. Marinca, R.-D. Ene**DUAL APPROXIMATE SOLUTIONS OF THE UNSTEADY VISCOUS FLOW OVER A SHRINKING CYLINDER WITH OPTIMAL HOMOTOPY ASYMPTOTIC METHOD** Adv. Math. Phys., 2014 (2014), p. 11 pages Special issue: Nonlinear Fluid Flow and Heat Transf. Article ID 417643.
16. V. Marinca, R.-D. Ene, B. Marinca**ANALYTIC APPROXIMATE SOLUTION FOR FALKNER–SKAN EQUATION** The Sci. World J., 2014 (2014), p. 22 pages Special issue: Dyn. Nonlinear Syst. Article ID 617453.
17. M. Turkyilmazoglu**AN EFFECTIVE APPROACH FOR APPROXIMATE ANALYTICAL SOLUTIONS OF THE DAMPED DUFFING EQUATION** Physica Scripta, 86 (1) (2012),
18. M. Turkyilmazoglu**THE AIRY EQUATION AND ITS ALTERNATIVE ANALYTIC SOLUTION** Physica Scripta, 86 (5) (2012),
19. D. Kumar, J. Singh, Sushila**APPLICATION OF HOMOTOPY ANALYSIS TRANSFORM METHOD TO FRACTIONAL BIOLOGICAL POPULATION MODEL** Rom. Rep. Phys., 65 (1) (2013), pp. 63-75
20. Ramswroop, J. Singh, D. Kumar**NUMERICAL STUDY FOR TIME-FRACTIONAL SCHRÖDINGER EQUATIONS ARISING IN QUANTUM MECHANICS** Nonlinear Eng. – Model. Appl., 3 (3) (2014), pp. 169-177
21. D. Kumar, J. Singh, S. Kumar, Sushila**NUMERICAL COMPUTATION OF KLEIN–GORDON EQUATIONS ARISING IN QUANTUM FIELD THEORY BY USING HOMOTOPY ANALYSIS TRANSFORM METHOD** Alexandria Eng. J., 53 (2014), pp. 469-474
22. D. Kumar, J. Singh, S. Kumar**ANALYTICAL STUDY FOR SINGULAR SYSTEM OF TRANSISTOR CIRCUITS** Alexandria Eng. J., 53 (2014), pp. 445-448
23. F. Mabood, W.A. Khan, A.I.M. Ismail**ANALYTICAL SOLUTION FOR RADIATION EFFECTS ON HEAT TRANSFER IN BLASIUS FLOW** Int. J. Mod. Eng. Sci., 2 (2) (2013), pp. 63-72
24. F. Mabood, W.A. Khan, A.I.M. Ismail**OPTIMAL HOMOTOPY ASYMPTOTIC METHOD FOR HEAT TRANSFER IN HOLLOW SPHERE WITH ROBIN BOUNDARY CONDITIONS**Heat Transf. – Asian Res., 43 (2) (2014), pp. 124-133
25. F. Mabood, W.A. Khan, A.I.M. Ismail**OPTIMAL HOMOTOPY ASYMPTOTIC METHOD FOR FLOW AND HEAT TRANSFER OF A VISCOELASTIC FLUID IN AN AXISYMMETRIC CHANNEL WITH A POROUS WALL** PLoS ONE, 8 (12) (2013),
26. F. Mabood, W.A. Khan**COMBINED ANALYTICAL-NUMERICAL SOLUTION FOR MHD VISCOUS FLOW OVER A STRETCHING SHEET** J. Comput. Eng. (2014), p. 7 pages

27. F. Mabood, W.A. Khan, M.J. Uddin, A.I.M. Ismail**OPTIMAL HOMOTOPY ASYMPTOTIC METHOD FOR MHD SLIPS FLOW OVER A RADIATING STRETCHING SHEET WITH HEAT TRANSFER** Far East J. Appl. Math., 90 (1) (2015), pp. 21-40
28. L. Elsgolts Differential Equations and the Calculus of Variations, Mir Publishers, Moscow (1980)
29. Ishak**MHD BOUNDARY LAYER FLOW DUE TO AN EXPONENTIALLY STRETCHING SHEET WITH RADIATION EFFECT** Sains Malays., 40 (2011), pp. 391-395

CHAPTER 14

On the Evolution of Consanguinities

Paul G. Bamberg

Department of Mathematics, Harvard University, Cambridge, Massachusetts, USA

ABSTRACT

"Consanguinity" is a gender-neutral term for "fraternity" or "sorority." Initially a consanguinity includes M male members and F female members. Each week a member, chosen at random, selects a new member, always of the same gender as the member making the selection. This model for evolution is isomorphic to the classic Pólya's urn. The male and female members play the same roles as the red and black balls in the urn, and the procedure for selecting a new member is equivalent to drawing a ball from the urn, then replacing it and adding a new ball of the same color. It is well known that for Pólya's urn, the proportion of red balls in the urn is a martingale. It follows that for a consanguinity, the proportion of the membership that is male is a martingale. Furthermore, being bounded, this martingale converges to a limit. For a martingale that is the sum of independent random variables, such as a symmetric random walk, there is also a well-known second-degree martingale from which the variance of the limiting distribution can be deduced. What the author discovered, in the process of solving his own examination problem, is that a similar martingale exists also for Pólya's urn, even though in this case the number of red balls is the sum of random variables that are not independent. This new martingale can be used to calculate the variance of the limiting distribution. Traditionally, the probability that r red balls will be drawn from Pólya's urn in n trials is derived by a rather tricky argument involving conditional probability. This article uses an obvious but overlooked simpler approach. Pólya's formula for the probability that m male members will be chosen in n weeks is derived, without any mention of conditional probability, by an elementary counting argument, and its limit is shown to be a beta distribution.

Keywords: Pólya's Urn, Martingale, Martingale Convergence, Beta Distribution

1. GENESIS OF THE PROBLEM

There is a long tradition of setting unsolved problems on mathematics examina- tions. The most famous example is Stokes's Theorem, a three-dimensional version of which appeared on the 1854 Smith's Prize Exam at Cambridge Uni- versity [1] , several years before a proof was published by Hankel.

The following problem appeared on the final examination in the author's course Mathematics 117, "Probability and Random Processes with Economic Applications," in December 2016.

"The initial membership of a newly founded MGSO (mixed gender social organization) is three women and one man. At the start of week n (starting with week 0), there are $n+4$ members, of whom M_n are male. A member is chosen at random, and he or she recruits a new member of the same gender as himself/herself.

Show that the fraction of the membership that is male, the random variable

$$\mu_n = \frac{M_n}{n+4},$$

is a martingale. State what property of this martingale guarantees that it converges almost surely, and describe the probability distribution for μ_∞."

An equivalent problem was stated and solved by Eggenberger and Pólya in 1923 [2] . Instead of a consanguinity with male and female members, they considered an urn which initially, for the special case in the exam problem, contained $R=1$ red balls and $S=3$ (S for "Schwarz") black balls. A ball was repeatedly chosen at random from the urn and then replaced, along with another ball of the same color.

Since the students had done examples based on Pólya's urn and had studied the martingale convergence theorem, everything was straightforward except for "describe the probability distribution for μ_∞," which the author of the exam question had not taken the trouble to work out!

This variant of Pólya's urn is of political interest at Harvard because of sanctions against single-gender social organizations that have been proposed for the 2017-18 academic year [3] . An existing single-gender organization can presumably avoid the sanctions by adding a single member of the opposite gender, and the proposed model for expanding the membership is not unreasonable.

The term "consanguinity" for an MGSO whose name consists of Greek capital letters was suggested to the author by Prof. Richard Thomas of the Harvard Classics Department. Instead of saying "X is my fraternity brother" or "Y is my sorority sister," one can say "Z is my consanguinity sibling."

None of the nine students who took the exam correctly guessed the describe the probability distribution for μ_∞. Perhaps James Clerk Maxwell, who tied for first on the 1854 Smith's Prize Exam, would have fared better.

Pólya's urn has developed since 1923 into an entire branch of mathematics. It is the subject of a recent book [4] . An excellent online resource that contains most of the key results is [5] .

The rest of the paper is organized as follows. The well-known proof that the proportion of the membership μ_n that is male is a convergent martingale is reviewed in Section 2 and is supplemented by a second martingale that involves M_n^2. In Section 3, the probability distribution for M_n is derived by an elemen- tary counting argument, and it is shown that for large n, the probability of adding m male members is proportional to $(1-m/n)^2$, consistent with the fact that a beta distribution with parameters 1 and 3 has the density function $3(1-x)^2$. In Section 4, it is shown that the expected number of weeks before a second male member is added is infinite. In Section 5, the general case of starting with M male members and F female members is analyzed.

2. MARTINGALE ANALYSIS

We first specify the sample space Ω and probability measure for an experi- ment in which a consanguinity evolves for n weeks. An element ω of this sample space is a sequence $(\omega_1, \omega_2, \cdots, \omega_n)$ where ω_1 specifies which of the four original members selects the member added in week 1, ω_2 specifies which of the five members present after week 1 selects the member added in week 2, and so forth. We specify not merely whether the new member added in week i is male or female, but also which existing member made the selection. Our probability model is then that each ω is equally likely.

Let the random variable M_n denote the number of males in the consanguinity after n weeks. M_n is a function on the sample space, whose value for each element ω can be determined by identifying the gender of each member who makes a selection. After n weeks the total membership is $n+4$. The ratio

$$\mu_n = \frac{M_n}{n+4}$$

is also a random variable, the proportion of the membership that is male after n members have been added.

We next review briefly the fundamentals of conditional expectation and martingales, a simplification of sections 8.2, 9.1 and 9.2 of the textbook used in Math117. [6]

Let Ω be a sample space, which in the case at hand is finite. Let A be an event (a subset of Ω), which in our case might be "after four weeks there are three male members." Then the conditional expectation $\mathbb{E}[X|A]$ is computed by restricting the sample space to the subset A and calculating expectation in the usual way for this restricted sample space.

The random variable M_n generates a finite partition of the sample space into events of the form $M_n = k$, i.e. "after n weeks, the consanguinity includes k male members." Since these are the only events of interest to us, we need make no mention of sigma-fields and can use the simplified notation

$$\mathbb{E}[X|M_n](\omega).$$

Since the number of males M_n and the proportional of males μ_n both generate the same partition of the sample space, "conditioning on M_n" and "conditioning on μ_n" are equivalent.

This conditional expectation defines a new random variable as follows: given an element ω of the sample space, we determine the value $M_n(\omega)$, then compute the expectation of X using only those elements ω that lead to that value.

We cite the three laws of conditional expectation, a simplified version of Proposition 8.8 from [6]

1) Taking out what is known (TOWIK):

   If X is constant on each level set of μ_n, then

$$\mathbb{E}\left[X \cdot Y \middle| \mu_n\right] = X\mathbb{E}\left[Y \middle| \mu_n\right] \tag{1}$$

2) Independence drops out (IDO):

   If X is independent of μ_n, then

$$\mathbb{E}\left[X \middle| \mu_n\right] = \mathbb{E}\left[X\right] \tag{2}$$

3) Tower Law:

   If $m < n$, then

$$\mathbb{E}\left[\mathbb{E}\left[X \middle| \mu_n\right] \middle| \mu_m\right] = \mathbb{E}\left[X \middle| \mu_m\right] \tag{3}$$

The term “martingale” was in use by roulette players long before it found its way into mathematics. Imagine that you are playing roulette in a fair casino where there is no zero on the wheel. If after n plays you have k chips and bet one chip on red, you are equally likely to end up with $k+1$ chips or $k-1$ chips, for an expectation of k chips. If random variable Y_n is the size of your stack of chips after n plays, then the sequence Y_n has the martingale property $\mathbb{E}\left[Y_{n+1} \middle| Y_n\right] = Y_n$.

For a consanguinity, the number of male members M_n, which can never decrease, is not a martingale. However, even someone who is unfamiliar with the term “martingale” is likely to guess correctly that the average number of male members is going to remain one-fourth, for the simple reason that the probability of choosing a new male member is equal to the probability that the new member is selected by a male, which in turn is equal to the proportion of the membership that is male.

More formally, we need to prove that the sequence of random variables μ_n has the martingale property: the expectation of μ_{n+1}, conditioned on μ_n, is equal to μ_n, or

$$\mathbb{E}\left[\mu_{n+1} \middle| \mu_n\right] = \mu_n.$$

First we consider the case $n=0$. The random variable μ_0 has the constant value $\frac{1}{4}$. For μ_1 there are two possible values. With probability $\frac{1}{4}$, a new male member is selected, and the proportion of males rises to $\frac{2}{5}$. With probability $\frac{3}{4}$, a new female member is selected, and the proportion of males drops to $\frac{1}{5}$. The expectation of μ_1 is therefore

$$\left(\frac{1}{4}\right)\left(\frac{2}{5}\right)+\left(\frac{3}{4}\right)\left(\frac{1}{5}\right)=\frac{2+3}{20}=\frac{1}{4}.$$

The general proof is scarcely more difficult. Introduce a new random variable X_n, the number of male members chosen in week n. This random variable has the value 1 with probability μ_n, the value 0 with probability $1-\mu_n$.

Clearly $M_{n+1} = M_n + X_n$, and since expectation is linear,

$$\mathbb{E}\left[M_{n+1} \middle| \mu_n\right] = \mathbb{E}\left[M_n \middle| \mu_n\right] + \mathbb{E}\left[X_n \middle| \mu_n\right]$$

For the first term, conditioning M_n on a function of itself changes nothing (TOWIK) (1). For the second term, the expectation of X_n is simply the probability that it equals 1 rather than 0, namely μ_n, so

$$\mathbb{E}\left[M_{n+1} \middle| \mu_n\right] = M_n + \mu_n = (n+4)\mu_n + \mu_n$$

Dividing by $n+5$ we find

$$\mathbb{E}\left[\mu_{n+1} \middle| \mu_n\right] = \mu_n,$$

which is the martingale property.

By the Tower Law(3), we can condition μ_{n+1} first on μ_n, then on μ_{n-1}, and eventually on μ_0 and conclude that

$$E\left[\mu_{n+1} \middle| \mu_0\right] = \mu_0 = \frac{1}{4}, \tag{4}$$

reaching a conclusion that requires half a page of computation in [2] , page 281.

This analysis appears in almost every introduction to martingales; for example, as exercise 27 on page 258 of [7] . What appears not to be well known is that there is also a second-degree martingale that makes the computation of variance equally easy.

The inspiration for this second martingale is a symmetric random walk, as described, for example, in Example 9.6 of [6] .

Let X_i denote the ith step in the walk: it is a random variable that has values +1 and -1, each with probability 1/2.

Let

$$Y_n = \sum_{i=1}^{n} X_i.$$

Then

$$\mathbb{E}\left[Y_{n+1} \middle| Y_n\right] = \mathbb{E}\left[\left(Y_n + X_{n+1}\right) \middle| Y_n\right] = \mathbb{E}\left[Y_n \middle| Y_n\right] + \mathbb{E}\left[X_{n+1} \middle| Y_n\right].$$

By TOWIK (1),

$$\mathbb{E}\left[Y_n \middle| Y_n\right] = Y_n.$$

By IDO (2),

$$\mathbb{E}\left[X_{n+1} \middle| Y_n\right] = \mathbb{E}\left[X_{n+1}\right] = 0,$$

and we conclude that

$$\mathbb{E}\left[Y_{n+1} \middle| Y_n\right] = Y_n.$$

For a second-degree martingale, define a new random variable, $Z_n = Y_n^2 - n$

Then

$$Z_{n+1}^2 = \left(Y_n + X_{n+1}\right)^2 - (n+1) = Y_n^2 + 2Y_n X_{n+1} + X_{n+1}^2 - n - 1.$$

Again condition on Y_n.

By TOWIK(1),

$$\mathbb{E}\left[Y_n^2 \middle| Y_n\right] = Y_n^2.$$

By IDO (2),

$$\mathbb{E}\left[X_{n+1} Y_n \middle| Y_n\right] = \mathbb{E}\left[X_{n+1} \middle| Y_n\right] = 0.$$

Finally, X_{n+1}^2 is the constant random variable 1.

So

$$\mathbb{E}\left[Z_{n+1} \middle| Y_n\right] = Y_n^2 + 1 - n - 1 = Z_n,$$

which is the martingale property.

In the case of the consanguinity we can write

$$M_n = M_0 + \sum_{i=1}^{n} X_i,$$

where X_i is the number of male members (0 or 1) added in week i. However, in this case X_{n+1} is no longer independent of M_n —indeed this lack of independence is precisely why Eggenberger and Pólya [2] found the urn problem interesting.

It took a bit of experimentation for the author to discover a second degree martingale even for the simple special case that appeared on the examination.

The martingale turns out to be

$$Z_n = \frac{1}{(n+4)(n+5)}\left(M_n^2 + M_n\right) \tag{5}$$

roof: consider

$$Z_{n+1} = \frac{1}{(n+5)(n+6)}\left(M_{n+1}^2 + M_{n+1}\right)$$

Since

$$M_{n+1} = M_n + X_{n+1},$$

$$Z_{n+1} = \frac{1}{(n+5)(n+6)}\left[M_n^2 + 2M_nX_{n+1} + X_{n+1}^2 + M_n + X_{n+1}\right].$$

This time it will be more difficult to deal with the second term, because X_{n+1} is not independent of M_n.

Again we condition everything on M_n. The key is that

$$\mathbb{E}\left[X_{n+1} \middle| M_n\right] = \mu_n = \frac{M_n}{n+4}.$$

By TOWIK (1),

$$\mathbb{E}\left[M_n^2 \middle| M_n\right] = M_n^2.$$

By TOWIK (1),

$$\mathbb{E}\left[M_n X_{n+1} \middle| M_n\right] = M_n \mathbb{E}\left[X_{n+1} \middle| M_n\right] = M_n \mu_n = \frac{M_n^2}{n+4}.$$

Since X_{n+1} equals 0 or 1, its square is equal to itself, and

$$\mathbb{E}\left[X_{n+1}^2 \middle| M_n\right] = \mathbb{E}\left[X_{n+1} \middle| M_n\right] = \mu_n = \frac{M_n}{n+4}.$$

Combining all the terms we find

$$\mathbb{E}\left[Z_{n+1} \middle| M_n\right] = \frac{1}{(n+5)(n+6)}\left(M_n^2\left(1+\frac{2}{n+4}\right) + M_n\left(1+\frac{2}{n+4}\right)\right)$$

or

$$\mathbb{E}\left[Z_{n+1} \middle| M_n\right] = \frac{1}{(n+5)(n+6)} \frac{n+6}{n+4}\left(M_n^2 + M_n\right) = Z_n,$$

which establishes the martingale property for the random variable Z_n (5). By the Tower Law (3), therefore,

$$\mathbb{E}\left[Z_n\right] = \mathbb{E}\left[Z_n \middle| M_0\right] = Z_0 = \frac{1}{20}(1+1) = \frac{1}{10}$$

It follows that

$$\frac{1}{(n+4)(n+5)}\left(\mathbb{E}\left[M_n^2\right] + \mathbb{E}\left[M_n\right]\right) = \frac{1}{10}$$

But we already know (4) that

$$\mathbb{E}\left[M_n\right] = \frac{n+4}{4}$$

and so

$$\mathbb{E}\left[M_n^2\right]=\frac{(n+4)(n+5)}{10}-\frac{n+4}{4}.$$

Thus the variance is

$$\mathrm{Var}M_n=\mathbb{E}\left[M_n^2\right]-\mathbb{E}\left[M_n\right]^2=\frac{3n(n+4)}{80}, \tag{6}$$

in agreement with the value that Eggenberger and Pólya calculate from the probability mass function [2] .

The Martingale Convergence Theorem, stated and proved, for example, in section 9.2 of [6] , asserts that if a sequence of random variables like μ_n is uniformly integrable, it converges almost surely to a random variable μ_∞. Since each μ_n is bounded below by 0 and above by 1, the requirement of uniform integrability is trivially satisfied.

Although we have still not determined the probability mass function, we now know enough to determine the expectation and variance of μ_∞. Since μ_n has an expectation of $\frac{1}{4}$ for all n, so does μ_∞. From the results that we have just obtained, combined with the fact that $\mu_n=\frac{M_n}{n+4}$, it follows that

$$\mathrm{Var}\mu_\infty=\lim_{n\to\infty}\frac{3n}{80(n+4)}=\frac{3}{80}$$

At this point we might conjecture an answer to the original exam question, "describe the probability distribution for μ_∞." Perhaps the limit is a beta distribution with parameters $\alpha=1$ and $\beta=3$, for which the density function is $f(x)=3(1-x)^2$. Any book on mathematical statistics, e.g. [8] has the formulas for the expectation and variance of the beta distribution. The expectation is

$$\frac{\alpha}{\alpha+\beta}=\frac{1}{4}$$

while the variance is

$$\frac{\alpha\beta}{(\alpha+\beta)^2(\alpha+\beta+1)}=\frac{3}{80}.$$

So the conjectured beta distribution has the right expectation and variance. Rather than attempt to compare higher moments, though, it will simpler just to work out the probability mass function and take its limit.

3. THE PROBABILITY MASS FUNCTION

The following derivation is equivalent to that of Eggenberger and Pólya [2] , but it is strictly an elementary counting argument, making no mention of the fact that the number of males M_n is the sum of a set of random variables which, although not independent, are nonetheless "exchangeable."

We have already identified a finite sample space whose elements all have equal probability. The number of elements in the sample space is the product of the number of equally likely choices that can be made in successive weeks: namely

$$4\times5\times6\cdots(3+n)=\frac{(n+3)!}{3!}.$$

Now we must count the number of elements of the sample space whose effect is to add m males and $f=n-m$ females to the membership.

For the first male added there is only one candidate for the member who did the choosing; for the second male added, there are two candidates, and so forth. Given a fixed set of m weeks in which males are chosen, there are in all $(m-1)!$ alternatives. Similarly, for the weeks in which females are chosen, there are $3\times4\times5\cdots(2+f)$ alternatives for the female choosers.

There are $\binom{n}{m}$ ways to select the set of m weeks in which a male is chosen. So the number of elements of the sample space that contribute to the event "m males are added" is

$$\binom{n}{m}(m-1)!\frac{(f+2)!}{2!}$$

The probability of adding m males is simply the proportion of elements in the sample space that contribute to this event: namely

$$\frac{\frac{n!}{m!(n-m)!}(m-1)!\frac{(n-m+2)!}{2!}}{\frac{(n+3)!}{3!}}=\frac{3(n-m)(n-m-1)}{n(n+1)(n+2)}=\frac{3\left(1-\frac{m}{n}\right)\left(1-\frac{m}{n}-\frac{1}{n}\right)}{n\left(1+\frac{1}{n}\right)\left(1+\frac{2}{n}\right)}. \tag{7}$$

This formula agrees with Equation (6) of Eggenberger and Pólya [2] , where the authors multiply conditional probabilities, then permute the numerators of the fractions that are multiplied.

We now set $\frac{m}{n}=x$ and take the limit as $n\to\infty$ by dropping all terms that approach 0, obtaining

$$\mathbb{P}(\mu_\infty=x)=\frac{3}{n}(1-x)^2$$

which corresponds to the density function for the beta distribution with parameters 1 and 3,

$$f(x)=3(1-x)^2. \tag{8}$$

4. WAITING FOR THE SECOND MALE MEMBER

The probability that only females are added over a period of n weeks is simply the ratio

$$\frac{3\times4\times5\cdots(2+n)}{4\times5\times6\cdots(3+n)}=\frac{3}{n+3}.$$

Suppose that the consanguinity agrees to pay a fine of 1 dollar for each week that its membership includes only one male. What is the expected amount of the fine?

Let 1_n be the indicator function of the event "in the first n weeks, all new members are female." The total fine is the random variable

$$F = 1_1 + 1_2 + 1_3 + \cdots.$$

For example, if the first male is added in week 3, the first two of these indicator functions equal 1, all the others are 0, and the fine is 2 dollars.

The expectation of the fine, by linearity of expectation, is

$$\mathbb{E}[F] = \mathbb{E}[1_1] + \mathbb{E}[1_2] + \mathbb{E}[1_3] + \cdots.$$

But the expectation of 1_n is just the probability of the event that only females are added over a period of n weeks, so

$$\mathbb{E}[F] = \sum_{n=1}^{\infty} \frac{3}{n+3}$$

This series is divergent, and the expected fine is infinite!

5. MARTINGALES FOR GENERAL STARTING CONDITIONS

If the consanguinity begins with N members, M males and F females, the proportion of male members is still a martingale:

$$\mu_n = \frac{M_n}{n+N}$$

Furthermore, the second-degree martingale is the simplest imaginable generalization: 4 gets replaced by N.

$$Z_n = \frac{1}{(n+N)(n+N+1)}\left(M_n^2 + M_n\right) \tag{9}$$

We repeat the proof, considering

$$Z_{n+1} = \frac{1}{(n+N+1)(n+N+2)}\left(M_{n+1}^2 + M_{n+1}\right)$$

Since

$$M_{n+1} = M_n + X_{n+1},$$

$$Z_{n+1} = \frac{1}{(n+N+1)(n+M+2)}\left[M_n^2 + 2M_nX_{n+1} + X_{n+1}^2 + M_n + X_{n+1}\right].$$

Again we condition everything on M_n and use

$$\mathbb{E}\left[X_{n+1}|M_n\right] = \mu_n = \frac{M_n}{n+N}.$$

By TOWIK (1),

$$\mathbb{E}\left[M_n^2|M_n\right] = M_n^2.$$

By TOWIK (1),

$$\mathbb{E}\left[M_nX_{n+1}|M_n\right] = M_n\mathbb{E}\left[X_{n+1}|M_n\right] = M_n\mu_n = \frac{M_n^2}{n+N}.$$

Since X_{n+1} equals 0 or 1, its square is equal to itself, and

$$\mathbb{E}\left[X_{n+1}^2|M_n\right] = \mathbb{E}\left[X_{n+1}|M_n\right] = \mu_n = \frac{M_n}{n+N}.$$

Combining all the terms we find

$$\mathbb{E}\left[Z_{n+1}\middle|M_n\right]=\frac{1}{(n+N+1)(n+N+2)}\left(M_n^2\left(1+\frac{2}{n+N}\right)+M_n\left(1+\frac{2}{n+N}\right)\right)$$

or

$$\mathbb{E}\left[Z_{n+1}\middle|M_n\right]=\frac{1}{(n+N+1)(n+N+2)}\frac{n+N+2}{n+N}\left(M_n^2+M_n\right)=Z_n,$$

which establishes the martingale property for Z_n (10).

With more than one initial male member, the calculation of variance becomes slightly more complicated, since Z_0 depends on the number of initial male members. By the Tower Law (3),

$$\mathbb{E}[Z_n]=\mathbb{E}\left[Z_n\middle|M_0\right]=Z_0=\frac{1}{N(N+1)}\left(M^2+M\right)=\frac{M(M+1)}{N(N+1)}$$

It follows that

$$\frac{1}{(n+N)(n+N+1)}\left(\mathbb{E}\left[M_n^2\right]+\mathbb{E}[M_n]\right)=\frac{M(M+1)}{N(N+1)}$$

But we already know that

$$\mathbb{E}[M_n]=\frac{(n+N)M}{N}$$

and so

$$\mathbb{E}\left[M_n^2\right]=\frac{(n+N)(n+N+1)M(M+1)}{N(N+1)}-\frac{(n+N)M}{N}$$
$$=\frac{(n+N)M}{N}\frac{n(M+1)+M(N+1)}{N+1}. \qquad (10)$$

Thus the variance is

$$\mathrm{Var}M_n = \mathbb{E}\left[M_n^2\right] - \mathbb{E}\left[M_n\right]^2 = \frac{(N-M)nM(n+N)}{N^2(N+1)}.$$

in agreement with the value that Eggenberger and Pólya [2] calculate from the probability mass function. Dividing by $n+N$ and taking the limit as the limit as $n \to \infty$, we find

$$\mathrm{Var}\mu_\infty = \frac{(N-M)M}{N^2(N+1)},$$

which is indeed the variance for a beta distribution with parameters

$$\alpha = M, \beta = F = N - M.$$

6. CONCLUSIONS

The original examination question has been solved at last.

- The proportion of male members, μ_n, is a martingale whose expectation is $\mathbb{E}(\mu_n) = 1/4$ (4) for all n.
- Being bounded below by 0 and above by 1, this martingale has a limit as $n \to \infty$.
- Without working out the distribution function for μ_n, a second martingale (5) can be used to show that the variance of the limiting distribution is $\mathrm{Var}\mu_\infty = 3/80$ (6), which is consistent with the conjecture that this limiting distribution is a beta distribution.
- The distribution of the number of male members (7) can be determined by an elementary counting argument, and its limit is a beta distribution(8).
- The newly found second-degree martingale generalizes easily (10) to arbitrary starting conditions.

ACKNOWLEDGEMENTS

I thank the Editor and the referee for their comments. I also thank Dean Rakesh Khurana, who made me aware of the issue of fraternities and sororities at Harvard, my student Morgan Breitmeyer, with whom I had enlightening discussions about sororities on campus, and my student Rebecca Jarvis, who reads German and was able to confirm my interpretation, initially done with the aid of Google Translate, of Eggenberger and Pólya's original article.

REFERENCES

1. From 1854: Smith's Prize Exam. http://www.clerkmaxwellfoundation. org/ Smiths PrizeExam_Stokes.pdf.
2. Eggenberger, F. and Pólya, G. (1923) über die Statistik verketteter Vorgange. Z. Angew. Math. Mech., 3, 279-289.
3. Letter on Single-Gender Social Organizations (2016) Cambridge, Mass. http://www. harvard.edu/president/news/2016/letter-on-single-gender-social-organizations.
4. Mahmoud, H. (2008) Polya Urn Models. Chapman and Hall/CRC.
5. Polya's Urn Process. http://www.math.uah.edu/stat/urn/Polya.html.
6. Dineen, S. (2013) Probability Theory in Finance. Graduate Studies in Mathematics, 70, 305p.
7. Grimmett, G. and Stirzaker, D. (2011) Probability and Random Processes. Oxford University Press, New York.
8. Chirara, L. and Hesterberg, T. (2011) Mathematical Statistics with Resampling and R. Wiley, Hoboken, New Jersey.

CHAPTER 15

Application of Simplex Method in the Radiotherapy Treatment

Thais R. Salvador[1], Silvia M. S. Carvalho[1], Mayk V. Coelho[2]

[1]*DFQM, Federal University of São Carlos, Sorocaba, São Paulo, Brazil*

[2]*Federal University of Alfenas-MG, Poços de Caldas, Minas Gerais, Brazil*

ABSTRACT

This work presents an application of the Simplex Method for solving an optimal planning problem for cancer treatment by radiotherapy. Linear Programming can aid the optimal planning for radiation therapy, where the concern is to apply a high enough radiation in the tumor while saving significantly healthy regions or critical organs.

Keywords: Linear Programming, Simplex Method, Radiotherapy

1. INTRODUCTION

Among the problems of global health, cancer has become one of the main subjects of research for different fields of knowledge. Many of these researches are conducted with the goal of developing and evaluating new forms of therapy or just new methodologies or treatment protocols. As there is a wide variety of algorithms for tomographic reconstruction and absorbed dose calculation in radiation therapy applications involving various mathematical techniques have been proposed and are still constantly improved [1] .

From a mathematical point of view, the challenge is to deliver a high radiation dose to the tumor, enough to eliminate them, and simultaneously minimize radiation in the surrounding regions, composed of healthy tissue, while minimizing the complications in these regions.

Among these techniques, linear programming stands out for its intense and wide application and the results obtained are considered promising. The objective of treatment is to eliminate tumor cells by radiation, and at the same time it seeks to prevent the destruction of surrounding healthy cells also affected by radiation. For planning the treatment with radiosurgery computerized tomography images are scanned and merged with magnetic resonance images obtained previously. From these images critical structures and tumor are located and then many computations may be performed to determine three-dimensional isocenters, isodose curves, dose to be prescribed, position, number and its weight conformal isocentric fields, among others. Such plan is prepared by the surgical team in order to obtain an adequate and optimal treatment for each case.

The treatment plans are commonly based on the desired dose levels for specific types of tumors and organs at risk in the irradiated region. These levels' dose is chosen so that high probability of cure is achieved, while the likelihood of complications in any organ is minimal risk. The aim of this paper is to show the optimal planning in radiotherapy looking for an optimal solution using the Simplex Method and its comparison with Interior Points Methods.

2. LINEAR PROGRAMMING IN RADIOTHERAPY

In recent decades, the impact of cancer in society got large attention, becoming a global concern, and in Brazil one of the top positions in the macro-regions mortality rates, along with the diseases of the circulatory system, external causes, among others. Cancer is a disease that begins when a genetic mutation in the DNA of the cell, where the control mechanism of normal tissue growth is changed. The methods for cancer treatment are: surgery, radiotherapy and chemotherapy.

The radiotherapy treatment aims elimination of cancerous cells through radiation or relief of symptoms and at the same time seeking to avoid destruction of surrounding healthy cells also affected by radiation. From a mathematical point of view, the challenge is to deliver a high dosage radiation into the tumor, sufficient for eliminating and stopping the growth of tumor cells, whilst minimizing the radiation neighboring regions composed of healthy tissue, while minimizing complications in those regions which are often critical.

At the beginning of the use of radiotherapy, was applied high doses of radiation, as well as the patient be able to afford. The dose constraint was generally established the tolerance of the skin [2] . There are two main types of radiation: brachytherapy and teletherapy. Give yourself will focus in this work to the second type. In this case, neither the radiation source nor device that emits, are in direct contact with the patient. Therefore, the radiation reaches beyond the tumor all organs and tissues that are in the way. In the brachytherapy radiation source is placed inside the patient in the region that would receive treatment. It is a kind of non-permanent radioactive implant [3] . Usually for the treatment using a beam with rectangular form, which passes through different types of patient body tissues, and depending on the attenuation and the depth of tissue, a large amount of energy is deposited at various points within the body. For radiotherapy treatment should take into account the depth dose distribution in the central axis and characterizing the radiation beam. To represent and to have the concept of the dose distribution and absorption throughout the volume and makes up the isodose curve, which are lines representing points of the same dose in a given arrangement of radiation beams. For the mathematical calculation, they are used clinical tumor data for each patient treated. These calculations can be performed by mathematical models through linear programming.

2.1 Mathematical Formulation

When cancer is diagnosed and there are medical indications for treatment by radiotherapy, various tests are performed on the patient, in order to learn about the location, form and tumor volume, as well as critical tissues in the region to be treated. The use of computerized tomography exams for reliable data collection and in the radiosurgery is required along with a computed tomography an magnetic resonance exam.

Based on these data, the dose to be received in the tumor and the volume to be irradiated may be prescribed by the doctor radiotherapist. Together with a physical, the radiation oncologist can then by the analysis of isodose curves can define the best type of treatment and the technique to be used. After obtaining the images by computed tomography or magnetic resonance imaging, the minimum dose to be applied to the tumor is prescribed as the maximum doses that critics and healthy tissue can receive. Through these images, it makes the selection of the anatomical structures of interest.

On the images obtained, the structures of interest can then be mapped (and critical lesion tissue) through a pixelation. Each pixel is related to an electron density and receives specific coordinates (i,j), which will be used in the creation of conformational planning or 3D treatment planning, as shown in Figure 1.

To illustrate, suppose a dimensional image m × n where each element's image is called pixel representing a part of the human body. Also, consider que each angle is composed of sub-rays. Modern processing systems are capable of performing complex combinations between these sub-beams. The number of angles is fixed such that are considered k angles locations to issue the main beam Radiation. Each one containing η sub-beams. Modern processing systems are able to perform combinations among these sub-beams to use them over a full arc, making planning using η sub-beams. This decomposition of the radiation beams is based on the theoretical and practical foundations of radiotherapy with modulated intensity. The geometry of a model using elementary beams, where $n = 2$, $p = 4$ and $\eta = 4$ is shown in Figure 2. The intervals between the angles $\frac{\pi}{4}$ and the initial angle is (usually) zero.

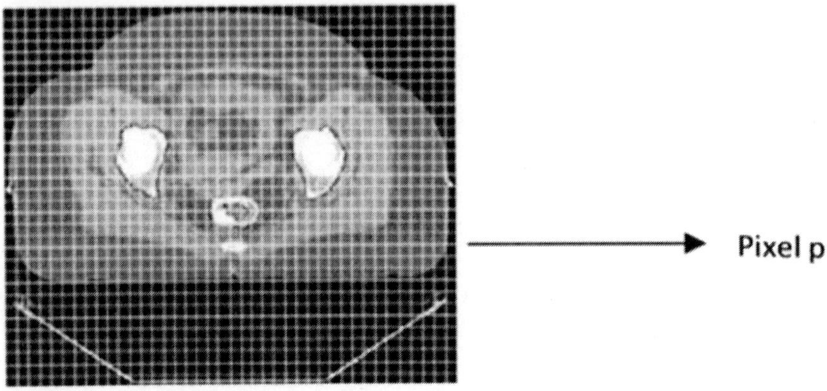

Figure 1. Image in pixels.

In the treatment planning, once defined the coordinates (i,j) of pixels, with $1 \le i \le \eta_1$, $1 \le j \le \eta_2$, one computes the total dose $D_{(i,j)}$ being received by the patient in each pixel coordinates (i,j). Thus, one can estimate the dose to be received in each patient's tissue (critical, healthy and tumor), the area to be treated worrying not to exceed the dose supported by healthy and critical tissues and a dose sufficient enough to eliminate the tumor.

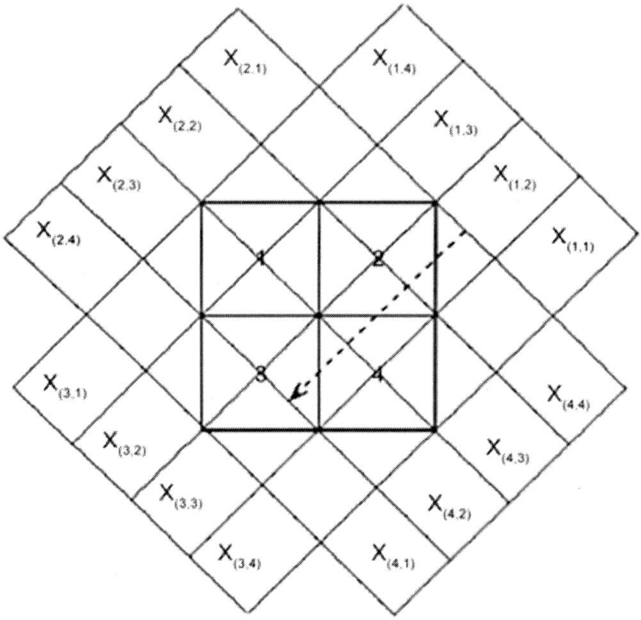

Figure 2. Geometry of an image.

This model can be represented by the following:

$$\begin{aligned} &\text{minimize } f = \sum_{i=1}^{m}\sum_{j=1}^{m} D_{ij} \\ &\text{subject to } D_{ij} = \sum_{p=1}^{n} w_p A_{ij}^{p} \ \forall (i,j) \\ &l \leq D_{ij} \ \ \forall (i,j) \in \Gamma \\ &w \geq 0, \end{aligned} \qquad (1.0)$$

where,

D_{ij} represents the total dose received by the patient in position (i,j);

A_{ij} represents the nominal dosage issued by the radius p in position (i,j);

w represents the weight of the dosage emitted by beams;

l represents the minimum dose to be received by the tumor;

Γ represents the subset of positions where the tumor is located;

n represents the number of rays;

m represents the total number of pixels.

It is possible to formulate a model with weights in each region of the patient privileged or penalize some areas the dosing quantity to be received. Given the delineation of structures, the treatment plan should be developed in order to manage the radiation dose on the tumor, thus seeking the homogeneity of the treatment plan, according to the determination of the dose prescription of regions of interest.

Defining the following subsets:

- R represents the subset of positions where critical tissue (healthy) or critical framework;
- N represents the subset of positions where the other healthy tissues.

It is possible to formulate a model with weights in each region of the patient privileging or penalizing some areas where the dosing quantity is received.

Given the delineation of structures, the treatment plan should be developed in order to manage the radiation dose to the tumor, thus seeking the homogeneity of the treatment plan, according to the determination of the dose prescription of regions of interest.

Holder [4] formulates some models of Linear Programming by weightings in each region of the patient's image, considering or privileging the type of tissue to be exposed to radiation doses.

As $x(a,i)$ the dose over the i-th subfeixe, to $i = 1, 2, \cdots, \eta$ a-th angle to $a = 1, 2, \cdots, k$ and $d_{(p,a,i)}$ the distance between the source positioned at an angle to sending subfeixe i (de dose $x_{(a,i)}$) and pixel p. The dose deposition in the pixel p due i-th subfeixe of the a-th angle $A_{(p,a,i)}$ is defined as:

$$A_{(p,a,i)} = Se^{-\mu d_{(p,a,i)}}, \tag{1.1}$$

The term S represents the geometric area of pixel p which receives the dose $x(a,i)$. With the component $A_{(p,a,i)}$ may be constructed the dose deposition of matrix A, where the rows in A are indexed by p and the columns are indexed by (a,i), as shown in the matrix (1.2), thus the matrix A has dimension $m \times k\eta$.

$$\begin{pmatrix} 0 & 0 & \frac{1}{2} & \frac{1}{2} & 0 & \frac{1}{2} & \frac{1}{2} & 0 & \frac{1}{2} & \frac{1}{2} & 0 & 0 & 0 & \frac{1}{2} & \frac{1}{2} & 0 \\ 0 & \frac{1}{2} & \frac{1}{2} & 0 & \frac{1}{2} & \frac{1}{2} & 0 & 0 & 0 & \frac{1}{2} & \frac{1}{2} & 0 & 0 & 0 & \frac{1}{2} & \frac{1}{2} \\ 0 & \frac{1}{2} & \frac{1}{2} & 0 & 0 & 0 & \frac{1}{2} & \frac{1}{2} & 0 & \frac{1}{2} & \frac{1}{2} & 0 & \frac{1}{2} & \frac{1}{2} & 0 & 0 \\ \frac{1}{2} & \frac{1}{2} & 0 & 0 & 0 & \frac{1}{2} & \frac{1}{2} & 0 & 0 & 0 & \frac{1}{2} & \frac{1}{2} & 0 & \frac{1}{2} & \frac{1}{2} & 0 \end{pmatrix} \tag{1.2}$$

Most research has used optimization models with linear constraints, with one of the two most obvious objective functions, maximizing the dose to the tumor and minimize the dose to the critical structure. Since this often leads maximization at high doses, other studies seek to maximize the minimum dose of the tumor or reduce the maximum dose of critical structure [5] .

The goals listed below indicate that this problem has many parameters to consider in deciding what would be desirable for a treatment plan:

- Transmitting a uniformly lethal dose in the tumor region;
- Transmitting radiation as small as possible in critical structure;
- Obtain a total dose as small as possible;
- Reducing the frequency of high doses outside the region of the tumor;
- Controlling the number of rays used in the treatment plan.

In search of uniform doses, Holder proposed a linear optimization model to assist in the optimal planning of radio surgery. This model has been widely cited in the literature by presenting good results in terms of conformal plans. So we opted for the study and implementation of this model, which will be discussed below.

Consider, m_T tumor pixel number, m_C the number of pixels of the critical structure and m_G the number of pixels of healthy tissue. Therefore the total number of pixels is given by $m = m_T + m_C + m_G$. Thus, in the model of Holder, the dose deposition matrix is divided into three parts:

$$A = \begin{bmatrix} A_T \\ A_C \\ A_G \end{bmatrix} \tag{1.3}$$

The rows of the dose deposition matrix are rearranged so that T is related to the set of pixels comprising the lesion or tumor, C to pixel set included the critical tissue or organs at risk and G to the pixel set of remaining healthy tissue.

Thus, the sectional image of computed tomography is converted into an array of pixels, which are classified to represent tumor and non-tumor regions. Due the division into pixels, the model that will be presented, the prescription dose is given in the form of vector and defined by four boundaries, with the following notation:

- u_t represents the upper limit vector to radiation in the tumor ($u_t \in R^{m_T}$);
- l_t represents the lower limit vector to the tumor radiation ($l_t \in R^{m_T}$);
- u_c represents the upper limit vector for radiation in the critical structure ($u_c \in R^{m_C}$);
- u_g represents the upper limit vector to radiation in the remaining healthy tissue ($u_g \in R^{m_G}$).

Obviously $0 < l_t \leq u_t, u_c \geq 0$ and $u_g \geq 0$ hold. If an uniform lethal dose is transmit ted to the tumor, the upper and lower limit to the tumor pixels are obtained through the established goals.

Assuming that the targets set for a cell are t_g, the values u_{ti} and l_{ti} they are generally (1 + e) t_g and (1 − e) t_g, respectively, where ε is the percentage of variation for tumor strength and uniformity is termed the tumor level. Typical values ε are found in the literature varying from 0.02 to 0.15.

The vector u_g represents the largest amount of radiation allowed for any pixel (healthy). Generally healthy tissues should not receive more than 10% of the dose established for the tumor. That is, $u_g = t_g(1+0.10)$. The rows of dose deposition array are rearranged in row a representing cancer regions critical structures and the remaining healthy tissue. This reordering is represented by the A_T, A_C and A_G as shown in (1.3).

The sub-beams that do not reach the tumor are removed by the elimination of the columns that have the zero vector in the corresponding column of submatrix A_T. Thus, without loss of generality let us assume that the A_T matrix does not have null columns. Therefore, $A \in R^{m \times n}$, $A_T \in R^{m_T \times n}$, $A_C \in R^{m_C \times n}$ and $A_G \in R^{m_G \times n}$.

3. LINEAR OPTIMIZATION PROBLEM

The model proposed by [4] allows a number of constraints on the amount of limitation of the dose in each tissue type. The objective function is represented by the weighted sum of three goals:

- $l^T t$, which measures how close the minimum dose the found plan apply in the tumor region;
- $u_c^T c$, which measuring the amount of radiation received by the prescribed critical region;
- $u_g^T g$, which measuring the amount of radiation prescribed gin other healthy tissues.

The positive scalar w weight the importance of formulating a plan to get the minimum dose in the tumor region, i.e., large values of w make $l^T t$ to be as small as possible. It would be desirable if there were a finite value for $w \geq 0$ such that the optimal value of the component $l^T t$ were zero which would ensure the tumor receives as much radiation level as possible.

The proposed model can be represented as follows:

$$\begin{aligned}
&\text{Minimize } wl^T t + u_c^T c + u_g^T \\
&\text{Subject to } l_t - Lt \leq A_T x \leq u_t \\
&\qquad A_C x \leq u_c + U_C c \\
&\qquad A_G x \leq u_g + U_G g \\
&\qquad 0 \leq Lt \leq l_t \\
&\qquad -u_c \leq U_C c \\
&\qquad U_G g \geq 0 \\
&\qquad x \geq 0.
\end{aligned} \tag{1.4}$$

where,

- w: positive scalar;
- x: dose of sub-beams entering the image to reach the pixel p, ($x \in R^n$);
- $t \in R^{m_T}, t \geq 0$;
- $c : c \in R^{m_C}$;

- $g \in R^{m_G}, g \geq 0$.

The constraints $l_t = Lt \leq A_T x, A_C x \leq u_c + U_C c$, and $A_G \leq u_g + U_G g$, are called elastic constraints, since their limits may vary according to the vectors t, c and g, respectively.

The matrices L, U_C and U_G define how to measure elasticity, and l, u_c and u_g control the penalty or reward relative to elasticity.

Fixed values l, u_c, u_g, L, U_C and U_G define a set of elastic functions. And these are incorporated for the following reasons:

1) The elastic constraint ensures that a set of elastic functions is always strictly feasible;
2) The difference of the limits on the elastic functions allow us to incorporate different treatment goals.

As previously discussed, the model (1.4) is a linear programming problem and can be solved by the simplex method. To solve this problem, means determining x^*, t^*, c^* and g^* that optimize the objective function and achieve the goals set the best way possible.

Thus, with the optimal dose of vector x is possible to compute the total dose of radiation, or total for each pixel. Given the optimal dose distribution in tissues it is possible to construct isodose curves.

4. SIMPLEX METHOD APPLIED TO THE TREATMENT MODEL FOR RADIOTHERAPY

As the image is in pixels, the dose limits should be divided by the number of pixels of the respective bodies, to ensure the uniformity of the dose [6] . Therefore, the objective function, the values dose limits will be $l = \frac{1}{m_T} e, u_c = \frac{1}{m_C} e$ and $u_g = \frac{1}{m_G} e$, where

$l \in R^{m_T}$, $u_C \in R^{m_C}$, $u_G \in R^{m_G}$ and e is unit vector.

Supposing, $A_T x = a$, and replacing bounded constraint $l_t - t \leq a \leq u_t$, as L = I, U_c = I and $U_g = I$, we obtain following problem:

$$
\begin{aligned}
&\text{Minimize } w\frac{1}{m_T}e^T t + \frac{1}{m_C}e^T c + \frac{1}{m_G}e^T g \\
&\text{Subject to } a + s_u = u_t \\
&\qquad a + t - s_l = l_t \\
&\qquad A_T x - a = 0 \\
&\qquad A_C x + s_c - c = 0 \\
&\qquad t + s_t = u_t \\
&\qquad (t, c, g, a, s_u, s_l, s_c, s_g, s_t) \geq 0.
\end{aligned}
\tag{1.5}
$$

Thus, we obtain the following constraint matrix:

$$
\begin{pmatrix}
I & 0 & 0 & 0 & 0 & I & 0 & 0 & 0 & 0 \\
I & 0 & I & 0 & 0 & 0 & -I & 0 & 0 & 0 \\
-I & A_T & 0 & 0 & 0 & 0 & 0 & 0 & 0 & 0 \\
0 & A_C & 0 & -I & 0 & 0 & 0 & I & 0 & 0 \\
0 & A_G & 0 & 0 & -I & 0 & 0 & 0 & I & 0 \\
0 & 0 & I & 0 & 0 & 0 & 0 & 0 & 0 & I
\end{pmatrix}
\cdot
\begin{pmatrix} a \\ x \\ t \\ c \\ g \\ s_u \\ s_l \\ s_c \\ s_g \\ s_t \end{pmatrix}
=
\begin{pmatrix} u_t \\ l_t \\ 0 \\ 0 \\ u_c \\ u_t \end{pmatrix}
\tag{1.6}
$$

5. COMPUTATIONAL RESULTS

This section will present and discuss some results obtained in the computational implementation of the mathematical model (1.4), aiming to help in optimal radiotherapy planning. The problem was solved in a Microcomputer-macbook pro, 2.5 GHz processor, i5, 4 GB memory, 1600 MHz, DDR3.

Following the described methodology, initially the cancer is diagnosed and indicated with radiation therapy. Thereafter, they are carried tomography computed or tumor sizing and location of the regions of interest (tumor, healthy and critical). Thus, the doctor determines the dose to be applied in the tumor (t_g).

The Image tomography and divided into pixels and are obtained as Geometric Positions of pixels locations relating to tissues Critics, healthy and tumor to be used in doses of Matrix Construction $A = [A_T A_C A_G]^T$ as defined in (1.3).

For the computational experiment, it is considered hypothetically, that the patient has a spherical tumor with a diameter of 15 mm, surrounded by critical tissue, exemplifying a spinal cord tumor, where there is difficulty in planning due to cancer was fully surrounded by a critical structure [7] .

In this case, it has been indicated a radiotherapy treatment with tumor dose of 80 Gy, where Gy represents the amount of absorbed ionizing radiation energy (or dose) for mass unit (1 Gray (Gy) = 1 J/kg). It was considered a percentage change $\varepsilon = 2\%$, and the critical structure can accommodate up to 40 Gy and healthy at most 60 Gy. As the positive scalar (w) considering the importance that the tumor receives the dose minimum, two values was considered for comparison, namely $w = 0.1$ and $w = 40$. Other bounds are given on the Table 1.

The considered lesion contains four pixels (2 × 2 pixels), the critical tissue around the lesion comprises 16 pixels (4 × 4 pixels). The developed planning system uses a grid of 16 × 16 pixels, with 4 angles from 0, $\frac{\pi}{2}$, π and $\frac{3\pi}{2}$ degrees, where each beam is composed by six sub-beams.

To improve the computational performance, we have considered only the region that tends to form hot pixels, which are pixels that receive radiation being considered healthy structure with 36 pixels. The pixels that will not receive radiation are called cold pixels [8] .

The 1.5 objective function is the sum of three goals: to achieve a high enough dosage to eliminate the cancer and not exceed the maximum dose in healthy tissues and critical. The Table 2summarizes the numerical results obtained on the value of the objective function, number of iterations and run time in seconds.

Analyzing the results in Table 2, we can see that the minimized function achieved a very low value, which is a good conformation of isodose curves, thus ensuring a safe treatment. The dose of excesses in critical and healthy regions and the dose deficit in the tumor region is almost zero, thus showing that the tumor received the dose required for elimination and the dose limit allowed for other regions was not exceeded.

Table 1. Values in the model relative the doses.

Used Value	Gy
u_t	81.6
l_t	78.4
u_c	40
u_g	60
t_g	80

Table 2. Numerical results in the objective function.

Variables	$W = 0.1$	$W = 40$
Objective function value	2.9553×10^{-6}	2.9800×10^{-6}
Iteration	38	38
Time	0.010718 s	0.011471 s
Tolerance	1.0×10^{-6}	1.0×10^{-6}
Excess dose in the tumor	0.1232×10^{-6}	0.1133×10^{-6}
Excess dose in critical tissue	0.1142×10^{-6}	0.1242×10^{-6}
Excess dose in healthy tissue	0.0045×10^{-6}	0.0045×10^{-6}

The weight w decides the importance of dose uniformity in the tumor. As we can see the results, if w is small, indicates that finding a treatment plan reach the dose lower limit of the tumor is not so important. Increasing w increases the chances that the treatment plan to reach an uniform dose and tumor need. Thus, higher values for w force excess dose in the tumor to be as minimal as possible.

5.1 Compared to the Interior Point Methods

At this moment, we compare the dual simplex method with interior points methods to three real problems.

Analysis report on the results obtained:

In Table 3 and Table 4 the computational time for the dual simplex method is slightly less when compared to the interior point method, unlike what happens with the larger problem 1.5 which is easily explained as the interior point method always has good results for large problems [9] , which does not occur with the Simplex method. In all other aspects the methods have the same performance.

Table 3. Problem 342 Pixels.

Method	Dual Simplex	IPM
Iteration	12	12
Time	4.535522×10^{-1} s	2.307250 s
Function objective	3.042952×10^{-9}	3.042952×10^{-9}
Excess dose in the tumor	4.153469×10^{-5}	4.153469×10^{-5}
Excess dose un critical tissue	7.304035×10^{-6}	7.304035×10^{-6}
Excess dose in healthy tissue	9.863899×10^{-8}	9.863899×10^{-8}

Table 4. Problem 747 Pixels.

Method	Dual Simplex	IPM
Iteration	13	13
Time	7.352247×10^{-1}	9.160095×10^{-1}
Function objective	3.000177×10^{-9}	3.000177×10^{-9}
Excess dose in the tumor	3.809950×10^{-5}	3.809950×10^{-5}
Excess dose un critical tissue	6.134144×10^{-6}	6.134144×10^{-6}
Excess dose in healthy tissue	7.212216×10^{-6}	7.212216×10^{-6}

Table 5. Problem 3718 Pixels.

Method	Dual Simplex	IPM
Iteration	24	24
Time	2.754705 s	2.738810 s
Function objective	1.258780×10^{-9}	1.258780×10^{-9}
Excess dose in the tumor	3.318770×10^{-5}	3.318770×10^{-5}
Excess dose un critical tissue	2.010262×10^{-6}	2.010262×10^{-6}
Excess dose in healthy tissue	4.522237×10^{-7}	4.522237×10^{-7}

In Table 5 the computational time for the interior point methods is smaller compared to the dual simplex method and this is common with increasing the dimension of the problem, since it is a robust method and very used in large problems.

6. CONCLUSIONS

This work presented an application of the simplex method in a cancer treatment plan for radiotherapy and a model based on linear programming for assistance in optimal planning. It can be seen that the mathematical model used in this paper can be a tool of great importance in the construction of optimized treatment plans; it provides a set of optimal solutions, which related to the treatment, may enable a high quality therapy [10] .

It is possible to consider the dual simplex method to ensure reliable treatment of medium and small problems, since their computational cost is lower than the interior point methods, but its performance is very similar.

ACKNOWLEDGEMENTS

This research was partially sponsored by the Foundation for the Support of Research of the State of São Paulo, (FAPESP) and the Brazilian Council for the Development of Science and Technology (CNPq) and Improvement of Higher Education Personnel (CAPES).

REFERENCES

1. Péricles Crisostomo de, S. (2008) Programação linear no planejamento do tratamento de cancerporradiocirurgia. Master's Thesis, Universidade Estadual Paulista (UNESP), São Paulo.
2. Rodrigo SartoreloSalemi, V. (2010) Programação Linear aplicada a criação de planejamentosotimizadosemradioterapia. Master's Thesis, Programa de Pós-Graduaçãoem Biometria, IB, UNESP, São Paulo.
3. David Moreira, F. (2009) Método de pontosinteriores no planejamentoótimo do tratamento de cancerporradioterapia. Master's Thesis, Universidade Estadual Paulista (UNESP), São Paulo.
4. Allen, H. (2003) Designing Radiotherapy Plans with Elastic Constraints and Interior Point Methods. Health Care Management Science, 6, 5-16.
5. Cecília Bollini Barboza, C. (2003) Planejamento do tratamentoporradioterapiaatravés de métodos de pontosinteriores. Master's Thesis, Dissertação de Mestrado, ICMC-USP-São Carlos.
6. Johns, H.E. and Cunningham, J.R. (1974) Physics of Radiology.

7. Barbosa, C. and Oliveira, A.R.L. (2006) Planejamento do tratamentoporradioterapia, através de métodos de pontosinteriores. Pesquisa Operacional.
8. Andrea Camila dos Santos, M. (2011) O método de pontosinteriores no planejamento da radioterapia. Universidade Estadual Paulista (UNESP), São Paulo.
9. Carvalho, S.M.S. and Oliveira, R. (2014) Predispatch of Hydroelectric Power Systems with Modifications in Network Topologies. Applied Mathematics, 5, 2271-2283.
10. Mageras, G.S. and Mohan, R. (1993) Application of Fast Simulated Annealing to Optimization of Conformal Radiation Treatments. Medical Physics, 20, 639-647.

CHAPTER 16

A Conceptual Numerical Model of the Wave Equation Using the Complex Variable Boundary Element Method

Bryce D. Wilkins, Theodore V. Hromadka, Randy Boucher

Department of Mathematics, United States Military Academy, West Point, NY, USA

ABSTRACT

In this work, a conceptual numerical solution of the two-dimensional wave partial differential equation (PDE) is developed by coupling the Complex Variable Boundary Element Method (CVBEM) and a generalized Fourier series. The technique described in this work is suitable for modeling initial-boundary value problems governed by the wave equation on a rectangular domain with Dirichlet boundary conditions and an initial condition that is equal on the boundary to the boundary conditions. The new numerical scheme is based on the standard approach of decomposing the global initial-boundary value pro- blem into a steady-state component and a time-dependent component. The steady-state component is governed by the Laplace PDE and is modeled with the CVBEM. The time-dependent component is governed by the wave PDE and is modeled using a generalized Fourier series. The approximate global solution is the sum of the CVBEM and generalized Fourier series approximations. The boundary conditions of the steady-state component are specified as the boundary conditions from the global BVP. The boundary conditions of the time-dependent component are specified to be identically zero. The initial condition of the time-dependent component is calculated as the difference between the global initial condition and the CVBEM approximation of the steady-state solution. Additionally, the generalized Fourier series approximation of the time-dependent component is fitted so as to approximately satisfy the derivative of the initial condition. It is shown that the strong formulation of the wave PDE is satisfied by the superposed approximate solutions of the time-dependent and steady-state components.

Keywords: Complex Variable Boundary Element Method (CVBEM), Partial Differential Equations (PDEs), Numerical Solution Techniques, Laplace Equation, Wave Equation

1. INTRODUCTION

In this work, the Complex Variable Boundary Element Method (CVBEM) pro- cedure is extended to modeling applications of the two-dimensional wave partial differential equation (PDE). The wave PDE, given by $c^2\left(u_{xx}+u_{yy}\right)=u_{tt}$, is com- monly used to model the propagation of waves in highly-stretched membranes that are uniformly dense with approximately constant tension. Consequently, the solution of the wave equation is important in disciplines such as physics, engineering, and other applied sciences that consider the propagation of such waves. The numerical methodology presented in this work uses a coupling of the CVBEM with a generalized Fourier series. The modeling outcome is novel since it is a function that satisfies the strong formulation of the wave PDE within the problem domain. The problem being considered here is

Domain:	$\Omega=\left\{(x,y)\in\mathbb{R}^2 : 0<x<L_1, 0<y<L_2\right\}$
PDE:	$c^2\left(u_{xx}+u_{yy}\right)=u_{tt}$
Boundary Conditions:	$u(x,y,t)=f(x,y) \quad \forall(x,y)\in\partial\Omega$
Initial Condition:	$u(x,y,0)=g(x,y)$
	$u_t(x,y,0)=h(x,y)$

The proposed solution technique is based on the standard approach of de- composing the global initial-boundary value problem into two components; namely, a steady-state component and a time-dependent component. The steady- state problem is given as

Domain:	$\Omega = \{(x,y) \in \mathbb{R}^2 : 0 < x < L_1, 0 < y < L_2\}$
PDE:	$\Delta u_1 = 0$
BoundaryConditions:	$u_1(x,y) = f(x,y) \quad \forall (x,y) \in \partial\Omega$

The time-dependent problem is given as

Domain :	$\Omega = \{(x,y) \in \mathbb{R}^2 : 0 < x < L_1, 0 < y < L_2\}$
PDE :	$c^2 \Delta u_2 = \frac{\partial^2 u_2}{\partial t^2}$
Boundary Conditions :	$u_2(x,y,t) = 0 \quad \forall (x,y) \in \partial\Omega$
Initial Condition :	$u_2(x,y,0) = g(x,y) - u_1(x,y)$
	$\frac{\partial}{\partial t} u_2(x,y,0) = h(x,y)$

The steady-state problem is modeled using the CVBEM [1] [2] [3] [4] , which produces an approximation of u_1 denoted by $\hat{u}_1$. The time-dependent problem is modeled by a generalized Fourier series, which produces an approximation of u_2 denoted by $\hat{u}_2$. The global approximation function is the sum $\hat{u}(x,y,t) = \hat{u}_1(x,y) + \hat{u}_2(x,y,t)$.

A necessary condition for using this methodology is that the initial condition of the global initial-boundary value problem must be equal on the boundary to the boundary conditions of the global BVP. That is, $g(x,y) = f(x,y)$ for all $(x,y) \in \partial\Omega$. This is a necessary condition because it ensures that the boundary conditions for the time-dependent problem will be identically zero. Therefore, a two-dimensional Fourier sine series in the spatial variables can be used to provide a particular solution to the time-dependent problem. It is noted that this methodology can still be used when this condition is not met, however, while the global approximation function might closely approximate the initial condition within the problem domain, it will not satisfy the initial condition on $\partial\Omega$ since the time-dependent problem would have non-homogeneous boundary conditions.

The CVBEM approximation function is both continuous and differentiable wherever the CVBEM basis functions are analytic. The approximate solution to the time-dependent component, given later in Equation (2), is also both continuous and differentiable. Therefore, the sum of the two approximate solutions is con- tinuous and differentiable wherever the CVBEM basis functions are analytic.

When the CVBEM basis functions are not analytic throughout the plane, special treatment of the resulting singularities or branch cuts is required in order to ensure that the CVBEM approximation function is at least analytic through- out the problem domain. The recent paper [5] examines several different types of CVBEM basis functions and includes a discussion about the treatment of basis functions that have branch cuts originating from nodes located exterior to the problem domain.

Suppose that the CVBEM basis functions are analytic throughout the problem domain (this can always be done by choosing entire functions, such as complex polynomials, as the CVBEM basis functions). Then, the global approximation function is continuous and differentiable throughout the problem domain.

Thus, computational estimates can be developed throughout the problem domain, and vector fields representing the gradient of the global approximation function can be developed throughout the problem domain.

These properties of CVBEM approximation functions are not shared with the usual real-variable domain numerical schemes such as the finite element method (FEM) and finite difference method (FDM), which develop point estimates at nodal points defined by meshing the problem domain and require interpolation to infill approximations of the potential function at points in the domain that are not nodes of the computational mesh.

The differentiability within the problem domain of the coupled CVBEM and generalized Fourier series approximation functions distinguishes this methodology from commonly-used domain methods such as the Finite Difference Method (FDM), Finite Element Method (FEM), and Finite Volume Method (FVM), which only estimate values of the solution at a finite number of nodes defined by a discretization (mesh) of the problem domain.

The purpose of this research is to the extend the well-known CVBEM modeling technique to an important hyperbolic PDE. The coupled CVBEM procedure that is presented in this work is significant because it applies some of the numerical advantages of solving Laplace's equation with a boundary element method to a solution for a PDE that does not have a boundary integral repre- sentation.

The result is a numerical technique that satisfies the strong formulation of the wave PDE. This is important because satisfying the strong formulation of a PDE is not generally a principle of the aforementioned domain discretization tech- niques, which only either satisfy the weak formulation of the wave equation or approximately solve the strong formulation of the wave equation.

2. STEADY-STATE COMPONENT AND THE COMPLEX VARIABLE BOUNDARY ELEMENT METHOD

As previously mentioned, the steady-state problem is governed by the Laplace equation. One numerical technique for approximating the solution to these problems is the Complex Variable Boundary Element Method, which has been the topic of numerous papers and books [6] [7] [8] [9] [10] . The CVBEM belongs to the general class of numerical methods known as boundary element methods (BEMs). As is the case with BEMs, the CVBEM relies on the use of a boundary integral equation in its formulation. The boundary integral equation for the CVBEM is the Cauchy integral formula.

Recent research related to the CVBEM has extended its use to modeling problems in three and higher spatial dimensions [11] as well as problems related to the diffusion equation [12] . Current efforts have been dedicated to improving the efficiency of the three-dimensional CVBEM as well as to extending the CVBEM to model applications of the time-dependent three-dimensional diffusion equation.

To solve the steady-state problem, the CVBEM is applied to the boundary conditions of the global BVP. The CVBEM formulation used in this particular work is based upon the use of complex polynomial basis functions, although any other family of analytic functions, including combinations of different families of analytic functions, could be used. It is noted that the choice of the analytic basis functions to be used in the CVBEM formulation may be motivated by the problem boundary geometry in order to simplify the modeling requirements and to better fit the boundary geometry.

The CVBEM approximation function is a linear combination of functions from the family of basis functions used in the CVBEM formulation. The coefficients are complex constants, and the basis functions are complex variable functions. Thus, the CVBEM approximation function has both a real and an imaginary part. The coefficients of the CVBEM linear combination are determined by collocating the real part of the CVBEM approximation function with the Dirichlet boundary conditions.

2.1 Review of CVBEM Modeling Approach

A CVBEM approximation function, $\hat{\omega}$, has the general form

$$\hat{\omega}(z)=\sum_{k=1}^{p}c_k g_k(z), \tag{1}$$

where c_k is the kth complex coefficient, $g_k(z)$ is the kth member of the family of CVBEM basis functions being used in the approximation, and p is the number of basis functions being used in the approximation.

The complex coefficients c_k, which are composed of two real-valued constants, each result in two degrees of freedom to be determined as part of the CVBEM modeling process. Thus, there are 2p degrees of freedom in a CVBEM approximation function, so it is necessary to know at least 2p boundary conditions in order to uniquely determine the coefficients of the CVBEM approximation function when using collocation.

Various techniques for determining the complex coefficients of the CVBEM linear combination include Fredholm integral types of formulations, minimizing the departure between boundary condition values and the CVBEM boundary values in a Hilbert space with least squares, and fitting the CVBEM approximation function exactly to the specified boundary conditions by using collocation [13] . In this paper, collocation is used to determine the coefficient values.

Provided that the basis functions used in the CVBEM formulation are analytic within the problem domain and along the boundary, the result of modeling with the CVBEM is the development of an approximation function that is analytic over the entire problem domain and boundary whose real part solves the Laplace equation and can be fitted to satisfy the prescribed Dirichlet boundary con- ditions.

3. TIME-DEPENDENT PROBLEM

The time-dependent component of the problem is modeled by a finite series of the form

$$\hat{u}_2(x,y,t)=\sum_{m=1}^{M}\sum_{n=1}^{N}\left[a_{m,n}\sin\left(\frac{m\pi x}{L_1}\right)\sin\left(\frac{n\pi y}{L_2}\right)\cos\left(\pi\sqrt{\frac{m^2}{L_1^2}+\frac{n^2}{L_2^2}}t\right)\right.$$
$$\left.+b_{m,n}\sin\left(\frac{m\pi x}{L_1}\right)\sin\left(\frac{n\pi y}{L_2}\right)\sin\left(\pi\sqrt{\frac{m^2}{L_1^2}+\frac{n^2}{L_2^2}}t\right)\right] \tag{2}$$

Notice that these functions satisfy the strong formulation of the wave PDE. Sine functions are used specifically for the spatial variables in the approximation function so as to satisfy the homogeneous boundary conditions of the time- dependent problem.

In this work, the coefficients $a_{m,n}$ are determined by collocation of the approximation function with points located in the interior of the problem do- main that are set to match the initial condition of the time-dependent problem. It is necessary to specify mn number of collocation points in order to calculate values for the $a_{m,n}$ coefficients. These initial condition collocation points are typically located so as to be reasonably uniformly spaced throughout the pro- blem domain (future research will assess advantages of defining collocation point locations non-uniformly). Additionally, the coefficients $b_{m,n}$ are deter- mined by collocation of the derivative of the approximation function with col- location points located in the problem domain that are set to match the de- rivative of the initial condition of the time-dependent problem. Again, it is necessary to specify mn number of collocation points in order to calculate values for the $b_{m,n}$ coefficients. These collocation points are also located so as to be reasonably uniformly spaced throughout the problem domain.

4. METHOD DEVELOPMENT

In this section, we do not consider the approximation functions $\hat{u}_1$ and $\hat{u}_2$. Rather, we use the exact functions u_1 and u_2 to demonstrate that the global problem can be decomposed in the manner that we have suggested. Suppose that the initial and boundary conditions are specified as the continuous functions $g(x,y)$ and $f(x,y)$, respectively. We say that the initial condition is con- sistent with the boundary conditions if $g(x,y)=f(x,y)$ for each boundary coordinate $(x,y)\in\partial\Omega$. Consequently, if there exists a boundary coordinate pair $(x,y)\in\partial\Omega$ such that $g(x,y)\neq f(x,y)$, we say that the initial condition is inconsistent with the boundary conditions.

4.1. A Consistent Initial Condition

A function u satisfies the two-dimensional wave equation if $c(u_{xx}+u_{yy})=u_{tt}$, or equivalently, if $c\Delta u=u_{tt}$. We can decompose the problem into a steady-state component and a time-dependent component. The solution to the steady-state problem is a function $u_1(x,y)$ such that $\Delta u_1=0$. That is, the solution to the steady-state problem is a function that satisfies the well-known Laplace partial differential equation as well as the boundary conditions from the global BVP. The solution to the time-dependent component of the problem is a function $u_2(x,y,t)$ that satisfies (1) the PDE $\Delta u_2 = \frac{\partial^2 u_2}{\partial t^2}$, (2) the homogeneous boundary conditions, and (3) the initial condition for the time-dependent component calculated as $g(x,y) - u_1(x,y)$. The global solution is obtained by adding the solutions to the steady-state and time-dependent problems, $u = u_1 + u_2$.

We shall show that an arbitrary term of Equation (2) satisfies the PDE $\Delta u_2 = \frac{\partial^2 u_2}{\partial t^2}$. This is sufficient to show that the linear combination in Equation (2) also satisfies the wave PDE.

$$\begin{aligned}
\frac{\partial^2 u_2}{\partial t^2} &= -a_{m,n}\pi^2\left(\frac{m^2}{L_1^2}+\frac{n^2}{L_2^2}\right)\sin\left(\frac{m\pi x}{L_1}\right)\sin\left(\frac{n\pi y}{L_2}\right)\cos\left(\pi\sqrt{\frac{m^2}{L_1^2}+\frac{n^2}{L_2^2}}t\right) \\
&\quad -b_{m,n}\pi^2\left(\frac{m^2}{L_1^2}+\frac{n^2}{L_2^2}\right)\sin\left(\frac{m\pi x}{L_1}\right)\sin\left(\frac{n\pi y}{L_2}\right)\sin\left(\pi\sqrt{\frac{m^2}{L_1^2}+\frac{n^2}{L_2^2}}t\right) \\
&= -a_{m,n}\left(\frac{m^2\pi^2}{L_1^2}\right)\sin\left(\frac{m\pi x}{L_1}\right)\sin\left(\frac{n\pi y}{L_2}\right)\cos\left(\pi\sqrt{\frac{m^2}{L_1^2}+\frac{n^2}{L_2^2}}t\right) \\
&\quad -b_{m,n}\left(\frac{m^2\pi^2}{L_1^2}\right)\sin\left(\frac{m\pi x}{L_1}\right)\sin\left(\frac{n\pi y}{L_2}\right)\sin\left(\pi\sqrt{\frac{m^2}{L_1^2}+\frac{n^2}{L_2^2}}t\right) \\
&\quad -a_{m,n}\left(\frac{n^2\pi^2}{L_2^2}\right)\sin\left(\frac{m\pi x}{L_1}\right)\sin\left(\frac{n\pi y}{L_2}\right)\cos\left(\pi\sqrt{\frac{m^2}{L_1^2}+\frac{n^2}{L_2^2}}t\right) \\
&\quad -b_{m,n}\left(\frac{n^2\pi^2}{L_2^2}\right)\sin\left(\frac{m\pi x}{L_1}\right)\sin\left(\frac{n\pi y}{L_2}\right)\sin\left(\pi\sqrt{\frac{m^2}{L_1^2}+\frac{n^2}{L_2^2}}t\right) \\
&= \Delta u_2
\end{aligned} \tag{3}$$

Since u_1 is a function of only x and y, it follows that $\frac{\partial^2 u_1}{\partial t^2} = 0$. Therefore, $u_{tt} = \frac{\partial^2 u_1}{\partial t^2} + \frac{\partial^2 u_2}{\partial t^2} = \frac{\partial^2 u_2}{\partial t^2}$. To see that u satisfies the governing PDE, observe that

$$\Delta u = \Delta(u_1 + u_2) = \Delta u_1 + \Delta u_2$$
$$= 0 + \frac{\partial^2 u_2}{\partial t^2} = \frac{\partial^2 u_2}{\partial t^2} = u_{tt} \qquad (4)$$

4.2. An Inconsistent Initial Condition

Now suppose $u(x, y, 0) \neq f(x, y)$ on Γ. When the boundary and initial conditions are only specified at a discrete number of points, this case is equivalent to saying that there exists a point (x_i, y_i) at which $u(x_i, y_i, 0)$ is specified, but $u(x_i, y_i, 0) \neq u_1(x_i, y_i)$. In this case, the initial condition is not equal on the boundary to the global boundary conditions, so the initial condition is said to be inconsistent. Since the initial condition of the global problem is inconsistent with the boundary conditions, the difference between the initial condition of the global problem and the CVBEM approximation of the steady-state solution would not be approximately zero on the boundary. Thus, use of the two-dimen- sional Fourier sine series, as done in Section 4.1, to approximate the initial condition of the time-dependent component would not be appropriate since the boundary conditions of this problem would not be approximately zero on the boundary. Problems with inconsistent initial conditions can still be modeled using this procedure, however, it is noted that they outcome of this procedure will not satisfy the initial condition on $\partial\Omega$.

5. GLOBAL SOLUTION

The global solution is achieved by summing the solutions to the steady-state and time-dependent problems. Since both of the approximation functions from the two problems are continuous throughout the two-dimensional plane (except for at any branch points or cuts associated with the basis functions used in the Complex Variable Boundary Element Method formulation), their sum, which represents the solution to the global problem is also continuous throughout the plane except for at any point where the CVBEM basis functions are not analytic. Consequently, computational results can be developed continuously throughout the two-dimensional plane except where the CVBEM basis functions have branch points or branch cuts.

5.1. Satisfying the Boundary Conditions

Substituting $c_k = \alpha_k + i\beta_k$ and $g_k(z) = \lambda_k(x,y) + i\mu_k(x,y)$ in Equation (1), we can derive a CVBEM approximation function of the form

$$\hat{\omega}(z) = \sum_{k=1}^{p} c_k g_k(z) = \sum_{k=1}^{p} (\alpha_k + i\beta_k)(\lambda_k(x,y) + i\mu_k(x,y)). \tag{5}$$

The real part, $\hat{u}_1$, of the CVBEM approximation function, which represents the potential function, is

$$\begin{aligned} \hat{u}_1(x,y) &= \alpha_1\lambda_1(x,y) - \beta_1\mu_1(x,y) + \alpha_2\lambda_2(x,y) - \beta_2\mu_2(x,y) \\ &\quad + \cdots + \alpha_p\lambda_p(x,y) - \beta_p\mu_p(x,y) \\ &= \sum_{k=1}^{p} (\alpha_k\lambda_k(x,y) - \beta_k\mu_k(x,y)). \end{aligned} \tag{6}$$

The boundary conditions for the steady-state problem are Dirichlet and are specified as the same boundary conditions as the global initial-BVP. This implies a system of 2p equations given by

$$\hat{u}_1(x_i,y_i) = \sum_{k=1}^{p} (\alpha_k\lambda_k(x_i,y_i) - \beta_k\mu_k(x_i,y_i)) = u_1(x_i,y_i) \tag{7}$$

for $i = 1,2,\cdots,2p$. Equivalently, in matrix form

$$[A_1]\{c\} = \{u_1\}, \tag{8}$$

where $\{u_1\}$ is a vector containing the specified potential boundary conditions from the global BVP, $[A_1]$ is the matrix obtained from evaluating the CVBEM basis functions at each of the collocation points, and $\{c\}$ is a vector containing the unknown coefficients $\alpha_1, \beta_1, \cdots, \alpha_p, \beta_p$. Once the coefficients are determined by solving the linear system in (8), they can be substituted back into Equation (6), which is the CVBEM approximation of the solution to the steady-state problem. The resulting function can be evaluated throughout the two-dimen- sional plane wherever the CVBEM basis functions are analytic.

In both two-dimensional and higher-dimensional problems, the use of a ran- dom number generator is sometimes employed to locate the collocation points approximately uniformly spatially on the problem boundary. However, they may also be located exactly uniformly spatially on the problem boundary, and the specific problem may dictate which approach is desirable.

Computational issues may arise depending on the choice of basis functions used in the CVBEM approximation function. For example, basis functions in-volving branch cuts, such as complex logarithms or reciprocals of complex monomials, among other types of functions, include considerations of positioning branch cuts to lie outside of the problem domain and boundary in order to enlarge the area of applicability of the CVBEM approximation. Procedures for handling these branch cuts have been well-documented in several papers and books including, [3] and the most recent book [4] and so are not repeated here.

5.2. Satisfying the Initial Condition

Since the exact solution to the steady-state problem is unknown, we need to use the CVBEM approximation of u_1 to determine the initial condition for the time-dependent problem. Thus, the initial condition of the time-dependent pro- blem is given by $u_2(x,y,0)=g(x,y)-\hat{u}_1(x,y)$. The coefficients $a_{m,n}$ are se- lected so as to satisfy

$$\hat{u}_2(x_i,y_i,0)=\sum_{m=1}^{M}\sum_{n=1}^{N}a_{m,n}\sin\left(\frac{m\pi x_i}{L_1}\right)\sin\left(\frac{n\pi y_i}{L_2}\right)=g(x_i,y_i)-\hat{u}_1(x_i,y_i) \tag{9}$$

at each of the initial condition collocation points. Equivalently, in matrix form

$$[A_2]\{\boldsymbol{a}\}=\{\boldsymbol{u}_2\}, \tag{10}$$

where $\{\boldsymbol{u}_2\}$ is a vector containing the specified initial conditions from the time-dependent problem, $[A_2]$is the $mn\times mn$ matrix obtained from evaluating $\sin\left(\frac{m\pi x}{L_1}\right)\sin\left(\frac{n\pi y}{L_2}\right)$ at each of the collocation points, and $\{\boldsymbol{a}\}$ is a vector containing the unknown coefficients $a_{m,n}$.

5.3. Satisfying the Derivative of the Initial Condition

The derivative of the initial condition is specified as $h(x,y)$. At $t=0$, the time derivative of the finite series approximation of the time-dependent component is

$$\frac{\partial}{\partial t}\hat{u}_2(x,y,0)=\sum_{m=1}^{M}\sum_{n=1}^{N}b_{m,n}\left(\pi\sqrt{\frac{m^2}{L_1^2}+\frac{n^2}{L_2^2}}\right)\sin\left(\frac{m\pi x}{L_1}\right)\sin\left(\frac{n\pi y}{L_2}\right). \tag{11}$$

The coefficients $b_{m,n}$ are selected so as to satisfy

$$\frac{\partial}{\partial t}\hat{u}_2(x_i,y_i,0)=\sum_{m=1}^{M}\sum_{n=1}^{N}b_{m,n}\left(\pi\sqrt{\frac{m^2}{L_1^2}+\frac{n^2}{L_2^2}}\right)\sin\left(\frac{m\pi x_i}{L_1}\right)\sin\left(\frac{n\pi y_i}{L_2}\right)=h(x_i,y_i). \tag{12}$$

at each of the collocation points for the derivative of the initial condition. Equivalently, in matrix form

$$[A_3]\{b\}=\{h\}, \tag{13}$$

where $\{h\}$ is a vector containing the value of the derivative of the initial condition at each of the collocation points, $[A_3]$ is the $mn\times mn$ matrix obtained from evaluating the expression

$$\left(\pi\sqrt{\frac{m^2}{L_1^2}+\frac{n^2}{L_2^2}}\right)\sin\left(\frac{m\pi x}{L_1}\right)\sin\left(\frac{n\pi y}{L_2}\right)$$

at each of the collocation points, and $\{b\}$ is a vector containing the unknown coefficients $b_{m,n}$.

The coefficients of the CVBEM approximation function are determined so that the CVBEM outcome approximately satisfies the specified Dirichlet boun- dary conditions of the global initial-boundary value problem. Further, the boun- dary conditions for the time-dependent component are identically zero on the boundary due to the use of the two-dimensional Fourier sine series for the spatial variables. Consequently, upon summing the approximate solutions from the two components, the value of the global approximation function on the boundary is exactly the value of the CVBEM outcome on on the boundary, which approximately satisfies the global boundary conditions.

The initial condition of the time-dependent component is calculated as the difference between the initial condition of the global initial-boundary value problem and the CVBEM approximation of the steady-state solution. Therefore, upon superposing the approximate solutions of the steady-state and time- dependent problems, the global initial condition is approximately satisfied. As indicated by the generalized Fourier series in Equation (2), there are two sets of coefficients to be determined; namely, the $a_{m,n}$ and $b_{m,n}$ coefficients. When evaluated at $t=0$, the only terms of the time-dependent approximation func- tion that are nonzero are those corresponding to the $a_{m,n}$ coefficients. Hence, the $a_{m,n}$ coefficients are determined so as to approximately satisfy the calcu- lated initial condition of the time-dependent component. Further, when evalu- ated at $t=0$, the only terms of the time derivative of the time-dependent appro- ximation function that are nonzero are those corresponding to the $b_{m,n}$ co- efficients. Thus, the $b_{m,n}$ coefficients are determined so as to approximately satisfy the specified derivative of the initial condition.

6. LIMITATIONS

To ensure that the boundary conditions of the global problem are satisfied, the steady-state solution is fitted to the global boundary conditions. Since the global boundary conditions are satisfied exclusively by the steady-state solution, it is necessary to prescribe homogeneous boundary conditions for the time-depen- dent problem. Therefore, only sine functions are used in the Fourier series approximation of the time-dependent problem. By using the Fourier sine series, the solution to the time-dependent problem is zero along the boundary of a rectangular domain of dimensions $L_1 \times L_2$. However, this is a property of the rectangular geometry of the domain and would not necessarily be true for other geometries. Thus, this technique is currently generally limited to problems with rectangular domain geometries or unions of rectangular subdomains. Additionally, this methodology is limited in that it will not satisfy the initial condition on the boundary if $g(x,y) \neq h(x,y)$ for all $(x,y) \in \partial\Omega$ since the boundary con- ditions for the transient component would no longer be homogeneous, which means that the two-dimensional Fourier sine series would not satisfy the boundary conditions of the transient problem.

7. CONCLUSIONS

The Complex Variable Boundary Element Method approach to solving partial differential equations is well-documented in other publications and the reader is referred to [4] for details regarding the mathematical foundations of the mo- deling approach. In this work, the Complex Variable Boundary Element Method is extended to modeling applications of the two-dimensional wave equations. The technique is based on resolving the global problem into two components, namely steady-state and time-dependent components, that are modeled sepa- rately. The solutions to each component are then summed, which yields a func- tion that satisfies the strong formulation of the wave PDE and approximately satisfies the boundary and initial conditions.

The steady-state component of the problem is solved by application of the CVBEM, and the time-dependent component of the problem is modeled by a truncated series of functions that would be used to solve the PDE analytically. The steady-state solution is fitted so as to satisfy the boundary conditions of the global problem. The boundary conditions of the time-dependent problem are specified to be zero continuously so that when the solutions to the steady-state and time-dependent components are added together, the value of the global solution at each boundary collocation point is as specified by the boundary con- ditions of the global problem. Although the current work focuses on the wave equation, the modeling approach presented applies to other PDE formulations as well as to problems in higher dimensions.

An alternative to the CVBEM is the recently-developed strong-form colloca- tion method known as the Singular Boundary Method (SBM), which uses a linear combination of the fundamental solution of the governing partial dif- ferential equation to approximate the field variables. In that sense, the SBM is a coupling of the boundary element method and the method of fundamental solu- tions. The SBM can easily be applied to problems with complex geometries and can be extended to model problems in three spatial dimensions. The reader is referred to [14] [15] [16] for more details.

REFERENCES

1. 1. Hromadka, T. and Guymon, G. (1984) The Complex Variable Boundary Element Method. International Journal of Numerical Methods Engineering.
2. Hromadka, T. and Lai, C. (1987) The Complex Variable Boundary Element Method. Springer-Verlag, New York.
3. Hromadka, T.V. and Whitley, R.J. (1998) Advances in the Complex Variable Boundary Element Method. Springer, New York.
4. Hromadka, T. and Whitley, R. (2014) Foundations of the Complex Variable Boundary Element Method. Springer, New York.
5. Wilkins, B.D., Hromadka, T., Whitley, R., Johnson, A.N., Boucher, R., Outing, D.A., Phillips, M.D., McInvale, H.D. and Horton, S. (2016) Assessment of Complex Variable Basis Functions in the Approximation of Ideal Fluid Flow Problems. International Journal of Computational Methods and Experimental Measurements.
6. Hromadka, T. and Whitley, R. (2005) Approximating Three-Dimensional Steady-State Potential Flow Problems Using Two-Dimensional Complex Polynomials. Engineering Analysis with Boundary Elements, 29, 190-194.
7. Dean, T. and Hromadka, T. (2009) A Collocation Cvbem Using Program Mathematica. Engineering Analysis with Boundary Elements, 34, 417-422.
8. Hromadka, T. and Whitley, R. (1996) A New Formulation for Developing CVBEM Approximation Functions. Engineering Analysis with Boundary Elements, 18, 39-41.
9. Dumir, P. and Kumar, R. (1993) Complex Variable Boundary Element Method for Torsion of Anisotropic Bars. Applied Mathematical Modelling, 17, 80-88.
10. Hromadka, T. and Pardoen, G. (1985) Application of the CVBEM to Non-Uniform st. Venant Torsion. Computer Methods in Applied Mechanics and Engineering, 53, 149-161.
11. Hromadka, T. (2002) A Multi-Dimensional Complex Variable Boundary Element Method. Topics in Engineering, Vol. 40, WIT Press, Billerica.
12. Wilkins, B.D., Greenberg, J., Redmond, B., Baily, A., Flowerday, N., Kratch, A., Hromadka, T.V., Boucher, R., McInvale, H.D. and Horton, S. (2016) An Unsteady Two-Dimensional Complex Variable Boundary Element Method (cvbem). International Journal of Computational Methods and Experimental Measurements.
13. Bohannon, A. and Hromadka, T. (2009) The Complex Polynomial Method with a Least-Squares Fit to Boundary Conditions. Engineering Analysis with Boundary Elements, 33, 1100-1102.
14. Chen, W. and Wang, F. (2016) Singular Boundary Method Using Time-Dependent Fundamental Solution for Transient Diffusion Problems. Engineering Analysis with Boundary Elements, 68, 115-123.
15. Li, J., Fu, Z. and Chen, W. (2016) Numerical Investigation on the Obliquely Incident Water Wave Passing through the Submerged Breakwater by Singular Boundary Method. Computers and Mathematics with Applications, 71, 381-390.
16. Qu, W., Chen, W. and Zheng, C. (2017) Diagonal form Fast Multipole Singular Boundary Method Applied to the Solution of High-Frequency Acoustic Radiation and Scattering. International Journal for Numerical Methods in Engineering.

CHAPTER 17

Analytical Solution for the Time-Dependent Emden-Fowler Type of Equations by Homotopy Analysis Method with Genetic Algorithm

Waleed Al-Hayani[1], Laheeb Alzubaidy[2], Ahmed Entesar[1]

[1]*Department of Mathematics, College of Computer Science and Mathematics, University of Mosul, Mosul, Iraq*

[2]*Department of Software Engineering, College of Computer Science and Mathematics, University of Mosul, Mosul, Iraq*

ABSTRACT

In this paper, Homotopy Analysis method with Genetic Algorithm is presented and used to obtain an analytical solution for the time-dependent Emden-Fowler type of equations and wave-type equation with singular behavior at $x = 0$. The advantage of this single global method employed to present a reliable framework is utilized to overcome the singularity behavior at the point $x = 0$ for both models. The method is demonstrated for a variety of problems in one and higher dimensional spaces where approximate-exact solutions are obtained. The results obtained in all cases show the reliability and the efficiency of this method.

Keywords: Homotopy Analysis Method, Genetic Algorithm, Emden-Fowler Equation, Wave-Type Equation, Adomian Polynomials, Noise Terms, Padé Approximants, Simpson Rule

1. INTRODUCTION

In many previous works of the diffusion of heat perpendicular to the surfaces of parallel planes are modeled by the heat equation

$$x^{-r}\left(x^{r}y_{x}\right)_{x}+af(x,t)g(y)+h(x,t)=y_{t},\ 0<x\leq L,\ 0<t<T,\ r>0, \tag{1}$$

or equivalently

$$y_{xx}+\frac{r}{x}y_{x}+af(x,t)g(y)+h(x,t)=y_{t},\ 0<x\leq L,\ 0<t<T,\ r>0, \tag{2}$$

where $f(x,t)g(y)+h(x,t)$ is the nonlinear heat source, $y(x,t)$ is the temperature, and t is the dimensionless time variable. For the steady-state case, and for $r=2$, $h(x,t)=0$, Equation (2) is the Emden-Fowler equation [1] [2] [3] given by

$$y''+\frac{2}{x}y'+af(x)g(y)=0,\ y(0)=y_{0},\ y'(0)=0, \tag{3}$$

where $f(x)$ and $g(y)$ are some given functions of x and y respectively. For $f(x)=1$ and $g(y)=y^{n}$, Equation (3) is the standard Lane-Emden equation that was used to model the thermal behavior of a spherical cloud of gas acting under the mutual attraction of its molecules [1] and subject to the classical laws of thermodynamics. For other special forms of $g(y)$, the well-known Lane- Emden equation was used to model several phenomena in mathematical physics and astrophysics such as the theory of stellar structure, the thermal behavior of a spherical cloud of gas, isothermal gas spheres, and theory of thermionic currents [1] [2] [3] . A substantial amount of work has been done on this type of problems for various structures of $g(y)$ in [1] - [7] .

On the other hand, the wave type of equations with singular behavior is given by

$$x^{-r}\left(x^{r}y_{x}\right)_{x}+af\left(x,t\right)g\left(y\right)+h\left(x,t\right)=y_{tt},\ 0<x\le L,\ 0<t<T,\ r>0, \tag{4}$$

or equivalently

$$y_{xx}+\frac{r}{x}y_{x}+af\left(x,t\right)g\left(y\right)+h\left(x,t\right)=y_{tt},\ 0<x\le L,\ 0<t<T,\ r>0, \tag{5}$$

will be examined as well, where $f\left(x,t\right)g\left(y\right)+h\left(x,t\right)$ is a nonlinear source, t is the dimensionless time variable, and $y(x,t)$ is the displacement of the wave at the position x and at time t. The singularity behavior that occurs at the point $x=0$ is the main difficulty in the analysis of Equations (2) and (5). Wazwaz [8] used the Adomian decomposition method (ADM) to get an analytical solution for the time-dependent Emden-Fowler type of equations and wave-type equation with singular behavior.

The main objective of this paper is to apply Homotopy Analysis Method (HAM) to obtain approximate-exact solutions for different models for the time- dependent Emden-Fowler type of equations and wave-type equation with singular behavior at $x=0$. While the VIM [9] [10] requires the determination of Lagrange multiplier in its computational algorithm, HAM is independent of any such requirements, HAM handles linear and nonlinear terms in a simple and straightforward manner without any additional requirements. Also, in this paper we apply Genetic Algorithm (GA) to obtain an approximate solution of the same equations.

In what follows, we give a brief review of Homotopy analysis method and Genetic algorith.

2. ANALYSIS OF THE HOMOTOPY ANALYSIS METHOD

To describe the basic ideas of the HAM, we consider the following differential equation:

$$N\left[y\left(x,t\right)\right]=k\left(x,t\right), \tag{6}$$

where N is a nonlinear operator, x and t denote the independent variables, $y(x,t)$ is an unknown function and $k(x,t)$ is a known analytic function. For simplicity, we ignore all boundary or initial conditions, which can be treated in the similar way. By means of generalizing the traditional Homotopy method, Liao [11] constructs the so called zero-order deformation equation

$$(1-q)L[\phi(x,t;q)-y_0(x,t)]=qh[N[\phi(x,t;q)]-k(x,t)], \quad (7)$$

where $q\in[0,1]$ is an embedding parameter, h is a nonzero auxiliary parameter, L is an auxiliary linear operator, $y_0(x,t)$ is an initial guess of $y(x,t)$ and $\phi(x,t;q)$ is an unknown function. It is important, that one has great freedom to choose auxiliary objects such as h and L in HAM. Obviously, when $q=0$ and $q=1$ it holds

$$\phi(x,t;0)=y_0(x,t), \quad \phi(x,t;1)=y(x,t), \quad (8)$$

respectively. Thus, as q increases from 0 to 1, the solution $\phi(x,t;q)$ varies from the initial guess $y_0(x,t)$ to the solution $y(x,t)$. Expanding $\phi(x,t;q)$ in Taylor series with respect to q, we have

$$\phi(x,t;q)=y_0(x,t)+\sum_{m=1}^{+\infty}y_m(x,t)q^m, \quad (9)$$

where

$$y_m(x,t)=\frac{1}{m!}\frac{\partial^m\phi(x,t;q)}{\partial q^m}\bigg|_{q=0}. \quad (10)$$

If the auxiliary linear operator, the initial guess, the auxiliary parameter h, and the auxiliary function are so properly chosen, the above series (9) converges at $q=1$, then we have

$$\phi(x,t;1) = y_0(x,t) + \sum_{m=1}^{+\infty} y_m(x,t), \tag{11}$$

which must be one of the solutions of the original nonlinear equation, as proved by Liao [11] . If $h=-1$, Equation (7) becomes

$$(1-q)L[\phi(x,t;q) - y_0(x,t)] + q[N[\phi(x,t;q)] - k(x,t)] = 0, \tag{12}$$

which is used mostly in the Homotopy perturbation method [12] .

According to Equation (10), the governing equation can be deduced from the zero-order deformation Equation (7). We define the vectors

$$\mathbf{y}_i = \{y_0(x,t), y_1(x,t), \cdots, y_i(x,t)\}. \tag{13}$$

Differentiating Equation (7) m times with respect to the embedding parameter q and then setting $q=0$ and finally dividing them by $m!$, we have the so- called mth-order deformation equation

$$L[y_m(x,t) - \chi_m y_{m-1}(x,t)] = hR_m(\mathbf{y}_{m-1}), \tag{14}$$

where

$$R_m(\mathbf{x}_{m-1}) = \frac{1}{(m-1)!} \left. \frac{\partial^{m-1}\{N[\phi(x,t;q)] - k(x,t)\}}{\partial q^{m-1}} \right|_{q=0}, \tag{15}$$

and

$$\chi_m = \begin{cases} 0, & m \le 1, \\ 1, & m > 1. \end{cases} \tag{16}$$

It should be emphasized that $y_m(x,t)$ $(m \ge 1)$ are governed by the linear Equation (14) with the linear boundary conditions that come from the original problem, which can be easily solved by symbolic computation softwares such as Maple and Mathematica.

3. GENETIC ALGORITHMS

Definition 1 Genetic Algorithms are search and optimization techniques based on Darwin's Principle of Natural Selection.

Definition 2 Genetic Algorithm Operators [13] [14]

The simplest form of genetic algorithm involves three types of operators: selection, crossover and mutation.

Selection: This operator selects chromosomes in the population for reproduction. The fitter the chromosome, the more times it is likely to be selected to reproduce.

Crossover: This operator randomly chooses a locus and exchanges the subsequences before and after that locus between two chromosomes to create two offspring. For example, the strings 10000100 and 11111111 could be crossed over after the third locus in each to produce the two offspring 10011111 and 11100100.

The crossover operator roughly mimics biological recombination between two single-chromosome (haploid) organisms.

Mutation: This operator randomly flips some of the bits in a chromosome. For example, the string 00000100 might be mutated in its second position to yield 01000100. Mutation can occur at each bit position in a string with some probability, usually very small (e.g., 0.001).

Algorithm 3 A Simple Genetic Algorithm [13] [14]

Given a clearly defined problem to be solved and a bit string representation for candidate solutions, a simple GA works as follows:

1) Start with a randomly generated population of n l-bit chromosomes (candidate solutions to a problem).
2) Calculate the fitness $f(x,t)$ of each chromosome x and t in the population.
3) Repeat the following steps until n offspring have been created:
    a) Select a pair of parent chromosomes from the current population, the probability of selection being an increasing function of fitness. Selection is done "with replacement," meaning that the same chromosome can be selected more than once to become a parent.

b) With probability pc (the "crossover probability" or "crossover rate"), cross over the pair at randomly chosen point (chosen with uniform probability) to form two offspring. If no crossover takes place, form two offspring that are exact copies of their respective parents. (Note that here the crossover rate is defined to be the probability that two parents will crossover in a single point. There are also "multi-point crossover" versions of the GA in which the crossover rate for a pair of parents is the number of points at which a crossover takes place).

c) Mutate the two offspring at each locus with probability pm (the "mutation probability" or "mutation rate"), and place the resulting chromosomes in the new population. If n is odd, one new population member can be discarded at random.

4) Replace the current population with the new population.

5) Go to step 2.

Also see the flow chart (Figure 1).

4. APPLICATIONS OF THE METHOD

In this work, we examine six distinct models with singular behavior at $x=0$, two linear time-dependent Lane-Emden type of equations, two linear models of wave-type equation, and two nonlinear singular models. To show the high accuracy of the solution results compared with the exact solution, we give the numerical results applying the Genetic Algorithm (GA), HAM$^{(n=5)}$, Padé approximants (PA) of order $[p/q]$, and the numerical solution with the Simpson rule (SIMP). Twenty points have been used in the Simpson rule. The computations associated with the examples were performed using a Maple 13 package with a precision of 20 dgits.

4.1. Time-Dependent Lane-Emden Type

Example 1. Firstly, let us consider the following linear homogeneous equation

$$y_{xx}+\frac{2}{x}y_x-\left(6+4x^2-\cos t\right)y=y_t, \tag{17}$$

subject to initial conditions

$$y(0,t)=e^{\sin t}, \quad y_x(0,t)=0. \tag{18}$$

To solve Equations (17)-(18) by means of the standard HAM, we choose the initial approximation

$$y_0(x,t)=e^{\sin t}, \tag{19}$$

and the linear operator

$$L[\phi(x,t;q)]=\frac{\partial^2\phi(x,t;q)}{\partial x^2}+\frac{2}{x}\frac{\partial\phi(x,t;q)}{\partial x}, \tag{20}$$

with the property

$$L\left[-\frac{c_1}{x}+c_2\right]=0, \tag{21}$$

where c_i $(i=1,2)$ are constants of integration. Furthermore, Equation (17) suggests that we define the nonlinear operator as

$$\begin{aligned}N[\phi(x,t;q)]=&\frac{\partial^2\phi(x,t;q)}{\partial x^2}+\frac{2}{x}\frac{\partial\phi(x,t;q)}{\partial x}\\&-(6+4x^2-\cos t)\phi(x,t;q)-\frac{\partial\phi(x,t;q)}{\partial t}.\end{aligned} \tag{22}$$

Using the above definition, we construct the zeroth-order deformation equation as in (7) and (8) and the mth-order deformation equation for $m\geq 1$ is

$$L[y_m(x,t)-\chi_m y_{m-1}(x,t)]=hR_m(y_{m-1}), \tag{23}$$

with the initial conditions

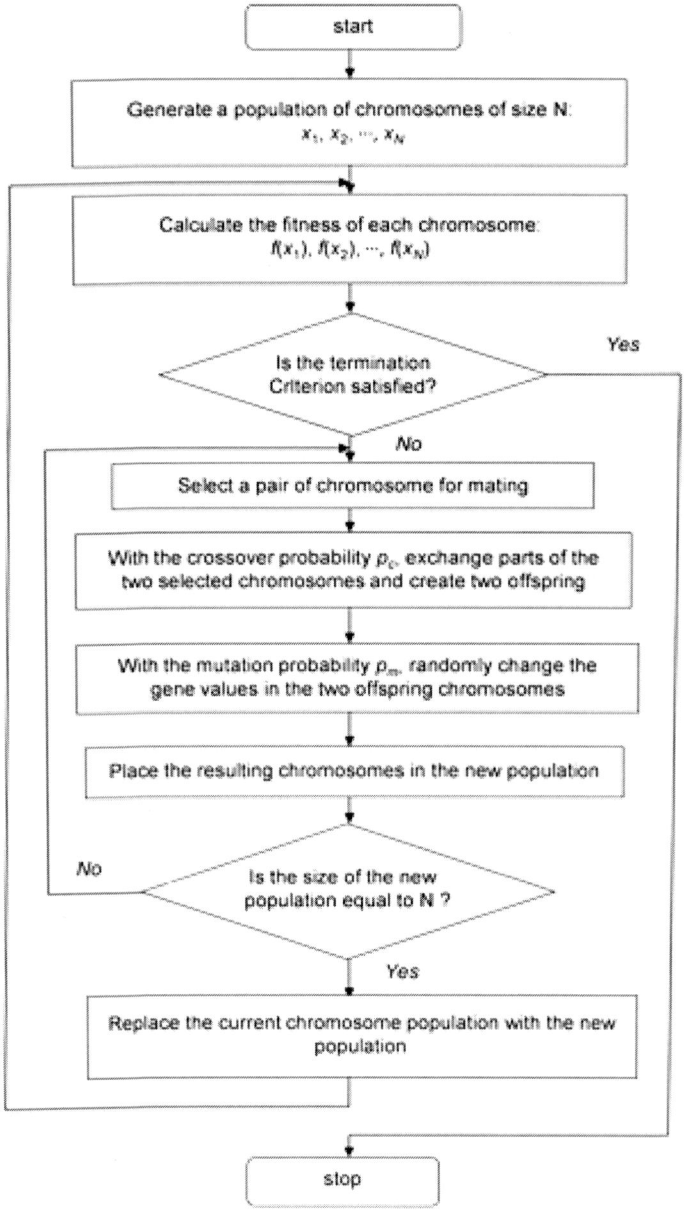

Figure 1. Flow Chart of Genetic Algorithm.

$$y_m(0,t)=0,\quad (y_m)_x(0,t)=0 \tag{24}$$

where

$$R_m(y_{m-1})=(y_{m-1})_{xx}+\frac{2}{x}(y_{m-1})_x-\left(6+4x^2-\cos t\right)y_{m-1}-(y_{m-1})_t. \tag{25}$$

Now, the solution of the mth-order deformation Equation (23) for $m\geq 1$ is

$$y_m(x,t)=\chi_m y_{m-1}(x,t)+h\int_0^x x^{-2}\int_0^x x^2 R_m(y_{m-1}). \tag{26}$$

We now successively obtain

$$\begin{aligned}
&y_1(x,t)=\left(-\frac{1}{5}hx^4-hx^2\right)e^{\sin t},\\
&y_2(x,t)=\left[\frac{1}{90}h^2x^8+\frac{13}{105}h^2x^6-\left(\frac{1}{5}h-\frac{1}{10}h^2\right)x^4-\left(h+h^2\right)x^2\right]e^{\sin t},\\
&y_3(x,t)=\left[-\frac{1}{3510}h^3x^{12}-\frac{59}{11550}h^3x^{10}+\left(\frac{1}{45}h^2-\frac{1}{210}h^3\right)x^8\right]e^{\sin t}\\
&+\left[\left(\frac{26}{105}h^2+\frac{43}{210}h^3\right)x^6-\frac{1}{5}\left(h-h^2-2h^3\right)x^4-\left(h+2h^2+h^3\right)x^2\right]e^{\sin t},\\
&\vdots
\end{aligned} \tag{27}$$

and so on, in this manner the rest of the iterations can be obtained. Thus, the approximate solution in a series form is given by

$$y(x,t)=y_0(x,t)+\sum_{m=1}^{+\infty}y_m(x,t), \tag{28}$$

Hence, the series solution when $h=-1$ is

$$y(x,t)=\left(1+x^2+\frac{1}{2!}x^4+\frac{1}{3!}x^6+\frac{1}{4!}x^8+\cdots\right)e^{\sin t}. \tag{29}$$

This series has the closed form as $n \to \infty$

$$y(x,t) = e^{x^2+\sin t}, \tag{30}$$

which is the exact solution of (17)-(18) compatible with ADM [8] .

Also, this example is solved by using GA as follows:

We'll choose six values of x between $(0,1)$ randomly, and converting them from the decimal format to the binary format

$$0.2 = \underbrace{0.001100}_{x_1}\underbrace{110011}_{t_1}$$

$$0.3 = \underbrace{0.010011}_{x_2}\underbrace{001100}_{t_2}$$

$$0.5 = \underbrace{0.100000}_{x_3}\underbrace{000000}_{t_3}$$

$$0.6 = \underbrace{0.100110}_{x_4}\underbrace{011000}_{t_4}$$

$$0.8 = \underbrace{0.110011}_{x_5}\underbrace{001100}_{t_5}$$

$$0.9 = \underbrace{0.111001}_{x_6}\underbrace{100100}_{t_6}$$

that is,

x_i	0.001100	0.010011	0.100000	0.100110	0.110011	0.111001
t_i	0.110011	0.001100	0.000000	0.011000	0.011000	0.100100

Now,

x_i	0.1875	0.2969	0.5000	0.5938	0.7969	0.8906
t_i	0.7969	0.1875	0.0000	0.3750	0.1875	0.5625
$f(x_i,t_i)$	2.1177	1.3159	1.2840	2.0521	2.2738	3.7675
$p(x_i,t_i)$	0.1653	0.1027	0.1002	0.1602	0.1775	0.2941

where $f(x_i,t_i)$ is the series solution of HAM given by (29) and

$$p(x_i,t_i)=\frac{f(x_i,t_i)}{\sum_{i=1}^{6} f(x_i,t_i)}$$

is the probability of each chromosomes with

$$\sum_{i=1}^{6} f(x_i,t_i)=12.8110.$$

We arrange the chromosomes an ascending order and then choose less than four chromosomes to find the best solution

$$\begin{aligned} x_{31} &= 0.100000, & t_{31} &= 0.000000 \\ x_{21} &= 0.010011, & t_{21} &= 0.001100 \\ x_{41} &= 0.100110, & t_{41} &= 0.011000 \\ x_{11} &= 0.001100, & t_{11} &= 0.110011 \end{aligned}$$

In this step of GA, crossover operation (two points) is done as follows:

$$\begin{aligned} cut_1 &= 2, & cut_2 &= 5 \\ x_{32} &= 0.100000, & t_{32} &= 0.001100 \\ x_{22} &= 0.010011, & t_{22} &= 0.000000 \\ x_{42} &= 0.101110, & t_{42} &= 0.010000 \\ x_{12} &= 0.000100, & t_{12} &= 0.111011 \end{aligned}$$

In the last step of GA a mutation operation (bit inverse, $m=3$) and then converting them from the binary format to the decimal format:

$$\begin{aligned} x_{33} &= 0.101000 = 0.6250, & t_{33} &= 0.000100 = 0.0625 \\ x_{23} &= 0.011011 = 0.4219, & t_{23} &= 0.001000 = 0.1250 \\ x_{43} &= 0.100110 = 0.5938, & t_{43} &= 0.011000 = 0.3750 \\ x_{13} &= 0.001100 = 0.1875, & t_{13} &= 0.110011 = 0.7969 \end{aligned}$$

The optimal solution is found after 51 generation to converge to the exact solution, where $x = 7.2959306E-04$ and $t = 2.2287430E-05$. After execute the Equation (29) many times by using GA as in Table 1 we found the optimal solution.

Table 1. Optimal solution of genetic algorithm for example 1.

x	t	$y_{Exact}(x, t)$	GA	$HAM(n = 5)$	PA [6/6]	SIMP
7.296E−04	2.229E−05	1.0000228	1.0000228	1.0000228	1.0000228	1.0000228
0.023	0.002	1.0025321	1.0023269	1.0025321	1.0025321	1.0025321
0.029	0.031	1.0323482	1.0328346	1.0323482	1.0323482	1.0323482
0.064	0.003	1.0071212	1.0074105	1.0071212	1.0071212	1.0071212
0.222	0.011	1.0621379	1.0623118	1.0621379	1.0621379	1.0621379

Example 2. We consider the following linear inhomogeneous equation,

$$y_{xx} + \frac{2}{x} y_x - \left(5 + 4x^2\right) y = y_t + 6 - 5x^2 - 4x^4, \tag{31}$$

subject to initial conditions

$$y(0,t) = e^t, \quad y_x(0,t) = 0. \tag{32}$$

To solve Equations (31)-(32) by means of the standard HAM, we choose the initial approximation

$$y_0(x,t) = e^t, \tag{33}$$

and the linear operator

$$L[\phi(x,t;q)] = \frac{\partial^2 \phi(x,t;q)}{\partial x^2} + \frac{2}{x} \frac{\partial \phi(x,t;q)}{\partial x}, \tag{34}$$

with the property

$$L\left[-\frac{c_1}{x}+c_2\right]=0, \tag{35}$$

where c_i $(i=1,2)$ are constants of integration. Furthermore, Equation (31) suggests that we define the nonlinear operator as

$$N[\phi(x,t;q)]=\frac{\partial^2\phi(x,t;q)}{\partial x^2}+\frac{2}{x}\frac{\partial\phi(x,t;q)}{\partial x}-(5+4x^2)\phi(x,t;q)$$
$$-\frac{\partial\phi(x,t;q)}{\partial t}-(6-5x^2-4x^4). \tag{36}$$

Using the above definition, we construct the zeroth-order deformation equation as in (7) and (8) and the mth-order deformation equation for $m\geq 1$ is

$$L[y_m(x,t)-\chi_m y_{m-1}(x,t)]=hR_m(y_{m-1}), \tag{37}$$

with the initial conditions

$$y_m(0,t)=0,\quad (y_m)_x(0,t)=0 \tag{38}$$

where

$$R_m(y_{m-1})=(y_{m-1})_{xx}+\frac{2}{x}(y_{m-1})_x-(5+4x^2)y_{m-1}$$
$$-(y_{m-1})_t-(1-\chi_m)(6-5x^2-4x^4). \tag{39}$$

Now, the solution of the mth-order deformation Equation (37) for $m\geq 1$ is

$$y_m(x,t)=\chi_m y_{m-1}(x,t)+h\int_0^x x^{-2}\int_0^x x^2 R_m(y_{m-1}). \tag{40}$$

We now successively obtain

$$y_1(x,t) = -\left(\frac{1}{5}x^4 + x^2\right)he^t + \left(\frac{2}{21}x^6 + \frac{1}{4}x^4 - x^2\right)h,$$
$$y_2(x,t) = \left(\frac{1}{90}h^2x^8 + \frac{13}{105}h^2x^6 + \frac{1}{10}h^2x^4 - \frac{1}{5}hx^4 - h^2x^2 - hx^2\right)e^t - \frac{4}{1155}h^2x^{10}$$
$$-\frac{31}{1512}h^2x^8 + \left(\frac{9}{56}h^2 + \frac{2}{21}h\right)x^6 + \left(\frac{1}{2}h^2 + \frac{1}{4}h\right)x^4 - \left(h^2 + h\right)x^2,$$
$$\vdots \quad (41)$$

and so on, in this manner the rest of the iterations can be obtained. Thus, the approximate solution in a series form is given by

$$y(x,t) = y_0(x,t) + \sum_{m=1}^{+\infty} y_m(x,t), \quad (42)$$

Hence, the series solution when $h = -1$ is given by

$$y(x,t) = x^2 + \left(1 + x^2 + \frac{1}{2!}x^4 + \frac{1}{3!}x^6 + \frac{1}{4!}x^8 + \cdots\right)e^t + \text{noise terms.} \quad (43)$$

This series has the closed form as $n \to \infty$

$$y(x,t) = x^2 + e^{x^2+t}, \quad (44)$$

which is the exact solution of the problem (31)-(32) compatible with ADM [8] . Notice that the noise terms that appear between various components vanish in the limit.

Using GA by the same procedure as in example 1, we get the optimal solution is found after 51 generation to converge to the exact solution, where x = 4.9489E−05 and $t = 7.6780E-05$. After execute the Equation (43) many times by using GA as in Table 2 we found the optimal solution.

4.2. Singular Wave-Type Equations

Example 3. Now, we consider the inhomogeneous singular wave-type equation,

$$y_{xx}+\frac{2}{x}y_x-\left(5+4x^2\right)y=y_{tt}+12x-5x^3-4x^5, \tag{45}$$

subject to initial conditions

$$y(0,t)=\mathrm{e}^{-t},\quad y_x(0,t)=0. \tag{46}$$

Table 2. Optimal solution of genetic algorithm for example 2.

x	t	$y_{Exact}(x,t)$	GA	HAM($n=5$)	PA [6/6]	SIMP
4.949E−05	7.678E−05	1.0000767	1.0000767	1.0000767	1.0000767	1.0000767
0.000	0.085	1.0887170	1.0882976	1.0887170	1.0887170	1.0887170
0.065	0.023	1.0318239	1.0322448	1.0323482	1.0323482	1.0323482
0.125	0.002	1.0334062	1.0333424	1.0334062	1.0334062	1.0334062
0.185	0.175	1.2669473	1.2667575	1.2669473	1.2669473	1.2669473

To solve Equations (45)-(46) by means of the standard HAM, we choose the initial approximation

$$y_0(x,t)=\mathrm{e}^{-t}, \tag{47}$$

and the linear operator

$$L\left[\phi(x,t;q)\right]=\frac{\partial^2\phi(x,t;q)}{\partial x^2}+\frac{2}{x}\frac{\partial\phi(x,t;q)}{\partial x}, \tag{48}$$

with the property

$$L\left[-\frac{c_1}{x}+c_2\right]=0, \tag{49}$$

where c_i $(i=1,2)$ are constants of integration. Furthermore, Equation (45) suggests that we define the nonlinear operator as

$$N\left[\phi(x,t;q)\right]=\frac{\partial^2\phi(x,t;q)}{\partial x^2}+\frac{2}{x}\frac{\partial\phi(x,t;q)}{\partial x}-\left(5+4x^2\right)\phi(x,t;q)$$
$$-\frac{\partial^2\phi(x,t;q)}{\partial t^2}-\left(12x-5x^3-4x\right).$$

Using the above definition, we construct the zeroth-order deformation equation as in (7) and (8) and the mth-order deformation equation for $m\geq 1$ is

$$L\left[y_m(x,t)-\chi_m y_{m-1}(x,t)\right]=hR_m\left(y_{m-1}\right), \tag{50}$$

with the initial conditions

$$y_m(0,t)=0,\quad \left(y_m\right)_x(0,t)=0 \tag{51}$$

where

$$R_m\left(y_{m-1}\right)=\left(y_{m-1}\right)_{xx}+\frac{2}{x}\left(y_{m-1}\right)_x-\left(5+4x^2\right)y_{m-1}$$
$$-\left(y_{m-1}\right)_{tt}-\left(1-\chi_m\right)\left(12x-5x^3-4x^4\right). \tag{52}$$

Now, the solution of the mth-order deformation Equation (50) for $m\geq 1$ is

$$y_m(x,t)=\chi_m y_{m-1}(x,t)+h\int_0^x x^{-2}\int_0^x x^2 R_m\left(y_{m-1}\right). \tag{53}$$

We now successively obtain

$$y_1(x,t)=-\left(\frac{1}{5}x^4+x^2\right)he^{-t}+\left(\frac{1}{14}x^7+\frac{1}{6}x^5-x^3\right)h,$$
$$y_2(x,t)=\left(\frac{1}{90}h^2x^8+\frac{13}{105}h^2x^6+\frac{1}{10}h^2x^4-\frac{1}{5}hx^4-h^2x^2-hx^2\right)e^{-t}-\frac{1}{462}h^2x^{11}$$
$$-\frac{43}{3780}h^2x^9+\left(\frac{43}{336}h^2+\frac{1}{14}h\right)x^7+\left(\frac{1}{3}h^2+\frac{1}{6}h\right)x^5-\left(h^2+h\right)x^3,$$
$$\vdots \tag{54}$$

and so on, in this manner the rest of the iterations can be obtained. Thus, the approximate solution in a series form is given by

$$y(x,t) = y_0(x,t) + \sum_{m=1}^{+\infty} y_m(x,t), \tag{55}$$

Hence, the series solution when $h=-1$ is given by

$$y(x,t) = x^3 + \left(1 + x^2 + \frac{1}{2!}x^4 + \frac{1}{3!}x^6 + \frac{1}{4!}x^8 + \cdots\right)e^{-t} + \text{noise terms.} \tag{56}$$

This series has the closed form as $n \to \infty$

$$y(x,t) = x^3 + e^{x^2 - t}, \tag{57}$$

which is the exact solution of the problem (45)-(46) compatible with ADM [8] . Notice that the noise terms that appear between various components vanish in the limit.

Using GA by the same procedure as in example 1, we get the optimal solution is found after 51 generation to converge to the exact solution, where $x=0$ and $t=0.879$. After execute the Equation (56) many times by using GA as in Table 3 we found the optimal solution

Example 4. We now examine the inhomogeneous wave-type equation,

$$y_{xx} + \frac{4}{x}y_x - \left(18x + 9x^4\right)y = y_{tt} - 2 - \left(18x + 9x^4\right)t^2, \tag{58}$$

subject to initial conditions

$$y(0,t) = 1 + t^2, \quad y_x(0,t) = 0. \tag{59}$$

To solve Equations (58)-(59) by means of the standard HAM, we choose the initial approximation

$$y_0(x,t) = 1 + t^2, \tag{60}$$

and the linear operator

$$L[\phi(x,t;q)]=\frac{\partial^2\phi(x,t;q)}{\partial x^2}+\frac{4}{x}\frac{\partial\phi(x,t;q)}{\partial x}, \tag{61}$$

with the property

$$L\left[-\frac{c_1}{3x}+c_2\right]=0, \tag{62}$$

where c_i $(i=1,2)$ are constants of integration. Furthermore, Equation (58) suggests that we define the nonlinear operator as

$$N[\phi(x,t;q)]=\frac{\partial^2\phi(x,t;q)}{\partial x^2}+\frac{4}{x}\frac{\partial\phi(x,t;q)}{\partial x}-(18x+9x^4)\phi(x,t;q)$$
$$-\frac{\partial^2\phi(x,t;q)}{\partial t^2}+(18x+9x^4)t^2+2. \tag{63}$$

Table 3. Optimal solution of genetic algorithm for example 3.

x	t	$y_{Exact}(x,t)$	GA	$HAM(n=5)$	PA [5/5]	SIMP
0.000	0.879	0.4151979	0.4150192	0.4151979	0.4151979	0.4151979
0.011	0.872	0.4181664	0.4183309	0.4181664	0.4181664	0.4181664
0.092	0.845	0.4339872	0.4338056	0.4339872	0.4339872	0.4339872
0.136	0.901	0.4400918	0.4161477	0.4400918	0.4400918	0.4400918
0.203	0.961	0.4069677	0.4071075	0.4069677	0.4069677	0.4069677

Using the above definition, we construct the zeroth-order deformation equation as in (7) and (8) and the mth-order deformation equation for $m\geq 1$ is

$$L[y_m(x,t)-\chi_m y_{m-1}(x,t)]=hR_m(y_{m-1}), \tag{64}$$

with the initial conditions

$$y_m(0,t)=0,\quad (y_m)_x(0,t)=0 \tag{65}$$

where

$$R_m(y_{m-1})=(y_{m-1})_{xx}+\frac{4}{x}(y_{m-1})_x-(18x+9x^4)y_{m-1}$$
$$-(y_{m-1})_{tt}+(1-\chi_m)\left[(18x+9x^4)t^2+2\right]. \tag{66}$$

Now, the solution of the mth-order deformation Equation (64) for $m \geq 1$ is

$$y_m(x,t)=\chi_m y_{m-1}(x,t)+h\int_0^x x^{-4}\int_0^x x^4 R_m(y_{m-1}). \tag{67}$$

We now successively obtain

$$y_1(x,t)=-\frac{1}{6}hx^6-hx^3,$$
$$y_2(x,t)=\frac{1}{120}h^2x^{12}+\frac{1}{9}h^2x^9+\frac{1}{6}(h^2-h)x^6-(h^2+h)x^3,$$
$$y_3(x,t)=-\frac{1}{5040}h^3x^{18}-\frac{23}{5400}h^3x^{15}-\left(\frac{1}{90}h^3-\frac{1}{60}h^2\right)x^{12}+\left(\frac{1}{6}h^3+\frac{2}{9}h^2\right)x^9$$
$$+\frac{1}{6}(3h^3+2h^2-h)x^6-(h^3+2h^2+h)x^3,$$
$$\vdots \tag{68}$$

and so on, in this manner the rest of the iterations can be obtained. Thus, the approximate solution in a series form is given by

$$y(x,t)=y_0(x,t)+\sum_{m=1}^{+\infty}y_m(x,t), \tag{69}$$

Thus, the series solution when $h=-1$ is given by

$$y(x,t)=t^2+\left(1+x^3+\frac{1}{2!}x^6+\frac{1}{3!}x^9+\frac{1}{4!}x^{12}+\cdots\right). \tag{70}$$

This series has the closed form as $n\to\infty$

$$y(x,t)=t^2+e^{x^3}, \tag{71}$$

which is the exact solution of the problem (58)-(59) compatible with ADM [8] .

Using GA by the same procedure as in example 1, we get the optimal solution is found after 51 generation to converge to the exact solution, where x = 4.6903E−06 and $t=1.1498E-06$. After execute the Equation (70) many times by using GA as in Table 4 we found the optimal solution

Table 4. Optimal solution of genetic algorithm for example 4.

x	t	$y_{Exact}(x,t)$	GA	$HAM(n=5)$	PA [6/6]	SIMP
4.693E−06	1.150E−06	1.0000000	1.0000000	1.0000000	1.0000000	1.0000000
0.000	0.102	1.0104040	1.0103960	1.0104040	1.0104040	1.0104040
0.058	0.035	1.0014201	1.0013941	1.0014201	1.0014201	1.0014201
0.164	0.079	1.0106616	1.0106261	1.0106616	1.0106616	1.0106616
0.173	0.001	1.0051921	1.0051844	1.0051921	1.0051921	1.0051921

4.3. Nonlinear Models

In what follows we close this analysis by examining two nonlinear time-dependent equations

Example 5. We now consider the nonlinear time-dependent equation,

$$y_{xx}+\frac{5}{x}y_x+\left(24t+16t^2x^2\right)e^y-2x^2e^{\frac{y}{2}}=y_t, \tag{72}$$

subject to initial conditions

$$y(0,t)=0, \quad y_x(0,t)=0. \tag{73}$$

To solve Equations (72)-(73) by means of the standard HAM, we choose the initial approximation

$$y_0(x,t)=0, \tag{74}$$

and the linear operator

$$L[\phi(x,t;q)]=\frac{\partial^2\phi(x,t;q)}{\partial x^2}+\frac{5}{x}\frac{\partial\phi(x,t;q)}{\partial x}, \tag{75}$$

with the property

$$L\left[-\frac{c_1}{4x}+c_2\right]=0, \tag{76}$$

where c_i $(i=1,2)$ are constants of integration. Furthermore, Equation (72) suggests that we define the nonlinear operator as

$$N[\phi(x,t;q)]=\frac{\partial^2\phi(x,t;q)}{\partial x^2}+\frac{5}{x}\frac{\partial\phi(x,t;q)}{\partial x}+\left(24t+16t^2x^2\right)e^{\phi(x,t;q)} -2x^2e^{\frac{\phi(x,t;q)}{2}}-\frac{\partial\phi(x,t;q)}{\partial t}. \tag{77}$$

Using the above definition, we construct the zeroth-order deformation equation as in (7) and (8) and the mth-order deformation equation for $m\geq 1$ is

$$L\left[y_m(x,t)-\chi_m y_{m-1}(x,t)\right]=hR_m(y_{m-1}), \tag{78}$$

with the initial conditions

$$y_m(0,t)=0,\quad (y_m)_x(0,t)=0 \tag{79}$$

where

$$R_m(y_{m-1})=(y_{m-1})_{xx}+\frac{5}{x}(y_{m-1})_x+\left(24t+16t^2x^2\right)A_{m-1}-2x^2B_{m-1}-(y_{m-1})_t. \tag{80}$$

For the nonlinear terms $e^{\phi(x,t;q)}=\sum_{m=1}^{\infty}A_{m-1}$ and $e^{\frac{\phi(x,t;q)}{2}}=\sum_{m=1}^{\infty}B_{m-1}$, the corresponding Adomian polynomials [15] [16] are:

$$\begin{aligned}
A_0&=e^{y_0},\\
A_1&=y_1e^{y_0},\\
A_2&=\left[y_2+\left(\frac{1}{2!}\right)y_1^2\right]e^{y_0},\\
A_3&=\left[y_3+y_1y_2+\left(\frac{1}{3!}\right)y_1^3\right]e^{y_0},\\
A_4&=\left[y_4+\left(\frac{1}{2!}\right)y_2^2+y_1y_3+\left(\frac{1}{2!}\right)y_1^2y_2+\left(\frac{1}{4!}\right)y_1^4\right]e^{y_0},\\
&\vdots
\end{aligned} \tag{81}$$

and

$$\begin{aligned}
B_0&=e^{y_0/2},\\
B_1&=\frac{1}{2}y_1e^{y_0/2},\\
B_2&=\left[\frac{1}{2}y_2+\frac{1}{4}\left(\frac{1}{2!}\right)y_1^2\right]e^{y_0/2},\\
B_3&=\left[\frac{1}{2}y_3+\frac{1}{4}y_1y_2+\frac{1}{8}\left(\frac{1}{3!}\right)y_1^3\right]e^{y_0/2},\\
B_4&=\left[\frac{1}{2}y_4+\frac{1}{4}\left(\frac{1}{2!}y_2^2+\frac{1}{4}y_1y_3\right)+\frac{1}{8}\left(\frac{1}{2!}\right)y_1^2y_2+\frac{1}{16}\left(\frac{1}{4!}\right)y_1^4\right]e^{y_0/2},\\
&\vdots
\end{aligned} \tag{82}$$

Now, the solution of the mth-order deformation Equation (78) for $m \geq 1$ is

$$y_m(x,t) = \chi_m y_{m-1}(x,t) + h\int_0^x x^{-5}\int_0^x x^5 R_m(y_{m-1}). \tag{83}$$

We now successively obtain

$$\begin{aligned} y_1(x,t) &= \frac{1}{2}ht^2x^4 - \frac{1}{16}hx^4 + 2htx^2, \\ y_2(x,t) &= \left(\frac{1}{12}h^2t^4 - \frac{1}{64}h^2t^2 + \frac{1}{1536}h^2\right)x^8 + \left(\frac{11}{15}h^2t^3 - \frac{3}{40}h^2t\right)x^6 \\ &\quad + \left(2h^2t^2 + \frac{1}{2}ht^2 - \frac{1}{8}h^2 - \frac{1}{16}h\right)x^4 + 2\left(h^2+h\right)tx^2, \\ &\vdots \end{aligned} \tag{84}$$

and so on, in this manner the rest of the iterations can be obtained. Thus, the approximate solution in a series form is given by

$$y(x,t) = y_0(x,t) + \sum_{m=1}^{+\infty} y_m(x,t), \tag{84}$$

Thus, the series solution when $h = -1$ is given by

$$y(x,t) = -2\left(tx^2 - \frac{1}{2}t^2x^4 + \frac{1}{3}t^3x^6 - \frac{1}{4}t^4x^8 + \cdots\right) + \text{noise terms.} \tag{85}$$

This series has the closed form as $n \to \infty$

$$y(x,t) = -2\ln\left(1 + tx^2\right), \tag{86}$$

which is the exact solution of the problem (72)-(73) compatible with ADM [8] . Notice that the noise terms that appear between various components vanish in the limit.

Using GA by the same procedure as in example 1, we get the optimal solution is found after 51 generation to converge to the exact solution, where $x=0.470$ and $t=0.895$. After execute the Equation (85) many times by using GA as in Table 5 we found the optimal solution

Example 6. Finally, we examine the nonlinear homogeneous equation,

$$y_{xx}+\frac{6}{x}y_x+\left(14t+x^4\right)y+4ty\ln y=y_{tt}, \tag{87}$$

subject to initial conditions

$$y(0,t)=1,\quad y_x(0,t)=0. \tag{88}$$

To solve Equations (87)-(88) by means of the standard HAM, we choose the initial approximation

$$y_0(x,t)=1, \tag{89}$$

and the linear operator

$$L[\phi(x,t;q)]=\frac{\partial^2\phi(x,t;q)}{\partial x^2}+\frac{6}{x}\frac{\partial\phi(x,t;q)}{\partial x}, \tag{90}$$

with the property

$$L\left[-\frac{c_1}{5x}+c_2\right]=0, \tag{91}$$

where c_i $(i=1,2)$ are constants of integration. Furthermore, Equation (87) suggests that we define the nonlinear operator as

$$N[\phi(x,t;q)]=\frac{\partial^2\phi(x,t;q)}{\partial x^2}+\frac{6}{x}\frac{\partial\phi(x,t;q)}{\partial x}+\left(14t+x^4\right)\phi(x,t;q)$$
$$+4t\phi(x,t;q)\ln\phi(x,t;q)-\frac{\partial^2\phi(x,t;q)}{\partial t^2}. \tag{92}$$

Table 5. Optimal solution of genetic algorithm for example 5.

x	t	$y_{Exact}(x, t)$	GA	$HAM(n = 5)$	PA [6/6]
0.470	0.895	−0.3608152	−0.3606302	−0.3603871	−0.3608131
0.841	0.813	−0.9085352	−0.8831179	−0.8170968	−0.9083255
0.866	0.878	−1.0117805	−0.9155520	−0.8245006	−1.0114234
0.946	0.728	−1.0033665	−0.9260635	−0.8376409	−1.0030238
0.974	0.672	−0.9863539	−0.9277153	−0.8436189	−0.9860390

Using the above definition, we construct the zeroth-order deformation equation as in (7) and (8) and the mth-order deformation equation for $m \geq 1$ is

$$L\left[y_m(x,t) - \chi_m y_{m-1}(x,t)\right] = hR_m(y_{m-1}), \tag{93}$$

with the initial conditions

$$y_m(0,t) = 0, \quad (y_m)_x(0,t) = 0 \tag{94}$$

where

$$R_m(y_{m-1}) = (y_{m-1})_{xx} + \frac{6}{x}(y_{m-1})_x + (14t + x^4)y_{m-1} + 4tA_{m-1} - (y_{m-1})_{tt}. \tag{95}$$

For the nonlinear term $\phi(x,t;q)\ln\phi(x,t;q) = \sum_{m=1}^{\infty} A_{m-1}$, the corresponding Adomian polynomials [15] [16] are:

$$
\begin{aligned}
A_0 &= y_0 \ln y_0, \\
A_1 &= (1+\ln y_0) y_1, \\
A_2 &= (1+\ln y_0) y_2 + \frac{1}{y_0}\left(\frac{1}{2!}\right) y_1^2, \\
A_3 &= (1+\ln y_0) y_3 + \frac{1}{y_0} y_1 y_2 - \frac{1}{y_0^2}\left(\frac{1}{3!}\right) y_1^3, \\
A_4 &= (1+\ln y_0) y_4 + \frac{1}{y_0}\left(\frac{1}{2!} y_2^2 + y_1 y_3\right) - \frac{1}{y_0^2}\left(\frac{1}{2!}\right) y_1^2 y_2 + \frac{2}{y_0^3}\left(\frac{1}{4!}\right) y_1^4, \\
&\vdots
\end{aligned}
\tag{96}
$$

Now, the solution of the mth-order deformation Equation (93) for $m \geq 1$ is

$$
y_m(x,t) = \chi_m y_{m-1}(x,t) + h\int_0^x x^{-6} \int_0^x x^6 R_m(y_{m-1}). \tag{97}
$$

We now successively obtain

$$
\begin{aligned}
y_1(x,t) &= \frac{1}{66} h x^6 + h t x^2, \\
y_2(x,t) &= \frac{1}{13464} h^2 x^{12} + \frac{7}{572} h^2 t x^8 + \frac{1}{66}(h^2+h) x^6 + \frac{1}{2} h^2 t^2 x^4 + (h^2+h) t x^2, \\
y_3(x,t) &= \frac{1}{5574096} h^3 x^{18} + \frac{965}{18290844} h^3 t x^{14} + \frac{1}{6732}(h^3+h^2) x^{12} \\
&\quad + \frac{67}{12870} h^3 t^2 x^{10} + \frac{7}{286}(h^3+h^2) t x^8 + \frac{1}{66}(11 h^3 t^3 + 2h^2 + h) x^6 \\
&\quad + (h^3+h^2) t^2 x^4 + (h^3 + 2h^2 + h) t x^2, \\
&\vdots
\end{aligned}
\tag{98}
$$

and so on, in this manner the rest of the iterations can be obtained. Thus, the approximate solution in a series form is given by

$$
y(x,t) = y_0(x,t) + \sum_{m=1}^{+\infty} y_m(x,t), \tag{99}
$$

Table 6. Optimal solution of genetic algorithm for example 6.

x	t	$y_{Exact}(x, t)$	GA	$HAM(n=5)$	PA [6/6]
0.270	0.935	0.8117150	0.8116540	0.8117157	0.8117155
0.478	0.913	1.0025321	1.0023269	1.0025321	1.0025321
0.855	0.954	0.4978791	0.4974576	0.4983086	0.4979100
0.903	0.973	0.4523075	0.4521125	0.4532268	0.4522629
0.981	0.938	0.4054755	0.4052946	0.4069536	0.4057609

Thus, the series solution when $h=-1$ is given by

$$y(x,t)=1-tx^2+\frac{1}{2!}t^2x^4-\frac{1}{3!}t^3x^6+\frac{1}{4!}t^4x^8+\cdots. \tag{100}$$

This series has the closed form as $n\to\infty$

$$y(x,t)=e^{-tx^2}, \tag{101}$$

which is the exact solution of the problem (87)-(88) compatible with ADM [8] .

Using GA by the same procedure as in example 1, we get the optimal solution is found after 51 generation to converge to the exact solution, where $x=0.270$ and $t=0.935$. After execute the Equation (100) many times by using GA as in Table 6 we found the optimal solution.

5. CONCLUSIONS

To our best knowledge, this is the first result on the application of the HAM with GA to solve models of the time-dependent Emden-Fowler type of equations and wave-type equation with singular behaviors. The HAM with GA has been successfully applied to solve models of the time-dependent Emden-Fowler type of equations and wave-type equation with singular behaviors. The HAM with GA has worked effectively to handle these models giving it a wider applicability. The proposed scheme of HAM has been applied directly without any need for transformation formulae or restrictive assumptions. The solution process of HAM is compatible with those methods in the literature providing analytical approximation such as ADM.

The approach of HAM with GA has been tested by employing the method to obtain approximate-exact solutions of six examples. The results obtained in all cases demonstrate the reliability and the efficiency of this method.

ACKNOWLEDGEMENTS

We would like to thank the referees for their careful review of our manuscript.

REFERENCES

1. Davis, H.T. (1962) Introduction to Nonlinear Differential and Integral Equations. Dover Publications, New York.
2. Chandrasekhar, S. (1967) Introduction to the Study of Stellar Structure. Dover Publications, New York.
3. Richardson, O.U. (1921) The Emission of Electricity from Hot Bodies. London.
4. Adomian, G., Rach, R. and Shawagfeh, N.T. (1995) On the Analytic Solution of Lane-Emden Equation, Found. Physics Letters, 8, 161-181.
5. Shawagfeh, N.T. (1993) Nonperturbative Approximate Solution for Lane-Emden Equation. Journal of Mathematical Physics, 34, 4364-4369.
6. Wazwaz, A.M. (2001) A New Method for Solving Differential Equations of the Lane-Emden Type. Applied Mathematics and Computation, 118, 287-310.
7. Wazwaz, A.M. (2005) Adomian Decomposition Method for a Reliable Treatment of the Emden-Fowler Equation. Applied Mathematics and Computation, 161, 543-560.
8. Wazwaz, A.M. (2005) Analytical Solution for the Time-Dependent Emden-Fowler Type of Equations by Adomian Decomposition Method. Applied Mathematics and Computation, 166, 638-651.
9. Yildirim, A. and Ozis, T. (2009) Solutions of Singular IVPs of Lane-Emden Type by the Variational Iteration Method. Nonlinear Analysis, 70, 2480-2484.
10. Dehghan, M. and Shakeri, F. (2008) Approximate Solution of a Differential Equation Arising in Astrophysics Using the Variational Iteration Method. New Astronomy, 13, 53-59.
11. Liao, S.J. (2003) Beyond Perturbation: Introduction to the Homotopy Analysis Method. Chapman and Hall, CRC Press, Boca Raton.
12. He, J.H. (2003) Homotopy Perturbation Method: A New Nonlinear Analytical Technique. Applied Mathematics and Computation, 135, 73-79.
13. Gen, M. and Cheng, R. (1997) Genetic Algorithms and Engineering Design. John Wiley & Sons, Hoboken.
14. Melanie, M. (1998) An Introduction to Genetic Algorithms. MIT Press, Cambridge.
15. Adomian, G. (1989) Nonlinear Stochastic Systems Theory and Applications to Physics. Kluwer Academic Publishers, Dordrecht.
16. Wazwaz, A.M. (2000) A New Algorithm for Calculating Adomian Polynomials for Nonlinear Operators. Applied Mathematics and Computation, 111, 53-69.

CHAPTER 18

A Study on Multipeutics

Jens Christian Larsen

Vanløse Alle 50 2. mf. tv, 2720 Vanløse, Copenhagen, Denmark

ABSTRACT

Multipeutics is the simultaneous application of $m \geq 4$ cancer treatments. $m = 4$ is quadrapeutics, which was invented by researchers at Rice University, Northeastern University, MD Anderson Cancer Centre and China Medical University, see [1] . Multipeutics is our idea. From section 6 Summary, it follows that multipeutics can be more potent than quadrapeutics by comparing these two mathematical models. The first two treatments in quadrapeutics are systemically administered nano gold particles G and lysosomal chemo therapeutic drug D. They form mixed clusters M primarily in cancer cells and can be excited by a laser pulse, the third treatment, to form plasmonic nanobubbles N. These nanobubbles can kill the cancer cells by mechanical impact. If they do not the chemo therapeutic drug can be released into the cytoplasm, which might be lethal to the cancer cell. The fourth treatment is x rays X and the cancer cells have been sensitized to x rays by the treatment. We present an ODE (ordinary differential equations) model of quadrapeutics and of multipeutics, which is quadrapeutics and $n \geq 1$ immune or chemo therapies. In the present paper we have found a polynomial p of degree at most $2(n + 3)$, such that a singular point $(C, D, G, M, N, I_1, \cdots, I_n)$ will have $p(M) = 0$ Here $I_1, \cdots, I_n$ are immune or chemo therapies. So this gives us candidates for singular points. Quadrapeutics is treated extensively. We find in theorem 3 a polynomium s of degree at most six in M such that a positive singular point (C, D, G, M, N) of the quadrapeutics system will have $s(M) = 0$. The main theorem of the present paper is the multipeutics theorem, saying that the more treatments we apply the lower the cancer burden, even if we take the doses of each treatment smaller. From the proof of this theorem, we can say, that quadrapeutics can outperform chemo radiation if the nanobubble kill rate k_{21} is sufficiently big. See also Figure 1 and Figure 2 and the text explaining them.

Keywords: Cancer, Mass Action Kinetic System, Quadrapeutics, Immune Therapy

1. INTRODUCTION

In [2] we introduced a discrete three dimensional model of cancer growth and a three dimensional ODE model of cancer growth. The variables are C cancer cells, GF growth factors (positive control) and growth inhibitors GI (negative control) in both models. In [3] we showed that there are cancer ODE models with oscillatory behaviour and in [4] we found a discrete mathematical model of C, GF, GI with a nontrivial attractor. In the same paper, we introduced mathematical models of the four phases of adoptive T cell therapy.

The first phase is the resection of the tumor and lymphodepletion with a chemo therapeutic drug. The second phase is the pre REP phase where the TILs (tumor infiltrating lymphocytes) from the patients' tumor are expanded with IL-2 treatment. The third phase is the REP phase of expanding and activating the TILs with IL-2 and anti CD3. The fourth phase is the injection of the expanded and activated TIL s back into the patient. We showed that this last phase of the treatment is bistable in the following sense. There is a positive sink and there exist initial conditions such that the cancer burden goes to infinity as $t \to +t^{+}$, where $]t^{-},t^{+}[\ t^{-} < 0, t^{+} > 0$ is the domain of definition of the maximal integral curve $c(t)$ of the corresponding vector field.

In the present paper, we propose to combine quadrapeutics and immune therapy.

We have written a paper [5] on fundamental concepts in dynamics intended for researchers with a background in medicine.

There are several important monographs related to the present paper, see [6] - [11] . [12] is a mathematical model of miRNAs. [13] [14] [15] are papers on quadrapeutics. [16] [17] are two papers of mine on cancer and mathematics.

2. THE MULTIPEUTICS SYSTEM

Consider the mass action kinetic system

$$C+N \to 0 \quad (1)$$

$$C+D \to 0 \quad (2)$$

$$D+G \to M \quad (3)$$

$$M \to N \quad (4)$$

$$N \to 0 \quad (5)$$

$$M \to 0 \quad (6)$$

$$C+X \to 0 \quad (7)$$

$$C \to 2C \quad (8)$$

$$C \leftrightarrows 0 \quad (9)$$

$$D \leftrightarrows 0 \quad (10)$$

$$G \leftrightarrows 0 \quad (11)$$

$$C+I_i \to 0 \quad i=1,\cdots,n \quad (12)$$

$$I_i \leftrightarrows 0 \quad (13)$$

$n \in \mathbb{N}_0$ ($n=0$ means that the reactions (12) and (13) are omitted). The complexes are $C(1)=C+N,\ C(2)=0,\ C(3)=C+D,\ C(4)=D+G,\ C(5)=M,$ $C(6)=N,\quad C(7)=C+X,\quad C(8)=C,\quad C(9)=2C,\quad C(10)=D,\quad C(11)=G,$

$C(10+2i)=C+I_i, \quad C(11+2i)=I_i, i=1,\cdots,n$ The variables here are C cancer cells, D lysosomal chemo therapeutic drug, G nanogold particles conjugated to an antibody panitumumab against the epidermal growth factor receptor EGFR, M mixed clusters in cancer cells, N nanobubbles and finally I_i immune or chemo therapies. X, the x rays, is not a variable but a parameter, a positive real number. (1) says that nanobubbles kill cancer cells and (2) that the chemo therapeutic drug kills cancer cells. (3) means that drug and nano gold particles form mixed clusters in cancer cells and normal cells and the cluster size is the largest in cancer cells. The mixed clusters generate nano bubbles (4), when excited with a laser pulse. The laser threshold pulse decreases with cluster size and since the cluster size is the largest in cancer cells it is mainly in cancer cells that plasmonic nanobubbles kill the host cell. (7) is the killing of cancer cells by x rays. (5) and (6) are decay rates for N and M. (8) says that cancer cells proliferate rapidly. (9), (10) and (11) give birth and decay rates for cancer cells, chemotherapeutic drug and nano gold particles, respectively. (12) and (13) are immune or chemo therapies. The rate constant in (7) is $\tilde{k}_{27}=k_{27}X.$ The ODEs are, see [18]

$$C'=-k_{21}C\cdot N-k_{23}C\cdot D+\left(k_{98}-k_{28}-k_{27}X\right)C+k_{82}-\sum_{i=1}^{n}k_{2,10+2i}C\cdot I_i \tag{14}$$

$$D'=-k_{23}C\cdot D-k_{54}D\cdot G+k_{10,2}-k_{2,10}D \tag{15}$$

$$G'=-k_{54}D\cdot G+k_{11,2}-k_{2,11}G \tag{16}$$

$$M'=k_{54}D\cdot G-\left(k_{65}+k_{25}\right)M \tag{17}$$

$$N'=-k_{21}C\cdot N+k_{65}M-k_{26}N \tag{18}$$

$$I_i'=-k_{2,10+2i}C\cdot I_i+k_{11+2i,2}-k_{2,11+2i}I_i \quad i=1,\cdots,n \tag{19}$$

The corresponding vector field is denoted f. We are going to consider subsystems of (14) to (19). First of all the chemo radiation system, the vector field in (14) and (15) with the rate constants in (16) to (19) equal to zero and the rest positive. This vector field is denoted

$$f_{cr} : \mathbb{R}^2 \to \mathbb{R}^2 \tag{20}$$

The quadrapeutics system (14) to (18) with the rate constants in (19) equal to zero and the rest positive. This vector field is denoted

$$f_q : \mathbb{R}^5 \to \mathbb{R}^5 \tag{21}$$

Finally the multipeutics systems: the system (14) to (19) with the rate constants in (19) equal to zero, when $i = n - j + 1, \cdots, n$ and the rest positive. This vector field is denoted

$$f_{n-j} : \mathbb{R}^{n-j+5} \to \mathbb{R}^{n-j+5} \tag{22}$$

$j \in \{0, \cdots, n\}$. Now introduce some important notation. A singular point of f_{cr} is denoted $c_{cr} = (C_*^{cr}, D_*^{cr})$. A singular point of f_q is denoted $c_q =$ $(C_*^q, D_*^q, G_*^q, M_*^q, N_*^q)$. A singular point of f_{n-j} is denoted

$$c_{n-j} = \left(C_*^{n-j}, D_*^{n-j}, G_*^{n-j}, M_*^{n-j}, N_*^{n-j}, I_*^{1,n-j}, \cdots, I_*^{n-j,n-j}\right) \tag{23}$$

We are going to compare the cancer burden for the different systems. In fact we shall show that there are stable equilibria of the different systems such that

$$C_*^n < C_*^{n-1} < \cdots < C_*^1 < C_*^q < C_*^{cr} \tag{24}$$

Looking from right to left, this says that the more treatments we apply the lower the cancer burden.

In Figure 1 we have plotted $C(t)$ versus t for the chemo radiation model and $C(t)$ versus t for the quadrapeutics model. All rate constants equal to 1, except $k_{21} = 10, k_{65} = 10, k_{11,2} = 20.$ Also $X = 1$. This supports the proof of Theorem 1 that if you make the nanobubble kill rate k_{21} sufficiently big, then quadrapeutics outperforms chemo radiation. I have iterated the Euler maps

$$H_{cr}(C,D)=(C,D)+hf_{cr}(C,D) \tag{25}$$

$$H_q(C,D,G,M,N)=(C,D,G,M,N)+hf_q(C,D,G,M,N) \tag{26}$$

where h is the step size and equal to 0.02. There are 1000 iterations. The numerical analysis indicated, that $C(t),D(t),G(t),M(t),N(t)$ for the quadrapeutics system converged, when t goes to infinity. And similarly for $C(t),D(t)$ of the chemo radiation model, see Figure 1 and Figure 2.

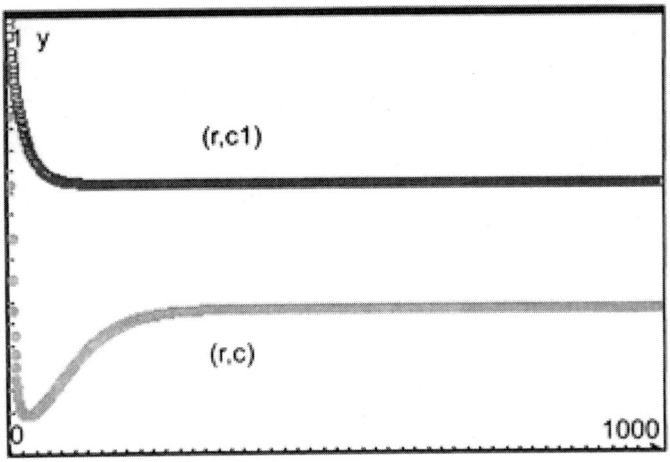

Figure 1. (r, c1) is C(t) versus t for the chemo radiation model and (r, c) is C(t) versus t for the quadrapeutics model.

Theorem 1. Assume, that $\hat{a} \triangleq k_{98} - k_{28} - k_{27}X < 0.$ There exist positive values of the rate constants k_{ij}, such that (24) holds and the singular points are all positive and stable.

Proof. Let K be defined by

$$k_1 = \left(k_{82}, k_{10,2}, k_{11,2}, \left\{k_{11+2i,2}\right\}_{i=1,\cdots,n}\right) \tag{27}$$

$$k_2 = \left(k_{21}, k_{23}, k_{98}, k_{28}, \tilde{k}_{27}, k_{54}, k_{2,10}, k_{2,11}, k_{65}, k_{25}, k_{26}, \left\{k_{2,10+2i}, k_{2,11+2i}\right\}_{i=1,\cdots,n}\right) \quad (28)$$

$$K = \left(k_1, k_2\right) \quad (29)$$

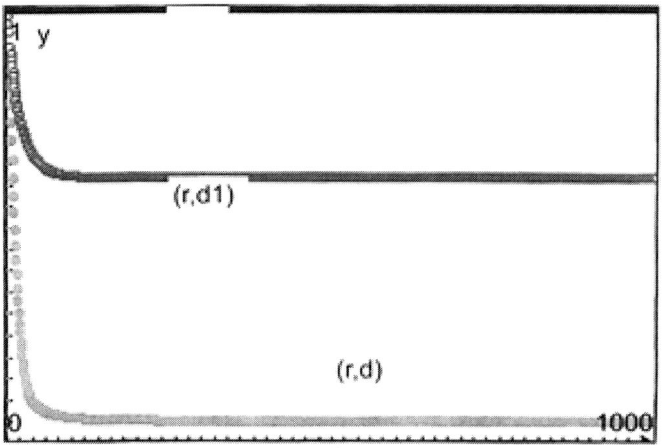

Figure 2. (r, d1) is D(t) versus t for the chemo radiation system and (r, d) is D(t) versus t for the quadrapeutics system.

Define

$$k_{82} = tk_{82}^{0} \quad (30)$$

$$k_{10,2} = tk_{10,2}^{0} \quad (31)$$

$$k_{11,2} = tk_{11,2}^{0} \quad (32)$$

$$k_{11+2i,2} = tk_{11+2i,2}^{0} \quad i = 1, \cdots, n \quad t \in \mathbb{R} \quad (33)$$

when $* = n$. Here

$$k_{82}^{0}, k_{10,2}^{0}, k_{11,2}^{0}, k_{11+2i,2}^{0} \in \mathbb{R}_{+} \qquad (34)$$

$i = 1, \cdots, n.$ Notice that $(0, \cdots, 0)$ is a singular point of f, when $t = 0$. The rate constants k_{ij} in k_2 are positive real numbers. Also note that $D_1 f_{*0}$ is an isomorphism.

Here

$$* = \begin{cases} cr \\ q \\ 1 \\ \vdots \\ n \end{cases} \qquad (35)$$

For instance in f

$$D_1 f_0 = \left\{ b_{ij} \right\}_{i,j=1,\cdots,n} \qquad (36)$$

is an isomorphism, where $b_{11} = \hat{a},\ b_{2,2} = -k_{2,10},\ b_{3,3} = -k_{2,11},\ b_{4,4} = -\left(k_{65} + k_{25}\right),$ $b_{5,5} = -k_{26},\ b_{5,4} = k_{65},\ b_{66} = -k_{2,13}, \cdots, b_{n+5,n+5} = -k_{2,11+2n},$ all other $b_{ij} = 0$.

So there exist C^k maps, $k \geq 5$,

$$c_* : V_* \subset \mathbb{R}^{p_*} \to \mathbb{R}^{s_*} \qquad (37)$$

such that

$$f_* \left(c_* \left(K_* \right), K_* \right) = 0 \quad c_* \left(\hat{K}_* \right) = \left(0, \cdots, 0 \right) \qquad (38)$$

by the implicit function theorem. $V_*, K_*, \hat{K}_*$ are defined below.

Also

$$p_{cr} = 7 \quad s_{cr} = 2 \qquad (39)$$

$$p_q = 14 \quad s_q = 5 \tag{40}$$

$$p_{n-j} = 14 + 3(n-j) \quad s_{n-j} = 5 + n - j \tag{41}$$

For $* = cr$ define

$$k_1^{cr} = (k_{82}, k_{10,2}) \tag{42}$$

and

$$k_2^{cr} = (k_{23}, k_{98}, k_{28}, \tilde{k}_{27}, k_{2,10}) \tag{43}$$

For $* = q$ we define

$$k_1^q = (k_{82}, k_{10,2}, k_{11,2}) \tag{44}$$

and

$$k_2^q = (k_2^{cr}, k_{21}, k_{2,11}, k_{54}, k_{65}, k_{25}, k_{26}) \tag{45}$$

Finally for the multipeutics system

$$k_1^{n-j} = (k_{82}, k_{10,2}, k_{11,2}, k_{11+2i,2i=1,\cdots,n-j}) \tag{46}$$

and

$$k_2^{n-j} = \left(k_2^q, \left(k_{2,10+2i}, k_{2,11+2i}\right)_{i=1,\cdots,n-j}\right) \tag{47}$$

Now let $\hat{K}_{cr} = (k_1^{cr}, k_2^{cr}), k_1^{cr} = 0$ and $\hat{K}_q = (k_1^q, k_2^q), k_1^q = 0,$ and finally $\hat{K}_{n-j} = (k_1^{n-j}, k_2^{n-j}), k_1^{n-j} = 0.$ V_* above is an open subset containing $\hat{K}_*$ in $\mathbb{R}^{p^*}$. Define $K_* = (k_1^*, k_2^*)$. Then

$$f_*\left(c_*\left(tk_1^{*,0},k_2^*\right),tk_1^{*,0},k_2^*\right)=0 \tag{48}$$

Define

$$d_*(t)=c_*\left(tk_1^{*,0},k_2^*\right) \tag{49}$$

For the chemo radiation model we let

$$d_*^G=0 \quad d_*^M=0 \quad d_*^N=0 \quad d_*^{I_i}=0 \tag{50}$$

$i=1,\cdots,n.$ For the quadrapeutics model we let $d_*^{I_i}=0, i=1,\cdots,n.$ And for the multipeutics model we let

$$d_*^{I_i}=0 \quad i=n-j+1,\cdots,n \quad j\in\{0,\cdots,n\} \tag{51}$$

Differentiate (48) with respect to t to get

$$D_1 f\left(\frac{\partial d_*}{\partial t}\right)+\frac{\partial f}{\partial t}=0 \tag{52}$$

Thus

$$\frac{\partial d_*^C}{\partial t}(0)=-\frac{1}{\hat{a}}k_{82}^0 \tag{53}$$

regardless of the system. We have

$$-k_{21}d_*^C d_*^N-k_{23}d_*^C d_*^D+\left(k_{98}-k_{28}-k_{27}X\right)d_*^C+k_{82}-\sum_{i=1}^{n}k_{2,10+2i}d_*^C d_*^{I_i}=0 \tag{54}$$

Differentiate this with respect to t to get

$$-k_{21}\left(\frac{\partial d_*^C}{\partial t}d_*^N+d_*^C\frac{\partial d_*^N}{\partial t}\right)-k_{23}\left(\frac{\partial d_*^C}{\partial t}d_*^D+d_*^C\frac{\partial d_*^D}{\partial t}\right) \tag{55}$$

$$+\hat{a}\frac{\partial d_*^C}{\partial t}+k_{82}^0-\sum_{i=1}^{n}k_{2,10+2i}\left(\frac{\partial d_*^C}{\partial t}d_*^{I_i}+d_*^C\frac{\partial d_*^{I_i}}{\partial t}\right)=0 \tag{56}$$

Also differentiate this equation with respect to t to find

$$-k_{21}\left(\frac{\partial^2 d_*^C}{\partial t^2}d_*^N+2\frac{\partial d_*^C}{\partial t}\frac{\partial d_*^N}{\partial t}+d_*^C\frac{\partial^2 d_*^N}{\partial t^2}\right) \tag{57}$$

$$-k_{23}\left(\frac{\partial^2 d_*^C}{\partial t^2}d_*^D+2\frac{\partial d_*^C}{\partial t}\frac{\partial d_*^D}{\partial t}+d_*^C\frac{\partial^2 d_*^D}{\partial t^2}\right) \tag{58}$$

$$-\sum_{i=1}^{n}k_{2,10+2i}\left(\frac{\partial^2 d_*^C}{\partial t^2}d_*^{I_i}+2\frac{\partial d_*^C}{\partial t}\frac{\partial d_*^{I_i}}{\partial t}+d_*^C\frac{\partial^2 d_*^{I_i}}{\partial t^2}\right) \tag{59}$$

$$+\hat{a}\frac{\partial^2 d_*^C}{\partial t^2}=0 \tag{60}$$

From this equation it follows that

$$\frac{\partial^2 d_{cr}^C}{\partial t^2}(0)=\frac{\partial^2 d_q^C}{\partial t^2}(0)>\frac{\partial^2 d_1^C}{\partial t^2}(0)>\cdots>\frac{\partial^2 d_n^C}{\partial t^2}(0) \tag{61}$$

anticipating that

$$d_*^N(0)=0 \quad \frac{\partial d_*^N}{\partial t}(0)=0 \tag{62}$$

Observe that

$$\frac{\partial d_*^C}{\partial t}(0) = -\frac{k_{82}^0}{\hat{a}} > 0 \quad \frac{\partial d_*^D}{\partial t}(0) = \frac{k_{10,2}^0}{k_{2,10}} > 0 \tag{63}$$

$$\frac{\partial d_*^G}{\partial t}(0) = \frac{k_{11,2}^0}{k_{2,11}} > 0 \quad \frac{\partial d_*^{I_i}}{\partial t}(0) = \frac{k_{11+2i,2}^0}{k_{2,11+2i}} > 0 \tag{64}$$

$i = 1,\cdots,n.$ The third equality holds for $* = q,1,\cdots,n.$ The fourth equality holds for the multipeutics system $f_{n-j}, i = 1,\cdots,n-j, j \in \{0,1,\cdots,n\}$.

By the fundamental theorem of calculus applied twice

$$d_*^C(t) = \int_0^t \left(\int_0^s \frac{\partial^2 d_*^C}{\partial t^2}(v)\mathrm{d}v + \frac{\partial d_*^C}{\partial t}(0) \right) \mathrm{d}s \tag{65}$$

So

$$d_q^C(t) > d_1^C(t) > \cdots > d_n^C(t) \tag{66}$$

To see the first inequality, observe that there exists a positive constant c, such that

$$d_q^C(t) - d_1^C(t) = \int_0^t \int_0^s \left(\frac{\partial^2 d_q^C}{\partial t^2}(v) - \frac{\partial^2 d_1^C}{\partial t^2}(v) \right) \mathrm{d}v\mathrm{d}s \geq \frac{1}{2}ct^2 > 0 \tag{67}$$

because a positive function on a compact interval $[0,\delta]$ assumes a minimum $c > 0$. The other inequalities follow in the same way. And this proves all the inequalities in (24) except $C_*^q < C_*^r$. Henceforth assume

$$k_{2,10+2i} = k_{11+2i,2} = k_{2,11+2i} = 0 \tag{68}$$

$i = 1, \cdots, n$. Finally differentiate (57) to (60) with respect to t to get

$$-3k_{21}\left(\frac{\partial^2 d_*^C}{\partial t^2}\frac{\partial d_*^N}{\partial t}+\frac{\partial d_*^C}{\partial t}\frac{\partial^2 d_*^N}{\partial t^2}\right) \tag{69}$$

$$-3k_{23}\left(\frac{\partial^2 d_*^C}{\partial t^2}\frac{\partial d_*^D}{\partial t}+\frac{\partial d_*^C}{\partial t}\frac{\partial^2 d_*^D}{\partial t^2}\right) \tag{70}$$

$$+\hat{a}\frac{\partial^3 d_*^C}{\partial t^3}=0 \tag{71}$$

when $t = 0$. But

$$\frac{\partial d_*^M}{\partial t}(0)=\frac{\partial d_*^N}{\partial t}(0)=0 \tag{72}$$

However when $* = q$

$$\frac{\partial^2 d_*^N}{\partial t^2}(0)>0 \tag{73}$$

To see this differentiate (17) twice with respect to t to find

$$\left(k_{65}+k_{25}\right)\frac{\partial^2 d_*^M}{\partial t^2}(0)=2k_{54}\frac{\partial d_*^D}{\partial t}(0)\frac{\partial d_*^G}{\partial t}(0)>0 \tag{74}$$

If we now differentiate (18) twice with respect to t we find

$$k_{26}\frac{\partial^2 d_*^N}{\partial t^2}(0)=k_{65}\frac{\partial^2 d_*^M}{\partial t^2}(0)>0 \tag{75}$$

which is what we sought to show.

Furthermore

$$\left(k_{65}+k_{25}\right)d_*^M\left(t\right)=k_{54}d_*^D\cdot d_*^G\left(t\right)>0 \tag{76}$$

and

$$d_*^N\left(t\right)=\frac{k_{65}d_*^M\left(t\right)}{k_{21}d_*^C\left(t\right)+k_{26}}>0 \tag{77}$$

when $t>0$ and small. Differentiate (15) twice with respect to t when $*=cr,q$ to get

$$-k_{2,10}\frac{\partial^2 d_*^D}{\partial t^2}=2k_{23}\frac{\partial d_*^C}{\partial t}\frac{\partial d_*^D}{\partial t}+2k_{54}\frac{\partial d_*^D}{\partial t}\frac{\partial d_*^G}{\partial t} \tag{78}$$

Now

$$\hat{a}\frac{\partial^2 d_*^C}{\partial t^2}=2k_{23}\frac{\partial d_*^C}{\partial t}\frac{\partial d_*^D}{\partial t} \tag{79}$$

Also

$$\hat{a}\frac{\partial^3 d_*^C}{\partial t^3}(0)=3k_{21}\left(\frac{\partial d_*^C}{\partial t}\frac{k_{65}k_{54}2}{k_{26}\left(k_{65}+k_{25}\right)}\frac{\partial d_*^D}{\partial t}\frac{\partial d_*^G}{\partial t}\right) \tag{80}$$

$$+3k_{23}\frac{1}{\hat{a}}\left(2k_{23}\frac{\partial d_*^C}{\partial t}\frac{\partial d_*^D}{\partial t}\right)\frac{\partial d_*^D}{\partial t} \tag{81}$$

$$-3k_{23}\frac{1}{k_{2,10}}\frac{\partial d_*^C}{\partial t}\left(2k_{23}\frac{\partial d_*^C}{\partial t}\frac{\partial d_*^D}{\partial t}+2k_{54}\frac{\partial d_*^D}{\partial t}\frac{\partial d_*^G}{\partial t}\right) \qquad (\quad 82)$$

where we have evaluated in $t=0$ on the right hand side. So if

$$3\frac{k_{21}k_{65}2}{k_{26}\left(k_{65}+k_{25}\right)}-\frac{3k_{32}2}{k_{2,10}}>0 \tag{83}$$

we have

$$\frac{\partial^3 d_{cr}^C}{\partial t^3}(t)>\frac{\partial^3 d_q^C}{\partial t^3}(t) \tag{84}$$

when $t=0$ and hence also for small $t>0$. The fundamental theorem of calculus applied three times gives

$$d_*^C(t)=\int_0^t\left(\int_0^s\left(\int_0^v\frac{\partial^3 d_*^C}{\partial t^3}(u)\,\mathrm{d}u+\frac{\partial^2 d_*^C}{\partial t^2}(0)\right)\mathrm{d}v+\frac{\partial d_*^C}{\partial t}(0)\right)\mathrm{d}s \tag{85}$$

So

$$d_{cr}^C(t)>d_q^C(t) \tag{86}$$

for small $t>0$ and the theorem follows in the same way as (67). Define the matrices

$$B=Df_{*_{c_*}}-\lambda\,\mathrm{id}\quad A=Df_{*_{c_*}} \tag{87}$$

and then write the definition of the determinant

$$\det B=\sum_{\sigma\in S^p}(-1)^{I(\sigma)}B_{1,\sigma(1)}\cdots B_{p,\sigma(p)} \tag{88}$$

where S^p is the set of permutations of $\{1,\cdots,p\}$ and $I(\sigma)=0$ if σ is even and $I(\sigma)=1$, when σ is odd. This formula shows immediately that

$$\det B=(-1)^p\lambda^p+\sum_{i=0}^{p-1}a_i\left(A_{11},\cdots,A_{pp}\right)\lambda^i \tag{89}$$

where a_i is a polynomial in $A_{11}, \cdots, A_{pp}$, and hence is smooth in $A_{11}, \cdots, A_{pp}$. You can prove this by induction. Now apply the continuous dependence of roots of a polynomial on its coefficients to show that $d_*(t)$ is stable for f_*, when t is small, see [19] . □

Remark. If the quadrapeutics chemo rate is

$$k_{10,2}^{q} = k_{10,2}^{cr} \tilde{\alpha} \quad \tilde{\alpha} \in \left]0,1\right[\tag{90}$$

and the x ray rate is

$$k_{27}^{q} X = k_{27}^{cr} X \tilde{\beta} \quad \tilde{\beta} \in \left]0,1\right[\tag{91}$$

where

$$k_{10,2}^{q} = t k_{10,2}^{q,0} \quad k_{10,2}^{cr} = t k_{10,2}^{cr,0} \tag{92}$$

while $\hat{a} = k_{98} - k_{28} - k_{27} X < 0.$ Here

$$k_{10,2}^{q,0}, k_{10,2}^{cr,0} \in \mathbb{R}_{+} \tag{93}$$

Then by taking k_{21} big we find

$$\frac{\partial^3 d_q^C}{\partial t^3}(0) < \frac{\partial^3 d_{cr}^C}{\partial t^3}(0) \tag{94}$$

since

$$\hat{a} \frac{\partial^3 d_q^C}{\partial t^3}(0) \to +\infty \tag{95}$$

as $k_{21} \to +\infty.$ Hence

$$d_q^C(t) < d_{cr}^C(t) \tag{96}$$

for small $t > 0$, when we impose (90) and (91), arguing as in the proof of theorem 1.

Experimentally this is what you see, that quadrapeutics can outperform chemo radiation even if we take the chemo radiation doses smaller, see [1] [13] [14] [15] .

Consider also the system (1) to (13) with

$$2C + X \to 0 \tag{97}$$

replacing (7). The ODE s are the same (15) to (19) except (14), which is

$$C' = -k_{21}C \cdot N - k_{23}C \cdot D + (k_{98} - k_{28})C + k_{82} - 2k_{27}X \cdot C^2 - \sum_{i=1}^{n} k_{2,10+2i}C \cdot I_i \tag{98}$$

We are now going to find candidates of the positive singular points of the different systems. We start with f and assume that all k_{ij} are positive. (19) gives

$$I_i = \frac{k_{11+2i,2}}{k_{2,10+2i}C + k_{2,11+2i}} \tag{99}$$

when $C > 0$ and (18) gives

$$N = \frac{k_{65}M}{k_{21}C + k_{26}} \tag{100}$$

(16) and (17) give

$$k_{11,2} - k_{2,11}G - (k_{65} + k_{25})M = 0 \tag{101}$$

or

$$G=\frac{1}{k_{2,11}}\left(k_{11,2}-\left(k_{65}+k_{25}\right)M\right)=\alpha M+\beta \tag{102}$$

Also

$$D=\frac{k_{10,2}}{k_{23}C+k_{54}G+k_{2,10}} \tag{103}$$

Insert these formulas in (14)

$$-k_{21}C\frac{k_{65}M}{k_{21}C+k_{26}}-k_{23}C\frac{k_{10,2}}{k_{23}C+k_{54}G+k_{2,10}} \tag{104}$$

$$+\left(k_{98}-k_{28}-k_{27}X\left(1+\delta\left(2C-1\right)\right)\right)C \tag{105}$$

$$-\sum_{i=1}^{n}k_{2,10+2i}C\frac{k_{11+2i,2}}{k_{2,10+2i}C+k_{2,11+2i}}+k_{82}=0 \tag{106}$$

$\delta = 0,1$. Insert the formula for D in $M' = 0$ to get

$$k_{54}G\frac{k_{10,2}}{k_{23}C+k_{54}G+k_{2,10}}-\left(k_{65}+k_{26}\right)M=0 \tag{107}$$

We can isolate C in this equation to find

$$C=\frac{aM^2+bM+c}{M} \tag{108}$$

where

$$a = \left(k_{65} + k_{25}\right)\frac{k_{54}}{k_{2,11}k_{23}} \tag{109}$$

and

$$b = -\frac{k_{54}k_{10,2}}{k_{23}k_{2,11}} - \left(\frac{k_{54}}{k_{2,11}k_{23}}k_{11,2} + \frac{k_{2,10}}{k_{23}}\right) \tag{110}$$

Finally

$$c = \frac{k_{54}k_{10,2}k_{11,2}}{k_{23}k_{2,11}\left(k_{65} + k_{25}\right)} \tag{111}$$

We now get by multiplying with the product of the denominators in (104), (105) and (106)

$$-k_{21}Ck_{65}M\left(k_{23}C + k_{54}G + k_{2,10}\right)\prod_{i=1}^{n}\left(k_{2,10+2i}C + k_{2,11+2i}\right) \tag{112}$$

$$-k_{23}Ck_{10,2}\left(k_{21}C + k_{26}\right)\prod_{i=1}^{n}\left(k_{2,10+2i}C + k_{2,11+2i}\right) \tag{113}$$

$$+\left(k_{82} + \left(k_{98} - k_{28} - k_{27}X\left(1 + \delta\left(2C - 1\right)\right)\right)C\right)\left(k_{21}C + k_{26}\right) \tag{114}$$

$$\cdot\left(k_{23}C + k_{54}G + k_{2,10}\right)\prod_{i=1}^{n}\left(k_{2,10+2i}C + k_{2,11+2i}\right) \tag{115}$$

$$-\sum_{i=1}^{n}k_{2,10+2i}Ck_{11+2i,2}\left(\prod_{j=1,j\neq i}^{n}\left(k_{2,10+2j}C + k_{2,11+2j}\right)\right) \tag{116}$$

$$\cdot\left(k_{23}C+k_{54}G+k_{2,10}\right)\left(k_{21}C+k_{26}\right)=0 \tag{117}$$

If we now insert the formulas for G,C and multiply with $M^{n+3},\delta=0$ or $M^{n+4},\delta=1,$ we get a polynomial p of degree at most $2(n+3)$ $(\delta=0)$ or $2(n+4)$ $(\delta=1)$ in M, such that a positive singular point

$$\left(C,D,G,M,N,I_1,\cdots,I_n\right) \tag{118}$$

will have $p(M)=0.$

Let $\delta=1.$ Then we have

Theorem 2. For $\delta=1,$ there exist rate constants $k_{ij}>0,$ such that (24) holds and all equilibria are positive and stable.

Proof. Now let

$$\hat{a}=k_{98}-k_{28} \tag{119}$$

and assume that it is negative. Then

$$\frac{\partial d_*^C}{\partial t}(0)=-\frac{k_{82}^0}{\hat{a}} \tag{120}$$

is the same for all treatments. But now the first coordinate of formula (48) becomes

$$-k_{21}d_*^C d_*^N-k_{23}d_*^C d_*^D+\hat{a}d_*^C-2k_{27}X\left(d_*^C\right)^2+k_{82}-\sum_{i=1}^{n}k_{2,10+2i}d_*^C d_*^{I_i}=0 \tag{121}$$

So

$$\hat{a}\frac{\partial^2 d_*^C}{\partial t^2}(0)=k_{21}\left(\frac{\partial^2 d_*^C}{\partial t^2}d_*^N+2\frac{\partial d_*^C}{\partial t}\frac{\partial d_*^N}{\partial t}+d_*^C\frac{\partial^2 d_*^N}{\partial t}\right) \tag{122}$$

$$+ k_{23}\left(\frac{\partial^2 d_*^C}{\partial t^2} d_*^D + 2\frac{\partial d_*^C}{\partial t}\frac{\partial d_*^D}{\partial t} + d_*^C \frac{\partial^2 d_*^D}{\partial t}\right) \tag{123}$$

$$+\sum_{i=1}^{n} k_{2,10+2i}\left(\frac{\partial^2 d_*^C}{\partial t^2} d_*^{I_i} + 2\frac{\partial d_*^C}{\partial t}\frac{\partial d_*^{I_i}}{\partial t} + d_*^C \frac{\partial^2 d_*^{I_i}}{\partial t}\right) \tag{124}$$

$$+ 4k_{27} X \left(\frac{\partial d_*^C}{\partial t}\right)^2 \tag{125}$$

and when $k_{2,10+2i} = k_{2,11+2i} = k_{11+2i,2} = 0, i = 1,\cdots,n.$

$$\hat{a}\frac{\partial^3 d_*^C}{\partial t^3}(0) = 3k_{21}\left(\frac{\partial^2 d_*^C}{\partial t^2}\frac{\partial d_*^N}{\partial t} + \frac{\partial d_*^C}{\partial t}\frac{\partial^2 d_*^N}{\partial t^2}\right) \tag{126}$$

$$+ 3k_{23}\left(\frac{\partial^2 d_*^C}{\partial t^2}\frac{\partial d_*^D}{\partial t} + \frac{\partial d_*^C}{\partial t}\frac{\partial^2 d_*^D}{\partial t^2}\right) \tag{127}$$

$$+12k_{27} X \frac{\partial^2 d_*^C}{\partial t^2}\frac{\partial d_*^C}{\partial t} \tag{128}$$

We can now take k_{21} big and argue as in theorem 1. □

3. THE CHEMO RADIATION MODEL

Consider the mass action kinetic system

$$C + D \rightarrow 0 \tag{129}$$

$$(1+\delta)C + X \rightarrow 0, \delta = 0,1 \tag{130}$$

$$C \leftrightarrows 2C \tag{131}$$

$$C \leftrightarrows 0 \tag{132}$$

$$D \leftrightarrows 0 \tag{133}$$

Here the complexes are $C(2)=0,\ C(3)=C+D,\ C(7)=(1+\delta)C+X,\ C(8)=C,$ $C(9)=2C,\ C(10)=D.$ With mass action kinetics the vector field is

$$f(C,D)=\begin{pmatrix} k_{82}-k_{23}C\cdot D+\left(k_{98}-k_{28}-k_{27}X\left(1+\delta(2C-1)\right)\right)C \\ -k_{23}C\cdot D+k_{10,2}-k_{2,10}D \end{pmatrix} \tag{134}$$

all $k_{ij}>0.$

This system with $\delta=0$ is similar to the reduced system from [17] .

$$GF \to C \tag{135}$$

$$C+GI \to 0 \tag{136}$$

$$C \to 2C \tag{137}$$

$$GF \leftrightarrows 0 \tag{138}$$

$$GI \leftrightarrows 0 \tag{139}$$

$$C \leftrightarrows 0 \tag{140}$$

where we have added the last reaction. The complexes are $C(1)=GF,$ $C(2)=C,\ C(3)=C+GI,\ C(4)=0,\ C(5)=2C,\ C(6)=GI.$ The vector field is

$$g(C,GF,GI)=\begin{pmatrix} k_{21}GF-k_{43}C\cdot GI+(k_{52}-k_{42})C+k_{24} \\ k_{14}-(k_{21}+k_{41})GF \\ -k_{43}C\cdot GI+k_{64}-k_{46}GI \end{pmatrix} \tag{141}$$

We can assume that GF is at equilibrium

$$GF=\frac{k_{14}}{k_{21}+k_{41}} \tag{142}$$

The reduced system is then

$$\hat{g}(C,GI)=\begin{pmatrix} k-k_{43}C\cdot GI+(k_{52}-k_{42})C \\ -k_{43}C\cdot GI+k_{64}-k_{46}GI \end{pmatrix} \tag{143}$$

$$k=\frac{k_{21}k_{14}}{k_{21}+k_{41}}+k_{24} \tag{144}$$

We shall find the singular points of the chemo radiation system. $D'=0$ gives

$$D=\frac{k_{10,2}}{k_{23}C+k_{2,10}} \tag{145}$$

when $C>0$ and then $C'=0$ amounts to

$$k_{82}-k_{23}C\frac{k_{10,2}}{k_{23}C+k_{2,10}}+\left(k_{98}-k_{28}-k_{27}X\left(1+\delta(2C-1)\right)\right)C=0 \tag{146}$$

which is equivalent to

$$k_{23}\left(k_{98}-k_{28}-k_{27}X\right)C^{2} \tag{147}$$

$$+\left(k_{2,10}\left(k_{98}-k_{28}-k_{27}X\right)+k_{23}k_{82}-k_{23}k_{10,2}\right)C+k_{82}k_{2,10}=0 \tag{148}$$

when $\delta = 0, C > 0$ and when $\delta = 1, C > 0$ it is equivalent to

$$-2k_{23}k_{27}XC^3+\left(\left(k_{98}-k_{28}\right)k_{23}-2k_{27}Xk_{2,10}\right)C^2 \tag{149}$$

$$+\left(k_{2,10}\left(k_{98}-k_{28}\right)+k_{82}k_{23}-k_{23}k_{10,2}\right)C+k_{82}k_{2,10}=0 \tag{150}$$

The linearization of f at a singular point $c_* = (C,D)$ is

$$A = Df_{c*} = \begin{pmatrix} -k_{23}D+\left(k_{98}-k_{28}-k_{27}X\right) & -k_{23}C \\ -k_{23}D & -k_{2,10}-k_{23}C \end{pmatrix} \tag{151}$$

when $\delta = 0$ and when $\delta = 1$ it is

$$\tilde{A} = Df_{c*} = \begin{pmatrix} -k_{23}D+\left(k_{98}-k_{28}-4k_{27}XC\right) & -k_{23}C \\ -k_{23}D & -k_{2,10}-k_{23}C \end{pmatrix} \tag{152}$$

If $\delta = 0$

$$\hat{a} = k_{98}-k_{28}-k_{27}X<0 \tag{153}$$

and (C,D) is a positive singular point, then

$$\sigma = \operatorname{trace} A<0 \quad \det A = -\hat{a}k_{23}C+k_{2,10}\left(k_{23}D-\hat{a}\right)>0 \tag{154}$$

hence this is a stable, positive singular point.

Consider the chemo radiation system where

$$k_{10,2}=23 \quad k_{98}=13 \quad X=1 \quad \delta=1 \tag{155}$$

and all other $k_{ij}=1$. Then the cubic polynomial giving candidates of singular points is

$$-2C^3+10C^2-10C+1=0 \tag{156}$$

and this gives three singular points

$$(C,D)=\begin{cases}(0.112,20.68)\\(1.21,10.41)\\(3.677,4.918)\end{cases} \tag{157}$$

where we have used that

$$D=\frac{k_{10,2}}{k_{23}C+k_{2,10}} \tag{158}$$

It is a simple matter to check that the first and last of these are stable equilibria and the middle is an unstable saddle by computing the trace and determinant of the linearization of f_q in the singular points. I have plotted a phase portrait for these values of the parameters in Figure 3.

4. THE QUADRAPEUTICS EQUILIBRIUM EQUATION

We shall find the quadrapeutics equilibrium equation too.

Theorem 3. Suppose $\delta=0$. There exists a polynomial of degree atmost six

$$s(M)=\sum_{i=0}^{6}q_iM^i \quad q_i\in\mathbb{R} \tag{159}$$

such that if (C,D,G,M,N) is a positive singular point of f_q then

$$s(M)=0 \tag{160}$$

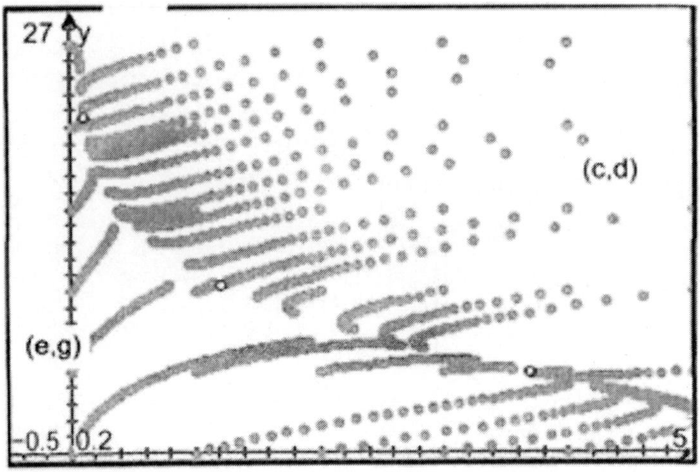

Figure 3. A tristable chemo radiation system. There are three singular points marked with a circle.

Proof. Isolate D, G, M, N from (15), (16), (17), (18) and insert it in $f_{q1}(C,D,G,M,N)=0$ to find

$$p(C,G,M) = -k_{21}Ck_{65}M\left(k_{23}C + k_{54}G + k_{2,10}\right) \tag{161}$$

$$-k_{23}Ck_{10,2}\left(k_{21}C + k_{26}\right) \tag{162}$$

$$+\left(k_{82} + \left(k_{98} - k_{28} - k_{27}X\left(1+\delta\left(2C-1\right)\right)\right)C\right) \tag{163}$$

$$\cdot\left(k_{21}C + k_{26}\right)\left(k_{23}C + k_{54}G + k_{2,10}\right) = 0 \tag{164}$$

where we have multiplied with

$$\left(k_{21}C + k_{26}\right)\left(k_{23}C + k_{54}G + k_{2,10}\right) \tag{165}$$

We shall now insert

$$C = \frac{aM^2 + bM + c}{M} \quad G = \alpha M + \beta \tag{166}$$

where

$$\alpha = -\frac{k_{65} + k_{25}}{k_{2,11}} \tag{167}$$

$$\beta = \frac{k_{11,2}}{k_{2,11}} \tag{168}$$

in the polynomial p. Compute

$$M^3 C^3 = \sum_{i=0}^{6} \tilde{a}_i M^i \tag{169}$$

where

$$\tilde{a}_6 = a^3 \tag{170}$$

$$\tilde{a}_5 = 3a^2 b \tag{171}$$

$$\tilde{a}_4 = 3\left(ab^2 + a^2 c\right) \tag{172}$$

$$\tilde{a}_3 = 6abc + b^3 \tag{173}$$

$$\tilde{a}_2 = 3\left(ac^2 + b^2 c\right) \tag{174}$$

$$\tilde{a}_1 = 3bc^2 \tag{175}$$

$$\tilde{a}_0 = c^3 \tag{176}$$

Now define

$$p_1(C,G,M) = M^3\left(-k_{21}Ck_{65}M\left(k_{23}C + k_{54}G + k_{2,10}\right)\right) \tag{177}$$

and this is

$$s_1(M) = p_1\left(\frac{aM^2 + bM + c}{M}, \alpha M + \beta, M\right) \tag{178}$$

Also put

$$p_2(C,G,M) = M^3\left(-k_{23}Ck_{10,2}\left(k_{21}C + k_{26}\right)\right) \tag{179}$$

and this is

$$s_2(M) = p_2\left(\frac{aM^2 + bM + c}{M}, \alpha M + \beta, M\right) \tag{180}$$

Finally set

$$p_3(C,G,M) = M^3\left(k_{82} + \hat{a}C\right)\left(k_{21}C + k_{26}\right)\left(k_{23}C + k_{54}G + k_{2,10}\right) \tag{181}$$

which is

$$s_3(M) = p_3\left(\frac{aM^2 + bM + c}{M}, \alpha M + \beta, M\right) \tag{182}$$

So

$$s(M) = s_1(M) + s_2(M) + s_3(M) \tag{183}$$

Now compute

$$s_1(M) = -M^2 k_{21} k_{65} k_{23}\left(a^2M^4 + b^2M^2 + c^2 + 2abM^3 + 2acM^2 + 2bcM\right) \quad (184)$$

$$- k_{21}\left(aM^2 + bM + c\right)k_{65}M^3\left(k_{54}\left(\alpha M + \beta\right)) + k_{2,10}\right) \quad (185)$$

$$= -k_{21}k_{65}k_{23}\left(a^2M^6 + b^2M^4 + c^2M^2 + 2abM^5 + 2acM^4 + 2bcM^3\right) \quad (186)$$

$$- k_{21}k_{65}\left(aM^2 + bM + c\right)M^3 k_{54}\alpha M$$
$$-k_{21}k_{65}\left(aM^2 + bM + c\right)M^3\left(k_{54}\beta + k_{2,10}\right) \quad (187)$$

Write

$$s_1(M) = \sum_{i=0}^{6} a_i M^i \quad (188)$$

Then by the previous computation

$$a_6 = -k_{21}ak_{65}\left(k_{23}a + k_{54}\alpha\right) \quad (189)$$

$$a_5 = -k_{21}bk_{65}\left(k_{23}a + k_{54}\alpha\right) - k_{21}ak_{65}\left(k_{54}\beta + k_{23}b + k_{2,10}\right) \quad (190)$$

$$a_4 = -k_{21}ck_{65}\left(k_{23}a + k_{54}\alpha\right) - k_{21}bk_{65}\left(k_{54}\beta + k_{23}b + k_{2,10}\right) - k_{21}k_{65}k_{23}ac \quad (191)$$

$$a_3 = -k_{21}ck_{65}\left(k_{54}\beta + k_{23}b + k_{2,10}\right) - k_{21}k_{65}k_{23}bc \quad (192)$$

$$a_2 = -k_{21}c^2k_{65}k_{23} \quad (193)$$

$$a_1 = 0 \quad (194)$$

$$a_0 = 0 \tag{195}$$

Now compute

$$s_2(M) = M^3\left(-k_{23}Ck_{10,2}\left(k_{21}C + k_{26}\right)\right) \tag{196}$$

$$= -M^3k_{23}k_{10,2}\left(k_{21}\frac{a^2M^4 + b^2M^2 + c^2 + 2abM^3 + 2acM^2 + 2bcM}{M^2}\right.$$

$$\left. + k_{26}\frac{aM^2 + bM + c}{M}\right) \tag{197}$$

$$= -k_{23}k_{10,2}k_{21}\left(a^2M^5 + b^2M^3 + c^2M + 2abM^4 + 2acM^3 + 2bcM^2\right) \tag{198}$$

$$- k_{23}k_{10,2}k_{26}\left(aM^4 + bM^3 + cM^2\right) \tag{199}$$

Write

$$s_2(M) = \sum_{i=0}^{6} b_i M^i \tag{200}$$

and by the computation above

$$b_6 = 0 \tag{201}$$

$$b_5 = -k_{23}a^2k_{10,2}k_{21} \tag{202}$$

$$b_4 = -k_{23}bk_{10,2}k_{21}a - k_{23}ak_{10,2}\left(k_{21}b + k_{26}\right) \tag{203}$$

$$b_3 = -2k_{23}k_{10,2}k_{21}ac - k_{23}k_{10,2}b\left(k_{21}b + k_{26}\right) \tag{204}$$

$$b_2 = -k_{23}ck_{10,2}\left(k_{21}b + k_{26}\right) - k_{23}bk_{10,2}k_{21}c \tag{205}$$

$$b_1 = -k_{23}c^2 k_{10,2} k_{21} \tag{206}$$

$$b_0 = 0 \tag{207}$$

Now consider

$$p_3(C,G,M) = k_{21}k_{23}\hat{a}C^3 + k_{21}k_{54}\hat{a}C^2G \tag{208}$$

$$+ k_{21}\hat{a}C^2 k_{2,10} + k_{26}k_{23}\hat{a}C^2 \tag{209}$$

$$+ k_{26}k_{54}G\hat{a}C + k_{26}k_{2,10}\hat{a}C \tag{210}$$

$$+ k_{82}k_{21}k_{23}C^2 + k_{82}k_{21}k_{54}CG \tag{211}$$

$$+ k_{82}k_{21}k_{2,10}C + k_{82}k_{26}k_{23}C \tag{212}$$

$$+ k_{82}k_{26}k_{54}G + k_{82}k_{26}k_{2,10} \tag{213}$$

Name these 12 summands $d_1,\cdots,d_{12}$. Now

$$M^3 d_1 = k_{21}k_{23}\hat{a}\sum_{i=0}^{6}\tilde{a}_i M^i \tag{214}$$

and

$$M^3 d_2 = k_{21}k_{54}\hat{a}\Big(\alpha\big(a^2M^6 + b^2M^4 + c^2M^2 + 2abM^5 + 2acM^4 + 2bcM^3\big) \tag{215}$$

$$+\beta\big(a^2M^5 + b^2M^3 + c^2M + 2abM^4 + 2acM^3 + 2bcM^2\big)\Big) \tag{216}$$

Also

$$M^3 d_3 = k_{21}\hat{a}k_{2,10}\left(a^2M^5 + b^2M^3 + c^2M + 2abM^4 + 2acM^3 + 2bcM^2\right) \tag{217}$$

Similarly

$$M^3 d_4 = k_{26}\hat{a}k_{23}\left(a^2M^5 + b^2M^3 + c^2M + 2abM^4 + 2acM^3 + 2bcM^2\right) \tag{218}$$

Then

$$M^3 d_5 = k_{26}k_{54}\hat{a}\left(\alpha aM^5 + \alpha bM^4 + c\alpha M^3 + \beta aM^4 + \beta bM^3 + \beta cM^2\right) \tag{219}$$

And

$$M^3 d_6 = k_{26}k_{2,10}\hat{a}\left(aM^4 + bM^3 + cM^2\right) \tag{220}$$

We also find

$$M^3 d_7 = k_{82}k_{21}k_{23}\left(a^2M^4 + b^2M^2 + c^2 + 2abM^3 + 2acM^2 + 2bcM\right)M \tag{221}$$

$$M^3 d_8 = k_{82}k_{21}k_{54}\left(\alpha\left(aM^5 + bM^4 + cM^3\right) + \beta\left(aM^4 + bM^3 + cM^2\right)\right) \tag{222}$$

Combine

$$M^3\left(d_9 + d_{10}\right) = \left(k_{82}k_{21}k_{2,10} + k_{82}k_{26}k_{23}\right)\left(aM^4 + bM^3 + cM^2\right) \tag{223}$$

The last two ds give

$$M^3 d_{11} = k_{82}k_{26}k_{54}\left(\alpha M^4 + \beta M^3\right) \tag{224}$$

$$M^3 d_{12} = k_{82}k_{26}k_{2,10}M^3 \tag{225}$$

Write

$$s_3(M)=\sum_{i=0}^{6} c_i M^i \tag{226}$$

Then we find from the above computations

$$c_6 = \hat{a}a^2 k_{21}\left(k_{23}a + k_{54}\alpha\right) \tag{227}$$

and

$$c_5 = k_{21}k_{23}\hat{a}3a^2 b + k_{21}k_{54}\hat{a}\left(\alpha 2ab + \beta a^2\right) + k_{21}\hat{a}k_{2,10}a^2 \tag{228}$$

$$+ k_{26}k_{23}\hat{a}a^2 + k_{82}k_{21}k_{23}a^2 + k_{82}k_{21}k_{54}a\alpha + k_{26}k_{54}\hat{a}\alpha a \tag{229}$$

Now collect terms

$$c_4 = k_{21}k_{23}\hat{a}3\left(ab^2 + a^2 c\right) + k_{21}k_{54}\hat{a}\alpha\left(b^2 + 2ac\right) \tag{230}$$

$$+ k_{21}k_{54}\hat{a}\beta 2ab + k_{21}\hat{a}k_{2,10}2ab + k_{26}k_{23}\hat{a}2ab \tag{231}$$

$$+ k_{26}k_{54}\hat{a}\left(\alpha b + \beta a\right) + k_{26}k_{2,10}\hat{a}a \tag{232}$$

$$+ k_{82}k_{21}k_{23}2ab + k_{82}k_{21}k_{54}\left(\alpha b + \beta a\right) \tag{233}$$

$$+\left(k_{82}k_{21}k_{2,10} + k_{82}k_{26}k_{23}\right)a + k_{82}k_{26}k_{54}\alpha \tag{234}$$

We can also find

$$c_3 = k_{21}k_{23}\hat{a}\left(6abc + b^3\right) + k_{21}k_{54}\hat{a}\left(2bc\alpha + \beta\left(b^2 + 2ac\right)\right) \tag{235}$$

$$+k_{21}\hat{a}k_{2,10}\left(b^2+2ac\right)+k_{26}k_{23}\hat{a}\left(b^2+2ac\right) \tag{236}$$

$$+k_{26}k_{54}\hat{a}\left(c\alpha+\beta b\right)+k_{26}k_{2,10}\hat{a}b \tag{237}$$

$$+k_{82}k_{21}k_{23}\left(b^2+2ac\right)+k_{82}k_{21}k_{54}\left(\alpha c+\beta b\right) \tag{238}$$

$$+\left(k_{82}k_{21}k_{2,10}+k_{82}k_{26}k_{23}\right)b+k_{82}k_{26}k_{54}\beta+k_{82}k_{26}k_{2,10} \tag{239}$$

Now

$$c_2=k_{21}k_{23}\hat{a}3\left(ac^2+b^2c\right)+k_{21}k_{54}\hat{a}\left(\alpha c^2+\beta 2bc\right) \tag{240}$$

$$+\left(k_{21}\hat{a}k_{2,10}+k_{26}k_{23}\hat{a}\right)2bc+k_{26}k_{54}\hat{a}\beta c \tag{241}$$

$$+k_{26}k_{2,10}\hat{a}c+k_{82}k_{21}k_{23}2bc \tag{242}$$

$$+k_{82}k_{21}k_{54}\beta c+\left(k_{82}k_{21}k_{2,10}+k_{82}k_{26}k_{23}\right)c \tag{243}$$

Also

$$c_1=k_{21}k_{23}\hat{a}3bc^2+k_{21}k_{54}\hat{a}c^2\beta+k_{21}\hat{a}k_{2,10}c^2 \tag{244}$$

$$+k_{82}k_{21}k_{23}c^2+k_{23}c^2\hat{a}k_{26} \tag{245}$$

Finally

$$c_0=k_{21}k_{23}\hat{a}c^3 \tag{246}$$

The quadrapeutics polynomial is

$$s(M)=\sum_{i=0}^{6}(a_i+b_i+c_i)M^i \tag{247}$$

If all rate constants $k_{ij}=1, X=1$ then

$$s(M)=2M^5-8M^4+10M^3-\frac{23}{4}M^2+\frac{3}{2}M-\frac{1}{8}=0 \tag{248}$$

This polynomium has three real roots

$$M=0.148638,\ M=\frac{1}{2},\ M=2.32234 \tag{249}$$

and two imaginary roots

$$0.514511\pm 0.312089i \tag{250}$$

There are thus two candidates of singular points ($M=\frac{1}{2}$ does not give a singular point).

$$(C,D,G,M,N)=\begin{cases}(0.661153,0.423634,0.702724,0.148638,0.089479)\\(1.85998,-1.27437,-3.64468,2.32234,0.812013)\end{cases} \tag{251}$$

and they are both singular points (numerical evidence). We have used, that

$$G=\alpha M+\beta \tag{252}$$

$$C=\frac{aM^2+bM+c}{M} \tag{253}$$

$$D=\frac{k_{10,2}}{k_{23}C+k_{54}G+k_{2,10}} \tag{254}$$

$$N = \frac{k_{65}M}{k_{21}C + k_{26}} \tag{255}$$

5. THE EXTENDED QUADRAPEUTICS MODEL

This is the system (1) to (11) and the reactions

$$\tilde{D} + C \to 0 \tag{256}$$

$$N + D \to \tilde{D} \tag{257}$$

$$\tilde{D} \to 0 \tag{258}$$

$$D \to \tilde{D} \tag{259}$$

taking into account, that plasmonic nanobubbles destroy the liposomes D and thus the chemo therapeutic drug $\tilde{D}$ is injected into the cytoplasm. But some of the liposomes decay, producing chemo therapeutic drug. The complexes here are $C(12) = N + D, C(13) = \tilde{D}, C(14) = \tilde{D} + C.$ The ODEs are

$$C' = -k_{21}C \cdot N - k_{23}C \cdot D + (k_{98} - k_{28} - k_{27}X)C + k_{82} - k_{2,14}\tilde{D} \cdot C \tag{260}$$

$$\tilde{D}' = -k_{2,14}\tilde{D} \cdot C + k_{13,12}N \cdot D - k_{2,13}\tilde{D} + k_{13,10}D \tag{261}$$

$$D' = -k_{23}C \cdot D - k_{54}D \cdot G + k_{10,2} - k_{2,10}D - k_{13,12}N \cdot D - k_{13,10}D \tag{262}$$

$$G' = -k_{54}D \cdot G + k_{11,2} - k_{2,11}G \tag{263}$$

$$M' = k_{54}D \cdot G - (k_{65} + k_{25})M \tag{264}$$

$$N' = -k_{21}C \cdot N + k_{65}M - k_{26}N - k_{13,12}N \cdot D \tag{265}$$

The vector field is denoted g_q. Denote by g_{cr} the vector field in (260), (261) and (262) with the rate constants in (263), (264) and (265) all equal to zero and the rest positive.

Theorem 4. There exist positive values of the rate constants such that

$$C_*^q < C_*^{cr} \tag{266}$$

holds and the singular points are both positive and stable.

Proof. Define for $* = q$

$$k_1 = \left(k_{82}, k_{10,2}, k_{11,2}\right) \tag{267}$$

$$k_2 = \left(k_{21}, k_{23}, k_{98}, k_{28}, \tilde{k}_{27}, k_{2,14}, k_{13,12}, k_{2,13}, k_{54}, k_{2,10}, k_{2,11}, k_{65}, k_{25}, k_{26}, k_{13,10}\right) \tag{268}$$

And for $* = cr$

$$k_1^{cr} = \left(k_{82}, k_{10,2}\right) \tag{269}$$

$$k_2^{cr} = \left(k_{23}, k_{98}, k_{28}, \tilde{k}_{27}, k_{2,14}, k_{2,13}, k_{2,10}, k_{13,10}\right) \tag{270}$$

Define

$$K_{cr} = \left(k_1^{cr}, k_2^{cr}\right) \quad K_q = \left(k_1^q, k_2^q\right) \tag{271}$$

and

$$\hat{K}_* = \left(k_1^*, k_2^*\right) \quad k_1^* = 0 \tag{272}$$

Use the implicit function theorem to find a mapping

$$c_q : V_q \subset \mathbb{R}^{18} \to \mathbb{R}^6 \tag{273}$$

and

$$c_{cr} : V_{cr} \subset \mathbb{R}^{10} \to \mathbb{R}^{3} \tag{274}$$

such that

$$g_{*}\left(c_{*}\left(K_{*}\right), K_{*}\right) = 0 \quad c_{*}\left(\hat{K}_{*}\right) = 0 \tag{275}$$

and define

$$d_{*}\left(t\right) = c_{*}\left(tk_{1}^{*,0}, k_{2}^{*}\right) \quad * = \begin{cases} cr \\ q \end{cases} \tag{276}$$

V_q, V_{cr} are open subsets of $\mathbb{R}^{18}, \mathbb{R}^{10}$ respectively. For $* = cr$ let $= d_{*}^{N} = 0.\ d_{*}^{G} = d_{*}^{M}$
By the implicit function theorem

$$d_{*}^{\tilde{D}}\left(0\right) = 0 \tag{277}$$

Differentiate (262) and (261) with respect to t to get

$$\frac{\partial d_{*}^{D}}{\partial t} = \frac{k_{10,2}^{0}}{k_{2,10} + k_{13,10}} \tag{278}$$

and

$$\frac{\partial d_{*}^{\tilde{D}}}{\partial t} = \frac{k_{10,2}^{0} k_{13,10}}{k_{2,13}\left(k_{2,10} + k_{13,10}\right)} \tag{279}$$

Now differentiate (261) twice with respect to t to find

$$k_{2,13} \frac{\partial^{2} d_{*}^{\tilde{D}}}{\partial t^{2}} = k_{13,10} \frac{\partial^{2} d_{*}^{D}}{\partial t^{2}} - 2k_{2,14} \frac{\partial d_{*}^{\tilde{D}}}{\partial t} \frac{\partial d_{*}^{C}}{\partial t} \tag{280}$$

and differentiate (262) with respect to t to get

$$\frac{\partial^2 d_*^D}{\partial t^2} = -\frac{2}{k_{2,10}+k_{13,10}}\left(k_{23}\frac{\partial d_*^C}{\partial t}\frac{\partial d_*^D}{\partial t}+k_{54}\frac{\partial d_*^D}{\partial t}\frac{\partial d_*^G}{\partial t}\right) \tag{281}$$

We also have

$$\frac{\partial d_*^C}{\partial t} = -\frac{k_{82}^0}{\hat{a}} \quad \frac{\partial d_*^G}{\partial t} = \frac{k_{11,2}^0}{k_{2,11}} \tag{282}$$

the last equality when $* = q$. We now get

$$\frac{\partial^2 d_*^C}{\partial t^2} = \frac{1}{\hat{a}}\left(2k_{23}\frac{\partial d_*^C}{\partial t}\frac{\partial d_*^D}{\partial t}+2k_{2,14}\frac{\partial d_*^{\bar{D}}}{\partial t}\frac{\partial d_*^C}{\partial t}+k_{21}\frac{\partial d_*^C}{\partial t}\frac{\partial d_*^N}{\partial t}\right) \tag{283}$$

so

$$\frac{\partial^2 d_{cr}^C}{\partial t^2}(0) = \frac{\partial^2 d_q^C}{\partial t^2}(0) \tag{284}$$

and

$$\frac{\partial^2 d_*^N}{\partial t^2} = \frac{k_{65}2k_{54}}{k_{26}\left(k_{65}+k_{25}\right)}\frac{\partial d_*^D}{\partial t}\frac{\partial d_*^G}{\partial t} \tag{285}$$

where we have used that

$$\frac{\partial d_*^M}{\partial t}(0) = 0 \tag{286}$$

hence

$$\frac{\partial d_*^N}{\partial t}(0) = 0 \tag{287}$$

Finally

$$\hat{a}\frac{\partial^3 d_*^C}{\partial t^3} = 3k_{23}\left(\frac{\partial^2 d_*^C}{\partial t^2}\frac{\partial d_*^D}{\partial t} + \frac{\partial d_*^C}{\partial t}\frac{\partial^2 d_*^D}{\partial t^2}\right) \tag{288}$$

$$+ 3k_{2,14}\left(\frac{\partial^2 d_*^{\tilde{D}}}{\partial t^2}\frac{\partial d_*^C}{\partial t} + \frac{\partial d_*^{\tilde{D}}}{\partial t}\frac{\partial^2 d_*^C}{\partial t^2}\right) \tag{289}$$

$$+ 3k_{21}\frac{\partial d_*^C}{\partial t}\frac{\partial^2 d_*^N}{\partial t^2} \tag{290}$$

and taking k_{21} big, this implies the theorem, arguing as in the proof of theorem 1. □

From $D' = 0$ isolate

$$C = \frac{1}{k_{23}D}\left(-k_{54}D \cdot G + k_{10,2} - \left(k_{2,10} + k_{13,10}\right)D - k_{13,12}N \cdot D\right) \tag{291}$$

and insert this in $N' = 0$ to get using

$$G = \alpha M + \beta \quad \alpha < 0 \quad \beta > 0 \tag{292}$$

and

$$D = \frac{-\alpha k_{2,11}M}{k_{54}\left(\alpha M + \beta\right)} \tag{293}$$

(this equation follows from (263)), that

$$\frac{k_{13,12}k_{21}}{k_{23}}N^2 + N\left(k_{54}\frac{k_{21}}{k_{23}}\left(\alpha M + \beta\right) + \frac{k_{21}k_{10,2}k_{54}\left(\alpha M + \beta\right)}{k_{23}k_{2,11}\alpha M}\right. \tag{294}$$

$$+k_{21}\frac{k_{2,10}+k_{13,10}}{k_{23}}-k_{26}+k_{13,12}\frac{k_{2,11}\alpha M}{k_{54}(\alpha M+\beta)}\Bigg)+k_{65}M=0 \tag{295}$$

$M>0, M\neq-\frac{\beta}{\alpha}.$ Let $\tilde{a}$ be the coefficient to N^2 and let $\tilde{b}(M)$ be the coefficient to N and $\tilde{c}(M)=k_{65}M.$ Then we get when

$$\Delta=\tilde{b}(M)^2-4\tilde{a}\tilde{c}(M)>0 \tag{296}$$

$$N^{\pm}(M)=\frac{-\tilde{b}(M)\pm\sqrt{\Delta}}{2\tilde{a}} \tag{297}$$

$\tilde{a}>0.$ Thus $D'=0$ gives

$$C^{\pm}=\frac{1}{k_{23}D(M)}\left(-k_{54}D\cdot G+k_{10,2}-(k_{2,10}+k_{13,10})D-k_{13,12}N^{\pm}\cdot D\right) \tag{298}$$

and

$$\tilde{D}^{\pm}=\frac{k_{13,12}N^{\pm}D+k_{13,10}D}{k_{2,14}C^{\pm}+k_{2,13}} \tag{299}$$

$$D=\frac{-k_{2,11}\alpha M}{k_{54}(\alpha M+\beta)} \tag{300}$$

So let

$$p_{\bullet}(M)=p_{\bullet}^{\pm}(M)=-k_{21}C^{\pm}N^{\pm}-k_{23}C^{\pm}D+\hat{a}C^{\pm}+k_{82}-k_{2,14}\tilde{D}^{\pm}C^{\pm} \tag{301}$$

If $(C,\tilde{D},D,G,M,N)$ is a positive singular point of g_q, then

$$p_{\bullet}^{\pm}(M)=0 \tag{302}$$

Define h by

$$\sqrt{1+x}-1=\int_0^1 \frac{\partial}{\partial s}\sqrt{1+xs}\,\mathrm{d}s \tag{303}$$

$$=x\int_0^1 \frac{1}{2\sqrt{1+xs}}\,\mathrm{d}s=xh(x) \tag{304}$$

where h is smooth and $x\in]-1,+\infty[$.

Proposition 5. The function

$$N^{-}(M)\triangleq -2\frac{\tilde{c}(M)}{\tilde{b}(M)}h\left(-\frac{4\tilde{a}\tilde{c}}{\tilde{b}^2}\right) \tag{305}$$

satisfies (294) and (295), when

$$-\frac{4\tilde{a}\tilde{c}}{\tilde{b}^2}\in]-1,+\infty[,\ \tilde{b}(M)<0 \tag{306}$$

Proof. If $\tilde{b}(M)<0$ then

$$N^{-}(M)=-\frac{2\tilde{c}(M)}{\tilde{b}(M)}h\left(-\frac{4\tilde{a}\tilde{c}(M)}{\tilde{b}(M)^2}\right) \tag{307}$$

is smooth and

$$N^{+}(M)=\frac{-\tilde{b}(M)+\sqrt{\Delta(M)}}{2\tilde{a}} \tag{308}$$

when $\tilde{a}>0$. If $\tilde{b}(M)>0$ then

$$N^{+}(M) = -\frac{2\tilde{c}(M)}{\tilde{b}(M)} h\left(-\frac{4\tilde{a}\tilde{c}(M)}{\tilde{b}(M)^2}\right) \tag{309}$$

is smooth and

$$N^{-}(M) = \frac{-\tilde{b}(M) - \sqrt{\Delta(M)}}{2\tilde{a}} \tag{310}$$

when $\tilde{a} > 0$. These are both negative, so $\tilde{b}(M) > 0$ is not interesting. □

Proposition 6. If $k_{13,12} = k_{2,13} = k_{2,14} = k_{13,10} = 0$ and (306) holds, then

$$s(M) = q(M) p_{\bullet}^{-}(M) \tag{311}$$

where

$$q(M) = M^3 (k_{21}C + k_{26})(k_{23}C + k_{54}G + k_{2,10}) \tag{312}$$

and

$$C = \frac{aM^2 + bM + c}{M} \tag{313}$$

$$G = \alpha M + \beta \tag{314}$$

Proof. We have

$$k_{23}C + k_{54}G + k_{2,10} \tag{315}$$

$$= k_{23}\frac{aM^2 + bM + c}{M} + k_{54}(\alpha M + \beta) + k_{2,10} \tag{316}$$

and here the terms with a, α cancel out to give

$$k_{23}b + k_{54}\beta + k_{2,10} + k_{23}\frac{c}{M} \tag{317}$$

$$= -\frac{k_{54}k_{10,2}}{k_{2,11}} + k_{23}\frac{c}{M} \tag{318}$$

So

$$\frac{k_{10,2}}{k_{23}C + k_{54}G + k_{2,10}} \tag{319}$$

$$= \frac{k_{10,2}}{-\frac{k_{54}k_{10,2}}{k_{2,11}} + k_{23}\frac{c}{M}} \tag{320}$$

which becomes

$$\frac{(k_{65} + k_{25})M}{\frac{k_{54}k_{11,2}}{k_{2,11}} - k_{54}\frac{k_{65} + k_{25}}{k_{2,11}}M} \tag{321}$$

$$= \frac{-k_{2,11}\alpha M}{k_{54}(\alpha M + \beta)} \tag{322}$$

We also can compute

$$\tilde{b}(M) = k_{54}\frac{k_{21}}{k_{23}}(\alpha M + \beta) + \frac{k_{21}k_{10,2}k_{54}}{k_{23}k_{2,11}}\frac{\alpha M + \beta}{\alpha M} \tag{323}$$

$$+k_{21}\frac{k_{2,10}}{k_{23}}-k_{26} \tag{324}$$

$$=k_{54}\frac{k_{21}}{k_{23}}\left(-\frac{k_{65}+k_{25}}{k_{2,11}}M+\frac{k_{11,2}}{k_{2,11}}\right) \tag{325}$$

$$+k_{21}\frac{k_{2,10}}{k_{23}}+\frac{k_{21}k_{10,2}k_{54}}{k_{23}k_{2,11}}-\frac{k_{21}k_{10,2}k_{54}}{Mk_{23}k_{2,11}}\frac{k_{11,2}}{k_{65}+k_{25}}-k_{26} \tag{326}$$

$$=-k_{21}\frac{aM^2+bM+c}{M}-k_{26}=-k_{21}C-k_{26} \tag{327}$$

Notice that

$$N^{-}(M)=-\frac{\tilde{c}(M)}{\tilde{b}(M)}=\frac{\tilde{c}(M)}{k_{21}C+k_{26}} \tag{328}$$

and

$$D=\frac{k_{10,2}}{k_{23}C+k_{54}G+k_{2,10}} \tag{329}$$

Now we get

$$p_{*}^{-}(M)q(M)=\left(-k_{21}C\cdot N^{-}-k_{23}C\cdot D+\left(k_{82}+\hat{a}C\right)\right)q(M) \tag{330}$$

$$=-k_{65}M\cdot C\left(k_{23}C+k_{54}G+k_{2,10}\right)M^{3} \tag{331}$$

$$-k_{23}C\left(k_{21}C+k_{26}\right)M^{3}+s_{3}(M)=s_{1}(M)+s_{2}(M)+s_{3}(M)=s(M) \tag{332}$$

If we assume there exists $M_i > 0$ such that

$$s(M_i) = 0 \tag{333}$$

$$\frac{\partial s}{\partial M}(M_i) \neq 0 \tag{334}$$

and $q(M_i) > 0$ (this is so if $G(M_i), C(M_i) > 0$). Then

$$p_*^{-}(M_i, k_0) = 0 \tag{335}$$

$$\frac{\partial p_*}{\partial M}(M_i, k_0) \neq 0 \tag{336}$$

where

$$k_0 = K_{q|k_{13,12}=k_{13,10}=k_{2,13}=k_{2,14}=k_{13,10}=0} \tag{337}$$

Then there exists a C^1 function

$$M_*^i : U \to \mathbb{R} \tag{338}$$

such that

$$p_*^{-}(M_*^i(k), k) = 0 \quad k \in U \tag{339}$$

$$M_*^i(k_0) = M_i \tag{340}$$

by the implicit function theorem. Here U is an open neighbourhood of k_0 in $\mathbb{R}^{18}$. So the roots of s are nearly zeroes of p_*^{-}, when

$$k_{2,14}, k_{13,12}, k_{2,13}, k_{13,10} \tag{341}$$

are small and (333) and (334) holds. The example in the end of section 4 applies here.

6. SUMMARY

Consider the extended quadrapeutics system and the additional reactions

$$L_a + C \to 0 \tag{342}$$

$$L_a \leftrightarrows 0 \tag{343}$$

Here the complexes are $C(15) = L_a + C, C(16) = L_a$, modelling adoptive T cell therapy, see [4] . L_a are activated and expanded tumor infiltrating lymphocytes. They are injected back into the patient. Then the differential equations are

$$\frac{\mathrm{d}C}{\mathrm{d}t} = -k_{21}C \cdot N - k_{23}C \cdot D + \hat{a}C + k_{82} - k_{2,14}\tilde{D} \cdot C - k_{2,15}L_a \cdot C \tag{344}$$

$$\tilde{D}' = -k_{2,14}\tilde{D} \cdot C + k_{13,12}N \cdot D - k_{2,13}\tilde{D} + k_{13,10}D \tag{345}$$

$$D' = -k_{23}C \cdot D - k_{54}D \cdot G + k_{10,2} - k_{2,10}D - k_{13,12}N \cdot D - k_{13,10}D \tag{346}$$

$$G' = -k_{54}D \cdot G + k_{11,2} - k_{2,11}G \tag{347}$$

$$M' = k_{54}D \cdot G - (k_{65} + k_{25})M \tag{348}$$

$$N' = -k_{21}C \cdot N + k_{65}M - k_{26}N - k_{13,12}N \cdot D \tag{349}$$

$$\frac{\mathrm{d}L_a}{\mathrm{d}t} = -k_{2,15}L_a \cdot C + k_{16,2} - k_{2,16}L_a \tag{350}$$

Denote this vector field g_q and let g_{cr} denote the vector field in (344), (345), (346) where the rate constants in (347), (348), (349), (350) are all zero and the rest positive. Use the implicit function theorem to find maps

$$c_{cr} : V_{cr} \subset \mathbb{R}^{10} \to \mathbb{R}^3 \tag{351}$$

and

$$c_q : V_q \subset \mathbb{R}^{21} \to \mathbb{R}^7 \tag{352}$$

such that

$$g_*\left(c_*\left(K_*\right), K_*\right) = 0 \quad c_*\left(\hat{K}_*\right) = 0 \tag{353}$$

Here

$$K_{cr} = \left(k_1^{cr}, k_2^{cr}\right) \tag{354}$$

is the same as in section 5. Also V_{cr}, V_q are open subsets of $\mathbb{R}^{10}$ and $\mathbb{R}^{21}$ respectively. And

$$k_1^q = \left(k_{82}, k_{10,2}, k_{11,2}, k_{16,2}\right) \tag{355}$$

Also

$$k_2^q = \left(k_{21}, k_{23}, k_{98}, k_{28}, \tilde{k}_{27}, k_{2,14}, k_{13,12}, k_{2,13}, k_{54}, k_{2,10}, k_{2,11}, k_{65}, k_{25}, \right. \tag{356}$$

$$\left. k_{26}, k_{2,15}, k_{2,16}, k_{13,10}\right) \tag{357}$$

Finally set

$$K_* = \left(k_1^*, k_2^*\right) \quad \hat{K}_* = \left(k_1^*, k_2^*\right) \quad k_1^* = 0 \tag{358}$$

Define

$$d_*(t) = c_*\left(tk_1^{*,0}, k_2^*\right) \tag{359}$$

Let

$$\hat{a} = k_{98} - k_{28} - k_{27}X \tag{360}$$

and assume that it is negative. Then

$$\hat{a}\frac{\partial^2 d_*^C}{\partial t^2} = 2k_{23}\frac{\partial d_*^C}{\partial t}\frac{\partial d_*^D}{\partial t} + 2k_{2,15}\frac{\partial d_*^{I_a}}{\partial t}\frac{\partial d_*^C}{\partial t} + 2k_{2,14}\frac{\partial d_*^{\tilde{D}}}{\partial t}\frac{\partial d_*^C}{\partial t} \tag{361}$$

But

$$\frac{\partial d_*^C}{\partial t} = -\frac{k_{82}^0}{\hat{a}} \tag{362}$$

$$\frac{\partial d_*^D}{\partial t} = \frac{k_{10,2}^0}{k_{2,10}} \tag{363}$$

$$\frac{\partial d_*^{I_a}}{\partial t} = \frac{k_{16,2}^0}{k_{2,16}} \tag{364}$$

$$\frac{\partial d_*^{\tilde{D}}}{\partial t} = \frac{k_{10,2}^0 k_{13,10}}{k_{2,13}\left(k_{2,10} + k_{13,10}\right)} \tag{365}$$

suggesting, that in this model multipeutics can be more potent than quadrapeutics.

In the present paper, we considered quadrapeutics and multipeutics cancer therapies. We proved the multipeutics theorem, stating that the more treatments we apply the smaller the cancer burden. We also found a polynomial of degree at most 6, giving candidates of singular points for the quadrapeutics system.

The cells of the immune system have a plasma membrane repair system and it turns out that this system is much more efficient in cancer cells than in normal cells. The two proteins S100A11 and Annexin A2 are involved in this plasma membrane repair system. They are commonly upregulated in cancer cells, see [20] . So we can ask: can cancer cells repair nanobubble injury.

REFERENCES

1. Lukianova-Hleb, E.Y., Kim, Y.S., Aryasomayajula, B., Boulikas, T., Phan, J., Hung, M.C., Torchillin, V.P., O'Neill, B.E. and Lapotko, D.O. (2015) Safety and Efficacy of Quadrapeutics versus Chemoradiation in Head and Neck Carcinoma Xenograft Model. American Journal of Cancer Research, 5, 3534-3547.
2. Larsen, J.C. (2017) Models of Cancer Growth. Journal of Applied Mathematics and Computing, 53, 613-645.
3. Larsen, J.C. (2016) Hopf Bifurcations in Cancer Models. JP Journal of Applied Mathematics, 14, 1-31.
4. Larsen, J.C. (2016) A Mathematical Model of Adoptive T Cell Therapy. JP Journal of Applied Mathematics, to Appear.
5. Larsen, J.C. (2016) Fundamental Concepts in Dynamics. Survey Article.
6. Adam, J.A. and Bellomo, N. (1997) A Survey of Models for Tumor-Induced Immune System Dynamics. Birkhauser, Boston.
7. Geha, R. and Notarangelo, L. (2012) Case Studies in Immunology. Garland Science, USA.
8. Marks, F., Klingmüller, U. and Müller-Decker, K. (2009) Cellular Signal Processing, Garland Science, USA.
9. Molina-Paris, C. and Lythe, G. (2011) Mathematical Models and Immune Cell Biology. Springer, New York.
10. Murphy, K. (2012) Immuno Biology. 8th Edition, Garland Science, USA.
11. Rees, R.C. (2014) Tumor Immunology and Immunotherapy. Oxford University Press, Oxford.
12. Cheng, F.H.C. and Agudas, B.D. (2014) A Mathematical Model of Bimodal Epigenetics Control of miR-193a in Ovarian Cancer Stem Cells. PLoS ONE, 9, e116050.
13. Lukianova-Hleb, E.Y. and Lapotko, D.O. (2015) Rapid Detection and Destruction of Squamous Cell Carcinoma of the Head and Neck by Nano-Quadrapeutics. Head Neck, 37, 1547-1555.
14. Lukianova-Hleb, E.Y., Ren, X., Sawant, R.R., Wu, X., Torchilin, V.P. and Lapotko, D.O. (2014) On-Demand Intracellular Amplification of Chemoradiation with Cancer-Specific Plasmonic Nanobubbles. Nature Medicine, 20, 778-784.
15. Lukianova-Hleb, E., Hu, Y., Latterini, L., Tarpani, L., Lee, S., Drezek, R.A., Hafner, J.H. and Lapotko, D.O. (2010) Plasmonic Nanobubbles as Transient Vapor Nanobubbles Generated around Plasmonic Nanoparticles. ACS Nano, 4, 2109-2123.
16. Larsen, J.C. (2016) The Bistability Theorem in a Model of Metastatic Cancer. Applied Mathematics, 7, 1183-1206.
17. Larsen, J.C. (2016) The Bistability Theorem in a Cancer Model. Preprint.
18. Horn, F. and Jackson, R. (1972) General Mass Action Kinetics. Archive for Rational Mechanics and Analysis, 47, 81-116.
19. Uherka, D.J. and Sergott, A.M. (1977) On the Continuous Dependence of the Roots of a Polynomial on Its Coefficients. The American Mathematical Monthly, 84, 368-370.
20. Nylandssted, J., et al. (2014) S100A11 Is Required for Efficient Plasma Membrane Repair and Survival of Invasive Cancer Cells. Nature Communications, 5, Article No. 3795.

INDEX

P

Q

R

S

T

V

W